CU00726231

Acknowledgements

A History of the Peninsular War. -Volume One 1807 -1800
by Sir Charles Oman

The Waterloo Companion by Mark Adkin

Wellington at War in the Peninsula 1808 -1814 - An Overview and
Guide by Ian C. Robertson

An Atlas of the Peninsular War by Ian Robertson

The 50th (Queen's Own) Regiment of Foot
– the 'Dirty Half Hundredth", so called because of the black facings and
cuffs on their uniforms. In this book, the narrative for 105th describes
their part in the battles of Rolica, Vimeiro and Corunna.

The 28th (North Gloucestershire) Regiment of Foot
– the "Slashers". The regiment earned its nickname in the American
War of Independence in 1775, when they slashed their way through
high grass with their bayonets to reach George Washington's troops.
The narrative for the 105th describes their role during the Retreat to
Corunna.

Novels by the same author
Worth Their Colours – the 105th in the Sicily Campaign of 1806 to
1808
A Question of Duty – a novel set within the war at sea 1809/1810

Published in 2013 by FeedARead.com Publishing

Copyright © - Martin McDowell 2013

First Edition

A CIP catalogue record for this title is available from the British
Library.

Contents

Chapter One
First Steps

Henry George Aloysius Carr; Captain, Light Company, 105th Foot, as he would introduce himself, sat rigid, being engulfed as he was, in a world of noise and tumult, his acute trepidation edging ever closer towards outright panic. He clung to any fixed point that presented itself, as he sat wedged between his Senior Lieutenant, Nathaniel Drake and the gunwale of a ship's longboat. This he found to be sturdy enough, but insufficiently high to engender any comforting peace of mind, as he anxiously judged the sights and sounds that came all too powerfully to his eyes, ears and skin. All around was a world filled with spray, surf and shouts, this acted out on a sea all too obviously angry at so many craft crowding and marring its surface, all with the temerity to chance the power of full Atlantic rollers, these being the last reminders of a violent August storm. The craft were purposed to land on a shore that could barely be seen through the murk of mist and spray and it seemed to all there afloat, riding perilously forward on the heaving waves, that their vessels were far too small to survive upon the white billows that reared up, intimidatingly beyond head height, to create a full surf that plunged onward, eager to crash against the narrow sands of this small fraction of the Portuguese coast.

Speech was impossible, so Carr contented himself by seeking what comfort he could against his state of peril, by transferring his right hand to inside his tunic, to either touch the latest letter from his secret beloved, Jane Perry, or finger her medallion, warm on his chest. This last had been given by her, equally secretly, as they marched off to a different war back in the summer of 1806, two years earlier, almost to the month. His thoughts landed momentarily on that indelible moment, but were soon brought back to reality as the longboat gave a sudden lurch to the left and he heard the feet of the sailor, standing in the stern, shift to give better purchase on the steering oar. He heard a shout from above him, but it was lost, unintelligible, in the roar of the surf, now alarmingly even nearer and larger, posing a threat even greater than before, to swamp their laden and all too vulnerable longboat. The oarsmen in the bows rowed frantically to keep their vessel stern on to the waves and Carr looked forward and leftwards to see the result of failure; a longboat had broached sideways in the surf and rolled over, the now useless oars appearing from the water as would the flailing legs of a stricken insect, to then be joined by the bobbing heads of the spilled occupants. A second malignant wave surged over the top of all, taking it from view.

Joe Pike took a gulp of seawater and opened his eyes to see a green and white world of water and froth. He opened his eyes wider in panic and gulped seawater again. He was sure he was sinking, his kit and weapon were dragging him down, but something, something unknown, was dragging him up. He instinctively kicked upwards and his face just broke the surface, just long enough for him to take a gulp of precious air, before the sea covered his face again, but he was being pulled to some form of safety. His head broke the surface again to see the face of his messmate, John Davey, Chosen Man, shouting some instruction, which he heard only on the second time.

"Hang on to this, boy. Don't let go. Get your arm behind it."

Davey pulled Joe to the turtle shape of the upturned boat and indicated the safety rope that ran in loops along its side. Joe Pike spat out water and gulped some air.

"John, where's Tom? Have you seen him?"

Davey spoke between his own gasps for air and between choking up water from his own lungs.

"No, don't know."

Davey, holding his own section of rope, seized the shoulder tab of another soldier and pulled him to the boat's side, just as he had for Pike.

"Keep hold of this, mate, don't let go. This boat's wood, 'twon't sink, and it'll get pushed in."

He paused as a wave cascaded over them, then he shook his head to clear it of water.

"Long as we hold onto it, we'll be alright. Grab any of the lads that you can."

Several, both soldiers and boat crew, had divined this for themselves, that this would be the most likely method to secure their own survival and so, they too, clung to the ropes and oars, which would at least keep them afloat. However, for Tom Miles himself, entombed within the upturned boat, his was a world of utter blackness and noise, the noise of panicking men, their screaming and their spluttering, also the surf pounding on the rolling hull above and the sound of foul language circulating in his own head. His evil temper gave weight to the words he shouted, easily heard, even above the pounding surf.

"Shuddup! Shuddup, the bloody lot of you."

Within this black world, what he felt holding him secure gave the weight of truth to his next shouted order, now that his angry shouts had silenced the panicking men trapped with him.

"Reach up, find a seat, anything you can. There's plenty above you can get your fingers round."

Water, imprisoned like themselves, slopped noisily against the sides, but Miles, his foul temper keeping his head clear enough to focus to good effect on the peril that they were in, had more to say.

"Check on anyone near you, get them to speak."

Tom Miles had no rank, but his infamous, belligerent character gained him immediate attention whenever he spoke. He heard mumbling from somewhere in the darkness, then a shout.

"There's Mick here, Tom, he's out, not movin', but not sunk"

"What's keepin' him up?"

A pause.

"His musket's caught under somewhere. The sling's holding him up. Lucky sod."

Miles ignored the judgment.

"Right, all of youse hold tight and we'll get pushed in. Soon as you feels sand under your feet, get under and out."

Then came another thought, its clarity again born of the anger with his own predicament.

"Not the beachside, t'other, then the boat won't be dumped on top of you. With ground under your feet, you'll get out. Just hold on and wait."

A crashing wave thrust the side of the boat against them out of the darkness, but it did, at least, mean that they were being pushed in. However, extra help was at hand. Sailors, on shore, had immediately seen the stricken longboat and lines were snaking out for John Davey and two others to secure them to the rope loops and soon they were being hauled in. The first to feel sand under his feet was the giant Ezekiel Saunders, the tallest in the Regiment, therefore unexpectedly, being so tall and muscular, a member of the Light Company. Simply by virtue of his surprising agility and prowess as a marksman, Saunders was not taken as a Grenadier.

"I can feel the bottom, lads, not too long now."

Now, so close to the shore, the surf had substantially eased and the incongruous shape of the upturned long boat had created a place of calm behind it. Soon all could feel sand beneath their feet and, having ducked under the longboat's side, were wading away and to the shore, their sodden kit hanging off them at all angles. Suddenly, from around the longboat, came a torrent of foul language. Davey looked at Joe Pike; he didn't need to look back as each waded ashore.

"I think Tom's alright."

Tom Miles had more to say.

"Where's that bloody sailor as couldn't keep this boat straight?"

Davey replied without looking back.

"You've probably just walked on top of him, not too many sailors've come up out with any of us."

Miles involuntarily lifted his feet to examine the water for human remains, but saw none and looked angrily at John Davey. However, Chosen Man Davey, being both a messmate and filemate with Tom Miles had a licence with him that no others had, even those of higher rank. This time he did turn to look at him.

"Now come up out of the water, you soft sod, and let's see what's missin', you, me, and the boy."

The boy, Joe Pike, the third filemate with Miles and Davey, had sagged down to his knees onto the wet sand, the now exhausted surf still lapping at his feet. Despite his young years and athletic build, his time in the cold Atlantic and his panic whilst gasping for breath, had taken its toll, but soon his innate strength asserted itself and he regained his feet. At this point both his section Lieutenant, Mr. Drake and his Company Captain, Mr. Carr, arrived on the scene. The distinctive blond figure of Joe Pike had told Carr that the overturned boat had involved members of his own Light Company, but he spoke immediately to John Davey.

"Casualties, Davey?"

John Davey came to the attention, at the "order arms", his soaked Baker rifle erect beside his left leg.

"Can't say, Sir, other than there's one, Michael Maguire, as've took a blow to his head, Sir. He's over there, Sir, with the other lads as was spilled out."

Davey's right arm jerked up to indicate the correct direction, just enough for the purpose, before returning to the attention. Carr gave the prone figure a quick glance, before turning to Drake.

"Ned, get Ellis to call the roll as soon as they're ashore, all that are going to get ashore that is."

Nathanial Drake saluted and ran off to find Company Sergeant Ellis. Carr took the moment to pull off his own boots and empty them of water, using the shoulder of the erect Davey for balance.

"Bad business, Davey"

"Sir."

"Thank that order from Major O'Hare, helping you to stay afloat. No packs nor knapsacks, just cartridge box, bayonet and an empty water canteen."

"Amen to that, Sir."

"Tell that to Chaplain Prudoe. He'll be pleased."

"I'll do that, Sir, when I see him."

Carr did not discern the note of doubt contained in Davey's voice, but, boots replaced, he regained his own feet and looked further up the beach. A parade of the Light Company was forming, with Sergeant Ethan Ellis smoothing out the folds of the Company Roll.

"Get your men up to Ellis, now, Davey. Let's see what we've got. Our orders are to get on picket as soon as we've landed, up over the dunes there. That's why we're first ashore."

Davey saluted and, as Carr sloshed away, he turned to his sodden and dishevelled shipmates.

"Right. All of us, up the beach. Ellis is calling the Roll."

He looked at the prone shape of Maguire, being ministered unto by Saunders.

"Dead?"

"No, John, but he's spark out. Don't know what damage've been done, can't tell."

"Well, bring him up with the rest of us, 'fore the tide washes him away."

For Saunders, that was a single-handed job. He seized Maguire's white crossbelts, hauled him to his feet, and then dropped him over his shoulder. Carrying both their weapons he strode up the beach, over the hard sand, to follow the others, all in various degrees of forlorn and dishevelment.

Ethan Ellis was not a large man, but many said that he had not been born, rather quarried and not "finished". Fierce of eye, with a countenance grim as wet granite, he ran the Light Company and, not surprisingly, he and Miles were natural enemies and the first thing he noticed was that Miles was without his shako.

"Miles! You'm without your headgear."

Miles, feeling hard done by and with some right on his side, replied in kind.

"Now just how much of a surprise is that, seein' as I've just been dumped out of a boat and spent nigh on ten minutes under'n"

Ellis had enough experience to know when to pick a fight with any soldier and let Miles retort go unanswered, at least regarding that issue, but he had a reserve.

"Then get yourself down to the water's edge and get one! There's any number swillin' about down there, and see if you can find one what's right, not the first what you comes to."

Miles shot a malignant look at Ellis, having been ordered to retrace the steps that he had just taken to get himself up there, but he returned to the water's edge to seek what he needed. There was little in the existence of Tom Miles that particularly mattered to him, apart from keeping alive, but one thing that did rise within him above the common was the fact that he was Light Infantry. In his mid-twenties, as far as he could calculate, he was naturally fit, wiry, cunning and aggressive. Within the persona that was Private Tom Miles, his status in an elite company was of no minor importance and so his search amongst the washed up kit, was for a shako with the distinctive Light Infantry hunting horn badge. One was found and eagerly seized, Miles hoping that it was his own, but it was too small. Annoyed and disgusted he threw it back into the water and was forced to try on several others of all three types, Line, Light and Grenadier, that were there, determined for good fit, for he was enough of a veteran soldier to know that kit had to be secure, or it would chafe and annoy over the long hard roads to come. One was found that did fit, but it had the plain George III badge with a large V in the centre, the shako of a Line Company of the 5th Foot. In foul temper he returned to his Company; Ellis having now just finished calling the roll.

Captain Carr awaited Ellis' report. He had heard very few names that were not answered and hope grew within him that the overturned boat had been but the one single mishap to befall his company. Ellis came to him and saluted.

"Three, Sir, not including Maguire, who may be beyond it, Sir."

Carr nodded.

"Thank you, Sergeant. Now get the men to present arms."

Ellis turned and bellowed the order. Five remained at the attention; they had lost their firearm. Carr walked forward to the first.

"Musket or Baker?"

"Musket, Sir."

Carr asked the same to each of the remaining four. The answer was the same, meaning that no man issued with a Baker rifle had lost their precious weapon, it being treasured as much more accurate than a musket and therefore a lifesaver in any open skirmish. More than half his Company now had one, provided at the personal expense of their Colonel, Bertram Lacey. Carr nodded.

"Sergeant, detail a Corporal to take these back and obtain five muskets, there must be some spares lying somewhere. Maguire's for one."

He then raised his voice.

"About face, skirmish order."

His command executed a smart "about face" and then broke up, to spread both left and right into their "files", which meant thirty columns of three men, five yards between the columns, five yards between the three ranks. Carr looked both left and right to check the alignment, feeling pleased that his men had completed the formation in the briefest of times. Lieutenant Drake was in front of Carr's Number 1 Section, on his right hand, Lieutenant Shakeshaft's Number 2 stood before that on his left. Shakeshaft was the Junior Lieutenant of the Company. And new! The closest file on Carr's immediate right was made up of Chosen Man Davey, with Joe Pike five yards behind him and Tom Miles, still incensed about his shako, five yards behind Pike in the final rank. Carr drew his sword and motioned his men forward. Motion only, because when advancing into the unknown, shouted orders were kept to a minimum and, in response, his command slogged their way up the soft sand of the featureless dune.

Nathaniel Drake looked across at his commanding Officer and remarked the contrast to himself. All throughout their easy cruise from Cork, even on the edge of the August storm, Henry Carr had been his usual indolent and casual self, talking in his usual matter-of-fact manner, even of his beloved Jane Perry. However, once in the presence of the enemy, either clearly perceived or but merely possible, there came a change, obvious and manifest. Out before his men, Carr resembled a creature of the hunt, awake and aware, his cold blue eyes casting everywhere, searching for any signs of a threat, either small or fully significant.

The Company's new Lieutenant, Richard Shakeshaft, was similarly looking before him with equal intensity. He was

The Company's new Lieutenant, Richard Shakeshaft, was similarly looking before him with equal intensity. He was young, not yet twenty, with an intelligent, active face, behind which existed a mind much inclined to overactive thought. At that moment, that mind had contrived the notion that ahead of him was enemy territory; in fact the very inch before the toecaps of his wet boots was enemy territory, thus he expected a French soldier to be behind every tree, wall and sagging fence within the next mile. Whilst behind him, first in their files, were Private Saunders and Private Byford, the latter of unknown background, but plainly, by his speech and the use he made of his leisure time, the nearest thing the Light Company had to what was known as a "Gentleman Ranker." The silence required of the parade ground had no place in skirmish order and so, now halted at the crest of the dunes, talking would not be a flogging offence. Saunders, leaning casually on his Baker rifle and, fully appreciating the "first time" of his Section Officer, felt the need to pile on the moment.

"Perfect place for a cavalry charge, eh, Byfe? French cavalry be the best in Europe, so they say."

Byford, fully appreciating what was happening in the mind of his immediate superior, felt no need to further add to what he knew must be his state of extreme anxiety, so he made no answer, just knowingly smiled. Shakeshaft craned his neck forward, the better to see what could be approaching, his overactive imagination now fully creative.

Carr, in the lead, had crested the dunes to see a dull, flat, plain from which a heat haze was beginning to rise and, beyond that, through the distortion, was a duller estuary, still awaiting the tide. On a promontory, where the river met the sea, was a collection of hovels and fishing boats, both in various degrees of dilapidation. He lifted his shako to block out the sun, revealing the two scars on his forehead, one from battle, the other from duelling, bisecting his left eyebrow. He spoke to himself, "Welcome to Portugal", then he spoke to each soldier either side of him.

"Out to twenty."

This would extend the gap between his files to twenty yards and thus guard four times the distance.

"Back files clean weapons."

As his commands were passed on and the files extended far out, his attention was suddenly drawn to the movement of horsemen to his right as they cantered towards him. He was pleased to see the front rank of Drake's section all present arms, a gesture of respect not lost on the horsemen, the leader of whom acknowledged all with a touch of his riding crop, up to the peak of his bi-corn hat. The group were definitely Staff and the leader, so Carr reasoned, was probably Wellesley himself; he certainly matched the descriptions that he had heard, a face on the thin side, aquiline nose and a lean athletic build. He wore no uniform, but a plain dark blue frock coat, another Wellesley characteristic, or so it was described. Whatever, they were not French and they were certainly superior in rank to himself. It was deeply probable that this was their General Commanding Officer, seen for the first time. The leading horseman reined in before Carr, at a comfortable speaking distance but not so close as to force Carr to awkwardly look up, a tedious ploy of so many General Officers. Carr waited for the General to open the conversation and it was not long in coming.

"Who are you?"

Now, from a General Officer, this was not an enquiry to make an introduction of a social nature, such as name and family, societal position or otherwise, it was more akin to an order to discover very straightforward military identity, personal names not being required. As his sword came to the present, Carr replied, as required.

"Light Company, 105th Foot. Here on picket. Sir."

The figure acknowledged the salute by again touching his hat with his riding crop, then he eased himself in the saddle, resting on the pommel. He was almost smiling, as though amused by something curious, but not altogether diverting. Carr lowered his sword and the Officer continued.

"Ah, Lacey's, if I'm not mistaken. 105th Foot. The Prince of Wales Own Wessex Regiment. Quite a mouthful."

"Sir."

"The heroes of Maida."

"Some people have been very kind, Sir."

"Including the Prince of Wales, it would seem."

"Sir."

The General nodded, then swung his horse around to give himself a view of both Carr and the country beyond.

"Keep a good eye to your front. As best you can."

The last words were added as the heat was making the distance more indiscernible and hazy by the minute.

"Cavalry, that's what I expect. Any kind of uniform on a horse and I want to know. The first thing you'll see is probably a glint from the sun on something. You'd agree?

"Sir."

"Were I Light Cavalry I would use those trees there, behind that white hut, to get as close as possible and get a good look. Probably dismounted."

He eased himself again in the saddle.

"I hope that helps."

"Yes Sir, it does."

Carr was both mildly surprised and pleased to receive such clear and knowledgeable orders from such as a Staff Officer.

"Have you a glass?"

"Yes Sir. A good Dolland."

The figure looked please and nodded.

"Now, retire your men back behind the crest. Always a good move, I find."

"Sir."

The horse was turned back to face the way they had come.

"I bid you good-day then, Captain."

"May I know your name, please, Sir?"

"Wellesley."

***

Captain Lord Charles Harvey Carravoy was somewhat anxious about his mount. He smelt of mutton fat and brick dust and could manage no more than a laborious plod. Carravoy and his Grenadiers of 105th Foot were finally disembarking and the Lord Charles could see no reason why his new boots, recently made for him and purchased from Batten's of Gloucester, should suffer a soaking in seawater, thus the nearest Grenadier was being employed to carry him ashore. A look to the right showed a picture similar to that created by himself, because his Junior Lieutenant, The Honourable Royston Marchman D'Villiers was also being born ashore by a labouring member of their Company. Carravoy's Grenadier had stopped, supposedly assuming that his efforts to preserve the dry feet of his Commanding Officer were over, but Carravoy looked down to see surf still making its last efforts to soak the bone-dry beach.

"Not yet, further."

The Grenadier took five more steps which then released him of his burden. Carravoy's feet met the sand and, with not a word to his severely fatigued and very wet Grenadier, he strode off to D'Villiers, now similarly ensconced on dry land.

"Royston, get the men assembled and get some kind of camp set up. I anticipate us marching off before the day's done. I'm off to find Lacey and O'Hare. Pass that onto Ameshurst, will you."

D'Villiers did no more than raise his finger to his temple; their equality in social class required no more. D'Villiers had transferred himself from Number Three Company, Commanded by Captain Heaviside, to fill a battle casualty in the Grenadiers and he now felt far more at ease with his situation, rather than, as he put it, to be "An altarboy for Holy Joe Heaviside". This description justified by the fact that the Captain of Number Three was a deeply devout Methodist. Although D'Villiers was the junior of his two Lieutenants, Carravoy tended to treat him as the senior, for the other, Simon Ameshurst, despite having shown himself in the past as a brave and resourceful Officer, was from a significantly lower social stratum, a gap too great either to bridge or even to be wished to. Lord Charles, exuded aristocracy from every pore, possessed of classic, finely chiselled good looks, every gesture and syllable spoke of sangfroid and self-regard. Also D'Villiers could equally be described as pleasing to the eye, were it not for a pinched peevishness that seemed to inform every expression and gesture. Thus, it was now the case, as Carravoy strode off, leaving D'Villiers to solve the problem, of where to place their Company in readiness to march off. Nowhere seemed to be absolutely right, but his conundrum was solved by his Officer companion, the aforesaid Lieutenant Ameshurst, who had spotted the arrival of the pair and felt it would be helpful to impart what he knew.

"Hello, Royston, not too damp I hope?"

Ameshurst was naturally cheerful, with a personality instinctively inclined towards being helpful. He was well regarded by his men and any would have willingly carried him ashore, but his uniform was wet almost to his waist.

"The Regiment's assembling but a little way further down the beach. A Headquarters has been set up, there's a small tent just up there, if you care to look."

He turned that way himself and counted.

"Third one along."

D'Villiers felt piqued that Ameshurst was privy to knowledge that he was not and should be. His reply was cold.

"Captain Carravoy told me pass on to you that he has gone off to look for the Colonel and Major O'Hare. He's of the opinion that we may be marching off this very day."

Ameshurst's countenance brightened further; he could be of additional help.

"Yes. I bumped into him just now and told him what I just told you."

He smiled, expecting something extra from the noble D'Villiers, but nothing came.

"Right, I'm off to get my lot organised. See you anon."

With that he touched his hat and disappeared up the beach, calling for a Sergeant, and using his full title.

"Sergeant Ridgway! Your help, if you please?"

D'Villiers scowled at such unnecessary, as he saw it, care in addressing someone of lower rank and then considered what he himself should do to move things forward. Many of the soldiers dropping off from the sides of beached longboats were 105th, very distinctive by the emerald green facing of their uniforms. He walked down the beach to the first group, an important group, for they were the Colour Party, but this made little difference to D'Villiers. They were all of lower rank to himself, including the two Ensigns, but he came first to a Colour Sergeant and forced himself to acknowledge that, from this particular individual, any information that came would be worth listening to.

"Sergeant."

Colour Sergeant Jedediah Deakin came to immediate attention. He had recognised the voice and straightened himself immediately, his musket beside his left leg. He knew that he was being addressed by

Lieutenant D'Villiers and his disdain for this particular Officer was as deep as could be, but this did not show as he stared rigidly to his front.

"Sir."

"Any Grenadiers further along from you?"

"No Sir. All beyond us is Number Three, Sir. The Colour Company. Sir."

Deakin thought it best to add the last piece of information, for he could not be to any degree certain that this particular Lieutenant would be aware, that, unusually for the British Army, in the 105th Number Three Company protected The Colours, therefore he, Deakin, was attached to that Company. D'Villiers did no more than nod, a gesture that Deakin could not even see as he stared straight ahead, then, without a word, the Honourable walked off in the direction of Captain Carravoy. Deakin stood stock still for a count of ten, then relaxed to look and see D'Villier's back, now some ten yards away and increasing. Deakin shook his head in disbelief at the utter ill manners of this Officer, for whom, if truth be told he felt the deepest and most thorough disdain, then he turned to his good and best friend, Corporal Tobias Halfway.

"Looks like we'n the last ashore, Toby, best get ours all on up the beach. The rest of the Regiment is up and over, musterin' for a rollcall, shouldn't wonder."

Deakin's friend of long, long, standing, looked along the beach and saw the last of the 105th Foot coming ashore, through the now exhausted surf.

"You'm right, Jed, I'll just mooch on up an' push 'em along a bit."

"Hold hard a mo', looks like we got company, of two sorts.

Deakin had looked both up and down the beach and had seen the Senior Company Sergeant, Obediah Hill approaching from one direction and their Captain, Joshua Heaviside, simultaneously approaching from the opposite. Whilst Hill was simply big and round, with pronounced ginger whiskers, Heaviside's build made him appear short and squat, when he was, in actuality, above average height. He was clean-shaven, at least for one hour of the day. The heavy jaw and grim mouth could justifiably be shaven twice each day and, with this being Noon, now showed seven hours of growth. Hill arrived first, as he had hurried to do so, and all three non-Commissioned Officers came to the attention together and waited for what each knew was inevitable.

"If the Lord delight in us, then He will bring us into this land, Numbers 14, verse eight".

It was Deakin who responded, as was usual, to a Heaviside Bible quote.

"Yes Sir. I'm sure the lads all see's it that way, Sir."

Heaviside looked at Deakin as though he had just received a response from the most devout of his Parishioners and gravely nodded, then he turned his mind to the task at hand. As usual, his orders were brief, to the point and reflected his taciturn attitude to all things temporal.

"Get the Company ready to move. I fancy we'll not be staying here long."

Three "Sirs" rang out in unison as Heaviside proceeded along the water's edge, considering how the scene chimed best with which appropriate passage of The Bible, his stocky figure being followed by the eyes of the two Sergeants and the Corporal, but not without affection. Heaviside, for all his lugubriousness and perpetual Bible quotations, was well liked by his men. Besides thoroughly knowing his trade, he had led them well in past times of battle and was always found alongside those of the front rank; which was easily good enough for them.

***

The two most Senior Officers of the 105th were at that moment in deep conversation, walking past their men setting up camp, subconsciously examining the state of each of their men as they came up and passed them by. Both were currently unconcerned with the military situation now pertaining, more reminiscing of campaigns past, for both were veteran Officers; Colonel Bertram Lacey and Major Padraigh O'Hare. Lacey was the taller, O'Hare more muscular, but both bore the marks on their faces and in their eyes of hard service, not least both carrying iron grey hair. Both were middle-aged and had been recalled from retirement to lead the 105th, when they were but a Battalion of Detachments and the threat of an invasion led by Napoleon was large and ominous. It was Lacey who began the conversation.

"Outside of my experience, this, Padraigh, but not yours I fancy. How does this compare?"

O'Hare had landed with Abercrombie in Egypt in the year 1801, whilst, contrastingly, almost all of Lacey's service had been in the American Wars. O'Hare answered in his usual lilting Irish tone.

"Sure, now wasn't this worse? The Med doesn't often boil up as this did for us today. Us in the 28th got ashore with barely a damp boot in the year one, but this was altogether different."

"What do you know of casualties?"

"The Lights lost some, being first ashore. Two, I believe, but the whole army, about a dozen, just under, I've heard. All men, no horses, and no guns, for which I'd fancy the General's grateful."

Lacey nodded.

"Where's Carr?"

O'Hare pointed to a section of dune further up towards the river mouth.

"That's his, all down this side of the crest, with him and his two, standing on top."

Lacey studied a while, seeing the three Company Officers, and he was also able to see that his Light Company was now stood down, almost all cleaning and examining their weapons. He then spoke further.

"Make sure he know's where we are. Rations and supplies should be arriving soon, along with the other kit you ordered left on ship. I'd like the men to eat soon. I wouldn't want Carr's to miss out, should Wellesley want to advance immediately. Such would surprise me not."

O'Hare nodded. Thorough preparation and attention to detail were Lacey's hallmarks and he, himself, had full patience with such concerns, knowing the importance of attending to even the smallest minutiae of running a battalion of infantry.

"I'll make sure, Sir, and also ensure that some is put by, if we are ordered to move off. It'll be no surprise if they're put out on picket for the column."

Lacey changed the subject.

"We'll be called by Fane soon."

"Yes Sir, I don't doubt at all that the Brigadier will have one or two things to say."

Lacey nodded whilst idly watching a group of Grenadiers wringing water from their stockings. He changed the subject again.

"What about "the followers"? Always the last."

Lacey was referring to the camp followers, the few, chosen by lots and their ability to be of use on campaign, that accompanied every Regiment. All were wives, either Common Law or legal Church and also the children of serving soldiers. No Officer had brought his wife, though each had the option. O'Hare looked at the Westering sun and the ocean beneath, the stretch nearest being still full of boats bringing guns, men, material and supplies ashore.

"They are the lowest of low priority, Sir. You know, just as much as myself, there could be a French army but an hour away. The army and its needs come first, and who could argue?'

Lacey made no reply, allowing his silence to speak his agreement.

***

Bridie Mulcahey, Mary O'Keefe, and Nelly Nicholls, the "wives" of Colour Sergeant Jed Deakin and Privates Joe Pike and Henry Nicholls respectively, sat in the dank gloom of their transport, the Chepstow Castle, their place within her closely matching one of the bleakest dungeons of her namesake. For the last pair, the Nicholls alone, their union was "Church blessed". Above them could be heard the rumbling sounds of barrels being rolled across the decking of the hold above and the shouts of sailors about their business unloading their ship's charges. Around the three women sat their children; Eirin, Patrick, Kevin and Sinead sat around Bridie, all the children of her previous husband, killed in battle. Sally, Trudie, and Violet sat around Nelly. Mary was newly attached to Joe and childless. Bridie's children looked upon Jed as their "uncle", which status he had enjoyed when the family was whole, but he had "taken on" the family when his messmate, Pat Mulcahey, was killed in Sicily, but they all acknowledged him as the head of their family. The children all sat in the dim half-light of the feeble candle, none complaining of their confinement, for all knew the uselessness of such grieving. Their Mothers were powerless to alter their situation by but one iota.

Bridie and Nelly were veterans of being shipped around to the various far-flung places where their husbands found themselves, but Mary not nearly so much, although she was Bridie's youngest sister. The two elder "wives" knew enough to be ready to move on the instant that they were called, thus they all had their few precious chattels bundled up or tied into convenient burdens, each of which could be carried by a child, which also applied to the family cooking pot and it's tripod. This precious item was in the charge of Bridie's eldest, the

21

seventeen year old Eirin, whilst that belonging to Nelly; she looked after herself.

Bridie was pleasant featured and well formed, despite the early onset of middle age, brought on by the privations of a life as the follower wife of a serving soldier. Any smile that formed on her mouth was quickly matched by the same light entering her eyes, both still bright green and complementing her dark brown hair. Jed loved her dearly despite their coming together as a matter of pure expediency and there were few in No. 3 Company that did not have a soft spot for Bridie Mulcahy. Nelly Nicholls was a horse of a different colour; in fact the word "horse" would not be misapplied. Her large Irish frame always seemed to be putting undue strain on her clothing and her face carried either a permanent scowl or a look of deep suspicion. Her grey hair, thin for so large a deportment and head, was pulled back, out of the way in a tight bun. Nelly Nicholls was built to last and to carry life's burdens for both her husband and her children. Few carried any affection for Nelly Nicholls, too many had felt the rough edge of her tongue, but she more than justified her inclusion in the Battalion and many had appreciated the weight of her stern common sense in time of need. Her and Bridie were firm friends, they relied on each other and supported each other as would two messmates on the field of battle. Nelly shared, more keenly than most, a common antipathy for Tom Miles.

Bridie's youngest, Sinead Mulcahy, looked at her mother, sat in the gloom, her patient arms wrapped around her precious bundle of possessions.

"I'm hungry, Ma."

Bridie's response was to smile indulgently and she reached into the soldier's knapsack that she carried down her right hand side. The tender hand produced two ship's biscuits and each was carefully broken in two and a half given to each of her children. At this prompting, Nelly did the same but having only three children, she offered the remaining half to Mary who shook her head and so it was returned to the haversack. Mary sat immersed in both the gloom of the room and the gloom of her mood. She was an Irish beauty of the fullest sort, dark chestnut hair with "green eyes and soft brown skin", words her husband Joe often used to describe her. Mary sat in depression engendered by anxiety, when they were apart, she could never stop worrying about her soldier Joe.

The door of their compartment opened and the head of an unknown sailor looked around.

"Not long now, ladies."

Then the head was gone and the door closed.

***

The short Portuguese evening had come and was going, the sun falling almost vertically into the horizon. Lacey and O'Hare sat outside their Headquarters tent, although from the languid attitude of each it could have been the terrace of a stately home, but such was far from the means of either. Some ale had come ashore and both gently imbibed from their pewter pint pots, both men plainly, as yet, untroubled by any orders or requirements. Brigadier Henry Fane had not called for his Senior Officers, the word being that Wellesley favoured caution until he knew more of the whereabouts of the French. Therefore Fane was fully occupied with attending Wellesley's own, higher order, Council of War. Such was the relationship between Lacey and O'Hare that neither had said a word since they sat to watch the descent of the sun into the now tranquil sea, each content to be lulled by the waves that now did no more than lap at a beach fully cleared of all detritus created by the perilous landing. What was still useful salvage was in a stores wagon, what was not, was stoking a cooking fire.

O'Hare's head turned seaward, away from regarding the multitude of campfires now giving life to the darkness fully along the length of the beach. He heard, before he saw, more longboats arriving, this being by way of female chatter coming across the dark water; neither raucous nor uncontrolled, but it was nevertheless reaching his ears above the sound of the exhausted waves. He saw a dozen points of light just out from the waters edge, denoting longboats that were bringing the followers ashore, their white oars somehow catching the light off the lingering surf. Soon their wide hulls bit softly into the fine sand and the sailors jumped down to aid their passengers' disembarkation. Being sailors of the sentimental kind each passenger was lifted from the bows and placed on the water's edge to stand and either wonder which way to go or to wait for the remainder of their "family". Soon all were on the sand and the longboats were out on the white of the surf and turning for the journey back. One woman, a shapeless bundle, made more incongruous by the bundles strapped about her, ascended the beach and approached Lacey and O'Hare. She spoke with a full German accent.

"May you excuse me, your Honours, but are you to know the place of the billets of the 97th The Queen's Own Germans?"

Lacey spoke immediately.

"Die Eigenen Deutsch der Königin. Sie sind zu unserem Recht, nicht zu weit, ich denke."

The reply came back, out of the dark.

"Vielen danke, mein Herr."

"Viel Gluck, gutige Dame."

The question came from his companion.

"German?"

Lacey took a deep breath, the memories were painful.

"In the American Wars, I had many Hessians under me."

O'Hare waited for more on that subject, but none came. He let the subject drop, he had no choice, for his own thoughts were soon interrupted.

"Ah sure, is that you up there, now, Colonel Sir, speakin' in some Heathen tongue?"

O'Hare heard Lacey break into a rolling laugh beside him, then fell to outright laughing himself. Neither could fail to recognise Nelly Nicholls and it was O'Hare who replied, matching the charm that only the Irish could command, including the normally taciturn Mrs. Nicholls, who could call it up as well as anyone, when the situation required.

"Mrs. Nicholls! Major O'Hare here. The top of this fine evening to youse, and to your fine family. 'Tis pleased we all are to see you safely here on shore and dry land. Thanks to the Good Lord."

"Thanks indeed to Him, and I thank you for your warm welcome, Major, but would you be knowin' the direction which we should take to find our husbands?"

O'Hare was about to answer when a lamp described a course down to the beach from behind and to the right. It stopped six yards from the seated Officers.

"Perhaps I can help, Sir. I saw the boats coming ashore and I divined their purpose. I can help these ladies, Sir."

Both Lacey and O'Hare immediately recognised the cultured voice of their ex-Cleric, Percival "Parson" Sedgwicke, once Vicar to his own Parish, then a convicted felon, then a "King's hard bargain" sentenced to serve in the Army, then a storesman, now Chaplain's Assistant. This time it was Lacey who answered.

"Sedgwicke?"

"Sir."

"You know the way?"

"Yes Sir, I think I have each Company placed, the Grenadiers are nearest, then each Company in order, then…."

"Yes, Sedgwicke, yes. Do the job, with my blessing. See these good ladies into the company of their good men."

The lamp quivered as Sedgwicke saluted in the dark and then it resumed its course to the waiting camp followers. Sedgwicke's voice had been recognised.

"Parson, me darlin'! Now there's kind to bring down a lamp and see us all safe and well and put amongst our dear husbands."

The lamp spoke.

"This way, ladies. The Grenadiers are first, and just up here."

The tramp of many feet across the sand and shingle spoke the end of the conversations and dim figures, little more than silhouettes against the fluorescent surf, passed across their Officers' front, but it was Bridie who spoke the last.

"Good night to you both, Sirs. God's blessing"

Lacey replied.

"God's Blessing on us all and a good night."

The sound of the small waves regained their pre-eminence and the short silence did not last long, soon broken by Lacey. The humour and kindliness of the encounter had passed.

"Rum fellow, that Sedgwicke!"

O'Hare felt more sympathetic.

"Odd cove he was, for sure, but the men like him, fish out of water even though he may be. He's found his place, grant him credit for that. And shouldn't our Chaplain be about a job such as that, helping the wives and children find their way?"

"For now I'll give Chaplain Prudoe the benefit of the doubt. Perhaps he's attending to the Spiritual needs of the men; that would take precedence, at least in my view."

O'Hare took a drink from his tankard; he had his doubts. However, at this moment, his doubts were unjust, for the Reverend Leviticus Prudoe and his good wife Beatrice were at that moment conducting a Service of Evensong for a substantial group of soldiers, a mixed congregation from a mixture of regiments. Their quiet singing could be heard from further down the beach, but they knew not to give Prudoe credit for it. Meanwhile, the three women had found their men and the joining together of each with their man was marked by three different degrees of affection, Mary flung herself on top of Joe, much to the amusement of Joe's messmates, Bridie sat close and snug up to Jed, and Nelly inquired of Henry if they'd had their rations and why wasn't there a better fire, 'Sure, wasn't there plenty of driftwood!'

The day was dying, nearly out. The British Army of Portugal was ashore, with its guns, horses, supplies and minimal casualties amongst the men. Soon but one tent showed a lantern through its thick canvas, the largest, Wellesley's headquarters, still debating the best course of action and digesting reports, English, Spanish, and Portuguese.

Guarding all atop the dune, Chosen Man John Davey held his spotless Baker rifle in the crook of his arm and pondered on his own circumstances within his able but uneducated mind, whilst allowing his poacher's hearing to subconsciously take over his sentry duties. He reflected on his own woman, Molly, her daughter Tilly and their son John, and one thought gladdened his musings. All three were safe back in England with his own mother and sisters, on a smallholding which they owned outright, every inch, bought with prize money that Davey had won in the fight ring. He looked out into the blackness of the Portuguese night, no point of light, not one, not even from the hovels on the point of the estuary. It gave him a chilling sense of foreboding, to stare out into the sepulchral stillness of the black night, black, like the beckoning of the grave. He heard a cry as of a baby in distress, but knew it to be a rabbit at the point of execution, a fox probably, they rarely made a quick kill. The thought further depressed his mood. He knew enough to place himself accurately on the surface of Europe and also enough to know that this particular piece of soldiering would only end when they marched through Paris and he knew enough to place that event, if at all, far, far into the future. Or, alternatively, it could end if they suffered severe defeat and he was killed, captured, or lucky enough to be evacuated. He eased his Baker in the crook of his arm and sighed.

The breath was barely out of his body when his alert hearing detected rattling pebbles over to his right and behind. The Baker came up and was cocked in one movement.

"Halt! Who goes there?"

At the same time as the reply, Captain Heaviside came out of the dark.

"Heaviside. 3rd Company. 105th Foot."

But Davey had not finished.

"Advance Captain Heaviside and be recognised."

Heaviside continued forward without a pause, despite the bayoneted muzzle pointed menacingly in his direction. Soon the threat turned to an immaculate "present arms".

"Captain Heaviside, Sir. All's well."

"Be ye steadfast, unmoveable, always abounding in the work of the Lord. First Corinthians, 15, verse 58."

Davey had little experience of Heaviside and his inevitable Bible quotes, but he knew one thing, that if an Officer of King George told you that black was blue, there was only one possible response.

"Yes Sir."

Heaviside returned the salute as formally as it was given, then continued on, into the silence beyond the dunes. He was carrying his small "prayer" bundle, his Bible, his plain crucifix and pictures of his wife and two children. Satisfied that he was beyond hearing, he sank to his knees and arranged the four items. He then began the prayer that he had been composing all day and, all the while, creatures of the night, scurrying home, gave no pause to hear the grave and sonorous intonations of Captain Joshua Heaviside praying to the God he loved, for the protection of the people he loved and the offering of his own life in the coming conflict if it would keep them safe. Now that the 105th had their own Chaplain, he could have joined him, the Chaplain that, coincidentally to Heaviside's lone ceremony, mumbled his own prayers to end the day beside the kneeling figure of his wife. However, the devout Heaviside had little time for the religion as practiced by Chaplain Prudoe, with its glorious vestments, choking incense, heavy serving vessels and writhing Christs upon their crosses. All appallingly Roman! He found the plain, empty space before him to be a superior, and more profound, church.

***

A marching army appears of a length that defies explanation and also strikes any observer as also surprising in its varied composition, it being made up from the widest possibilities of army life. First, the Light

Cavalry, vigorous and deeply suspicious, riding hither and thither, investigating anything that could be the first signs of the enemy, these perhaps waiting in position or, more likely, out scouting, as they were themselves. Then comes the Light Infantry screen, in this case the 95th Rifles, green clad and every one a lean, hard eyed man, walking with a confident and easy gait. They were the best the army had and they knew it. Next came the infantry, the Line Battalions, endless columns of fours, then, as the final ingredient of what may be described as the active component, the artillery. However, the whole of this length was then doubled by supply wagons, in this case an Irish Wagon Train, then the camp followers, these latter achieving their progress in marked contrast to their menfolk held within the formality of their Regimental columns further at the front. The followers walked as if in holiday mood, always noisy and always animated, with various bunches of women and children, forming and dissolving according to the changes in the needs of the moment and the topics of conversation.

For the Line Battalions, if they were formally marching in step and possibly with a band also playing, this depended on where they were. If they were in open country, then the men were allowed to walk at their ease, required merely to hold their place, with the fifes and drums back somewhere in the rear of their column. If the army neared a town or large village, then, as often as not, the men would be called to order, the colours uncased and the bandsmen advanced. A bit of show for the locals never came amiss, for perhaps some recruits may even be gained, taken in by the brazen colours, the uplifting music and the smart uniforms.

Thus in this way, in a barren countryside, this moving army displayed the track of its progress; its serpentine length disappearing back into the dust, the distance, and the shimmering heat of a Portuguese summer. Wellesley had not ordered a march for the day following their landing, nor even the day after that. Having landed on the 1st August, the march South to Lisbon did not begin until the 5th, four days later. Lacey and O'Hare did meet their Brigadier, Henry Fane, but not in any formal manner. He, a gentle, depressed Scotsman, had taken the trouble to call on them and issued orders to be ready to march at two day's notice. This was met with pleasant surprise, for it meant a more permanent camp, therefore a few extra comforts, extra to that afforded by the requirement of perhaps moving within the hour. Fane felt able to pass on, that Wellesley was in deep disagreement with the Portuguese; they were insisting on the British joining them far inland before beginning the march South, for there the Portuguese Generals felt safer, the reason being that further inland there were many good defensible positions, where their customary annihilation at the hands of the French, at least, could be avoided. Fane added his firm opinion on the Portuguese, this being that they considered the French to be invincible. Wellesley, in his turn, was insisting on marching South close to the coast, where they had the security of their own fleet, always there for a hurried evacuation, were one needed. French numbers were unknown; all that was known was that they were against the veteran Marshal Junot and that his subordinates Generals Dellaborde and Loison were closest to them, their numbers also unknown. Lacey and O'Hare felt grateful to have been made aware of the full picture, but their minds soon turned to the micro business of managing their battalion, in particular, that there were possibly insufficient spare flints.

Therefore, it was not until the 5th that they had set off South, close to the shore. Their fleet was in sight, but the marching was hard going, over soft sand in the heat of late Summer. Portugal in early August presented herself in a garb not so very different from that of June and early July, that being hot, brown and dusty, all covered by the slow accumulation of the latter, which dulled all colour, the dust being picked up by the summer winds and deposited on any convenient surface. Perhaps April and May would seem different, the full green of Spring showing in the fields, the bright blossom of the orange groves and the more delicate, but equally welcome, blossom of the olive groves.

Fane's Brigade was in the lead, for his contained the army's two Rifle Battalions. The one leading was the 5th Battalion of the 60th Rifles, formed in column on the road, whilst four Companies of 95th were out before the army in skirmish order. The 105th brought up the rear of their Brigade and on each flank was their own picket screen made up from their own Light Company, Nathaniel Drake's 1 Section placed inland, whilst Dick Shakeshaft's 2, was placed off to seaward. Ellis marched with Shakeshaft, Carr with Drake, and Carr was thoroughly out of sorts. He was hot and tired and was just now considering removing his jacket for extra comfort, but he knew that it would be viewed as appalling "bad form" for a Captain to succumb to such informality. The one point that had made him feel better was the squelch of his boots, he had filled them with water at the last stream and so his feet at least felt cooler, but even the comfort of that was beginning to pale as the soaked skin was beginning to blister. He resolved that, at the next stop, he would change his socks but soak his shirt if he could. Thus was his mind occupied.

If any mind was occupied by military affairs it was Drake's, him contemplating the seemingly pointless role that they played on the flanks of the army, because a thick screen of cavalry was very visible inland from where they were; groups of riders appearing and disappearing from and behind the olive trees and thick gorse. Finally, the thought would out.

"I do feel somewhat superfluous here, Henry. Our noble horse, out there, cover any sneak approach, surely?"

Carr's own thoughts automatically addressed the question and the answer came without hesitation.

"If Johnny does show, it's our job to hold him up, especially if it involves some column, sneaking, as you put it, out of a valley on our flank, like that one."

He pointed to a shallow gap in their own enveloping hills.

"The cavalry warn us, and we delay their approach, that's our role; us, the 60th and 95th, that is."

Drake dropped the subject, but, cheerful and garrulous as ever, he soon found another.

"Have you written to Jane?"

If, by raising this topic he hoped to cheer up his Commanding Officer, his hopes were misplaced, for Carr's demeanour sank further.

"No. At least, not since we landed. You, to Cecily?"

"Yes. I left one behind when we marched."

Carr nodded, merely to acknowledge, but his face remained grim.

"I got one off when we left Ireland, in response to one from her, and it was not a pleasant one; very businesslike."

Drake looked at his friend, but said nothing. If Carr wanted to say more, he would, but Drake was not going to press for more and, after some dark looks passaging Carr's face and the working of his jaw, more detail arrived.

"Her Father's pestering her to marry! He's made a few suggestions, even. It's giving her some disquiet."

Drake looked shocked and looked hard at Carr.

"But she says she's resisting, surely? I mean, you two; you practically have an understanding, words of fire, tablets of stone, sort of thing."

Carr nodded.

"Perhaps, yes, perhaps, but it wasn't mentioned."

Drake's reply was instant.

"Which leaves it up to you! You are the suitor, you should plead your case."

He paused to allow the words to be absorbed.

"She is waiting for your proposal, being a woman of quality she can do no other thing!"

He looked again at Carr, then continued.

"It's true. You know what a hopeless dullard you are when it comes to dealings with the fairer sex. Take it from me, one who knows, being betrothed and set on a clear and unswerving matrimonial course. The thing lies with you!"

Carr turned to his companion in arms, his brow knitted in puzzlement.

"You really think so?"

Drake nodded and Carr continued.

"That I should write pleading undying affection and wonderful prospects? That sort of thing?"

"That sort of thing exactly."

Carr worked his mouth and bit his lower lip.

"Well, the affection bit, I can manage, quite easily, but…."

Drake interrupted.

"With the right words."

Carr nodded.

"With the right words, yes, but prospects? A Captain in the 105th, unknown but unto God and the Prince of Wales."

"Well, that's a start! And a good one, I'm sure we can make something out of that, at least enough to persuade her to mount a spirited resistance to anyone laying any degree of siege to her affections, especially anyone at the request of her Father, our esteemed "used to be" General that once commanded over us. You do, if you cared to take advantage of it, have a head start over them all."

Carr worked his mouth some more, before finally managing a conclusion.

"Next billet, then, we'll get to work?"

"More like I'll get to work and you'll busy yourself with the pen!"

Carr nodded and smiled for the first time that day. However, smiles had very much been the order of the day on the far side of the 105th 's column, where Lieutenant Dick Shakeshaft's section was holding picket between the army and the coast. He was in the most buoyant of spirits, for this was real soldiering, independent, out and away, detached from the army, his men properly posted and looking the part. This was a description which he readily applied to himself, because all about him was as required for the full effect, sword drawn and held over his right shoulder, jacket fully buttoned, shako badge gleaming, as were his boots, at least the upper half. His veteran Sergeant, Ethan Ellis walked beside him, this not being exactly the best of disciplinary form, as Ellis saw it, but he had little choice, Shakeshaft had placed himself there, beside him. To Ellis, it made little difference at all, his task was "on picket". He carried his Baker in the crook of his arm, that being the only static part of him, for his eyes were in total contrast, casting everywhere, trusting nothing. His over twenty years of soldiering, in all conditions possible, provided the fund of knowledge from which came the automatic answers to Shakeshaft's almost incessant questioning. Having queried Ellis' dispositions of the men and found nothing to argue over, his latest batch of questions concerned the possibility of meeting the French.

"What if the French were over that ridge, Sergeant, that one there?"

Ellis had already seen the ridge and was instead examining a big gap in the cavalry screen out to his right.

"If both armies know their business, Sir, and I think we can safely say that about the French, there's no possibility of one bumping into the other, sort of blind, like."

He took a breath, not quite short of exasperated, but this teenager would be giving orders that he would have to obey. The more this "rank beginner" knew, the safer they would be.

"If they were over that ridge, we'd have to trust the cavalry to see enough and see it soon enough. There'd be some musketry and on hearing that, we'd all stop dead, wait for orders, but expect an attack."

"Would it be cavalry?"

"Most likely, Sir. That's the order I'd be waiting for, Sir, "form square".

Shakeshaft remained silent, thinking on the answer. Ellis used the pause to check on the men and found all was well. He nodded his head, pleased, all were at their correct spacing, all looking out to their right and so he passed judgment, the highest praise he ever gave, "good lads, these".

Meanwhile, at the head of the main column, back on the coastal track, the conversation was taking a similar vein. Ensigns Rushby and Neape, their Colours now being carried by their Sergeants, were asking questions of the NCO's around them, these being Colour Sergeants Deakin and Bennet, Corporal Halfway and Sergeant Major Cyrus Gibney. The last named was amongst the largest men in the Regiment, he could meet, with no need to incline his head, a level gaze from even such as Private Ezekiel Saunders. Whilst Sergeant Hill seemed to have all parts round and bulbous, from whichever direction you looked, Gibney looked large, solid, fierce and capable. A dour Yorkshireman, his soldierlike demeanour was completed by dark whiskers which descended in clean lines from in front of his ears, to then swoop up from his lower cheeks to meet magnificently above his upper lip and below his prominent nose. Sergeant Major Gibney was the solid foundation of the whole battalion, none would question his capabilities as a soldier, nor questioned that Gibney would endure, no matter what the fortunes of war would send their way.

Gibney, for his part, allowed himself little social interaction with any of the men, but he was "Old Norfolk", a survivor of a shipwreck which created many "castaways" who helped form the 105th in its very beginnings. This also applied to Deakin and Halfway, and so, as often as not, on the march Gibney took his place close to these two. However, it was Deakin who was providing most of the answers, the latest question being whether or not there was going to be a battle, this from Neape, their youngest Ensign, not yet full grown nor started shaving, as Deakin would describe him, by way of judgment.

"I wouldn't bet against it, Sir. There's been occasions, as I've heard, of armies landin', then goin' back aboard if things looks a mite too perilous, if you take's my meanin', Sir, but my feelin' is that our General won't be shy of takin' on the French, an' they'll be anxious to make an end to us, here in their Manor, on their piece of territory, if you take's my meanin', Sir."

Ensign Rushby then spoke up, in conversational tone. He had been sketching, all the while, the fearsome figure of Sergeant Major Gibney.

"And don't forget the Portuguese! We've come here to support them. It'll not look good if we land, march about a bit, then go back on our ships, leaving them to face the Johnnies alone. Shameful, to say the least, I'd say."

What visible agreement could be seen, came from Neape, sagely nodding his head, whilst Deakin, Gibney and Halfway, looked gravely ahead. Battles were not outside their experience and never anything to be looked forward to. Rushby, also, had his own memories and flexed his left shoulder, with a wound scar to both front and rear, but he had more to say.

"Then there's Parliament. Think of the questions that would be asked, if we more or less arrived down here, padded about a bit, then slunk off and away."

More nodding from Neape, but nothing from the remainder of his audience. However, Rushby was hoping for a contribution from Gibney. None came, so he asked for it and when it came, it finished the conversation.

"Gi' it a week, an' we'll be muzzle to muzzle, never fear."

Silence fell, filled only by the sound of their marching feet, all subconsciously in step. Deakin eased his musket and reached for his water bottle, his thoughts flying back down the column to the followers. If this march was taxing for him, what would it be like for them? He had his worries, but that was nothing unusual.

***

All the women, collectively described as "followers" had hitched up their skirts, the easier to walk without pushing at the heavy cloth and the better to keep their legs cool. All sported army issue hose and army boots, these being more like a clog than a boot, but none concerned themselves over exposing their legs to view. There were no men in immediate proximity and it would have made no difference, for all were well used to barrack life where due deference and concern for privacy existed only in their absence. All had their packs and bundles placed as carefully as any veteran soldier and all had their own water, including the children, who, in addition, all had their own bundle to carry, well

34

supported by strapping across their shoulders. Bridie Mulcahy and Nelly Nicholls walked side by side, Mary O'Keefe a little behind with the children. Conversation, even for these, had tailed off. The heat and the soft sand underfoot was taking its toll and so each looked forward to the next halt, because the next should be a major stop, enough to brew some tea and make some porridge. None had any kind of timepiece, but Nelly, being a veteran, had marked the possible time for a break by identifying a feature in front, some way ahead, about two hours marching further on. She felt the need to check and so she turned to the wagon, following immediately behind the children and stepped out, to shout beyond the pair of horses.

"Are we due a stop yet, Parson? Can y'tell, you having the clock?"

Percival Sedgwicke pulled out from his pocket the one relic of his once comfortable life as a Country Vicar, respected and unassailable in a remote Parish. He studied the elegant hands that slowly traversed the elaborate surface beneath the glass, then, after some elementary maths, made pronouncement.

"One hour and perhaps a quarter, or more, beyond that, Mrs. Nicholls."

The answer, including a fraction, meant little to Nelly; better to have heard the reply, "Not soon, but not too long". A whole "one" was inside her understanding, the concept of a "quarter", was beyond, but his tone sounded cheery enough, conveying that it would not be horribly far into the future. She looked at Bridie.

"Not long then."

"No, not long."

Sedgwicke eased his fingers around the reins that led down to the team that pulled their wagon. Beside him was Chaplain Prudoe and behind them both, in the shade, Prudoe's wife, Beatrice. Chaplain Prudoe could best be described as a "soft" looking man, pudgy from any direction; face, jowls and fingers, this confirmed by a full waist that bulged against his jacket. By appearance he was mid-forties, older than he actually was, and the growing jowls beside his face had turned what had begun years ago as a look of simple seriousness into a look of stern disapproval. He looked comfortable, but somehow at odds with all around him, as though all he encountered, should naturally be a copy of himself, but rebelliously was not. Sedgwicke, by contrast, was a figure best described as thin and wispy. Whilst Prudoe's hair was hidden under a soft forage cap, Sedgwicke sat bare headed, his thin, grey hair, tied tight in a regulation soldier's queue at the back of his head, whilst at the front sat his narrow face, always beset by a look of concern, but he was learning to make the best of things.

Prudoe was in a quandary as to how to address his Chaplain's Assistant. He knew Sedgwicke's background, that he was an educated man, but he was a defrocked Priest, and, above all, a mere ranker. However, he did not doubt that he was a devout Christian, a "Man of God", and as such, this elevated him, however slightly, above the mere chaff. Also, another factor was that Sedgwicke had been on campaign before, whilst for Prudoe this was all into the future and very unique. On the voyage down Sedgwicke had been a source of good advice, all deferentially and politely spoken, and so some form of relationship had formed between them, but, at that moment, the heat and the hard seat had ended any social intercourse almost before it had begun. However, with the strife of conflict very possible within the not too distant future, Prudoe began the questioning that had been increasingly preying on his mind since they landed.

"In battle, our role is to comfort the dying and administer last rites to the dead, that's well understood, but you've been in battle yourself have you not?"

Sedgwicke gave the reins a flick, being interrogated as if a mere underling always grated upon his own self-esteem, even much reduced as it was.

"Yes, I have."

A pause.

"Sir."

"What else could we be called upon to do?"

Sedgwicke's annoyance was heightened by a question that he judged as wholly facile, born of pure ignorance which an ounce of common sense could dispel. They were part of an army, that fights battles.

"Supplying the line, Sir. Water, ammunition, anything that may be needed."

Sedgwicke flicked the rein again.

"In a fierce conflict, that will take priority over any spiritual concerns for the wounded and dying."

A pause.

"Sir!"

Sedgwicke's tone of finality killed off that particular topic, but the sight of a church on a hill off to their left, reminded Prudoe of an ambition he had awarded to himself upon their embarkation, now weeks back.

"I hope to make a study of local churches, Private."

Prudoe had never used Sedgwicke's name since they had first met and the latter found this dimunition hard to cope with, especially with the instinctive status which he gave himself, despite this now being but a mere remnant of his earlier life. However, by now he had fully acknowledged that he was, indeed, a mere Private and that Prudoe was an Officer. Sedgwicke gave himself the time to form an answer.

"I'd prefer if you'd not ask me to join you, Sir. I count myself as very much Low Church and these Roman interiors are too far against the creed and customs of the Faith that I personally adhere to."

So eloquent a reply left Prudoe in a quandary. He could order Sedgwicke to accompany him, as his servant, but on the other hand he was faced with religious conscience, which he felt bound to respect. However, being an Anglican himself, this somewhat Puritan viewpoint was one that thoroughly jarred with him.

"I, myself, see no problem in worshipping our Lord Jesus in the best way our means allow. I find that your plain and poor interiors do little to convey the Glory of God, but I will respect your beliefs, even if I detect little foundation in them."

Sedgwicke looked straight ahead and said nothing, he had already pushed at the boundaries of due deference to rank, but his quandary was solved by Nelly Nicholls again turning around.

"How much longer, now, Parson, darlin'?"

Sedgwicke, with the sense of time possessed of a trained mind, had no need to extract his watch for another consultation.

"One hour, Mrs. Nicholls."

Nelly nodded her head and faced the direction of her march. "One hour" gave her a better idea of the time left, but her landmark, the one of her own that measured the hours between stops, was now giving her a far more practical notion of the time before they could, indeed, halt, rest and eat.

***

The army had slowly increased its distance from the coast, their route diverging away inland and the sea was now but a suggestion on their right hand side, indicated by the view now being practically flat, no significant upland, unlike the scene off to their left, which was one of impressive hills framed against distant mountains. The sinuous track of their slow advance had progressed steadily for the week now passed, but, if the reassurance of the fleet on the ocean was now missing, at least they were now on a decent road, where a foot placed forward did not sink down into a surface more sand than soil.

Far out on the left, inland of the column, the 105th Light Company formed the far left of the advanced picket line. Inside them were the four Companies of the 95th Rifles completing the picket screen. Carr had his men advancing in their files, making three ranks, covering a good 600 yards, for he had ordered 20 yards between each of the 30 files. He had placed himself out on the far left, accompanied by Nathaniel Drake and thus they were the most inland of all the infantry of the army, only the cavalry screen was out beyond them, still guarding their most vulnerable inland side, these horsemen being seen, then not seen, as trees and hillocks hid them from view. The day was again searingly hot, the heat building through the hours immediately after the Noon zenith of the uncaring sun, it, as always indifferent to occurrences below, remaining brilliant and dazzling, tracking its casual course off towards the Western horizon. The heat gave rise to the question most frequent in the minds in all in the Light Company; whether or not they could afford another swallow of precious water before their stand-down. Carr, as much to appease his boredom as for any other reason, pulled his Dolland glass from his waistband and extended it for a slow traverse of the country immediately to his front. He stopped as something white came into the lens and this soon proved to be buildings; they were approaching a town. Carr closed the glass and turned to Drake.

"We're approaching some town."

He raised his arm to point down into the valley ahead that harboured the road.

"Down there, in the centre, they may not have seen it. Do you think we should send a runner?"

Drake continued to stare ahead.

"Somewhat superfluous, I feel. Looks to me as though our cavalry are doing, have done, what they are out there to do."

Carr looked forward and couldn't miss what Drake had seen, a group of horse from their cavalry screen, namely the 20th Light

Dragoons, who were cantering diagonally across their front, down towards Fane and his Staff, these riding the road immediately behind the Riflemen still patrolling forward.

"Nevertheless, we need to be sure they're not riding back because one of them has cast a shoe or something. Send a runner."

The sarcasm was not lost on Drake and he grinned momentarily, before turning to the file immediately to his right. The leading man was Tom Miles.

"Miles. Get down to our Brigadier and make sure they know that we are approaching a town."

Tom Miles' mouth closed into an angry line at the task he had been given, but he was enough of a soldier to know that such a message had something missing.

"How far off, Sir?"

Drake raised his eyebrows and nodded.

"Good question. Tell them under a mile."

Miles hoisted his rifle sling onto his shoulder and grasped the barrel just up from the flintlock, this to steady the weapon, as he trotted off, not happy; this job would need a run of something over half a mile, in the heat. Mouthing obscenities at those he passed, all much amused at the discomfited Miles, he took himself down to his Brigadier. He was even less amused when he saw cavalry arriving first and knew full well that they were on the same errand, rendering his own wholly pointless. He came to the command party, stood in the road before them, brought himself to the attention and saluted. He was but a little out of breath, for running came easily to Tom Miles. He also knew the exact form in which such a message should be delivered.

"Beg pardon, Sir, but Captain Carr, 105th Foot, sends his compliments, Sir, and asks to report that he can see a town, about a mile ahead. Sir."

General Fane, appreciating the task just undertaken, had not the heart to tell this soldier that that which he was saying, Fane already knew, so he decided to send the messenger back with a message of his own.

"Very good, Private. So, return and inform your Captain that what he can see is called Leiria. Tell him, from me, that there is a road running inland from it and that he is to take his men onto that road and set up a picket to the East, using his whole Company. In whichever suitable place he chooses. Also tell him that the cavalry are being called in."

Fane reached into his waistcoat pocket.

"That for your trouble."

The coin was tossed accurately forward and Miles needed to do little other than to reach out his left hand to catch it. His face changed to one that almost conveyed pleasure at this gift from an Officer, one with a broad Scottish accent, which flavoured his voice with something approaching drone of a bagpipe. Miles saluted, which was returned, then he set off on his run, this time uphill, but in much better mood, if such could ever be ascribed to the character of Tom Miles.

Less than an hour later, the Light Company were stepping off the brown, dried, grass of the fields onto the white dust of a road that ran East and inland, running with little deviation, off over the hillside for some 500 yards. It found its way over using its own cutting, through an avenue of stray olive trees and drought stunted oak. Carr looked both ways. To his right the fields dropped further downhill, whilst to his left the ridge created a skyline with what looked like a building not too far off the road. His mind was soon made up, because, as the far left picket of the army, they should be higher rather than lower. He turned left himself and indicated for his men to follow. The cutting had been worn into the chalk by countless generations of feet and hooves, so Drake, and also Shakeshaft, at the suggestion from Ellis, sent several men up to the top of the banks on each side to maintain a better lookout all around.

They were nearing the summit when a dense column of horsemen came over the ridge, easily 50 strong. They must have seen them, but they had slowed not at all. The shout of "cavalry" came from many that were ahead and Carr, taking no chances, gave an instant order.

"Cavalry! Take post. Sunken road defence!"

His Company had practiced this several times, that of meeting cavalry riding down a road. All his men climbed the banks, each section lining the top of their own bank, four ranks deep, two facing in towards the road, two outwards, as if within a square. Carr drew his sword and remained in the road. He heard the orders being given to load and fix and soon the bank ahead and either side of him was bristling with bright steel bayonets at the end of muskets held "en garde". Carr waited and a small smile gradually spread across his face. These horsemen were from their picket, but he was going to make the point that no cavalry should have the temerity to simply ride up, en masse, without a word, through any picket line, particularly one commanded by him. Almost at that instant, their Commander saw the bayonets and the formed up infantry. He hauled his horse to a halt, the sudden stop causing confusion further back as horses and men careered into each other. Carr walked forward, looking at a face he thought he recognised and a few more paces confirmed that he did. He spoke in a leisurely drawl, his voice thick with irony.

"Halt! Who goes there?"

The Commander had recovered his composure and reinforced it with anger and indignation, but he had no choice but to answer the challenge.

"Picket! 20th Light Dragoons. Captain Tavender. And you?"

Now Carr really did smile, in his indulgent, even insulting, manner, which annoyed so many who believed unambiguous deference to be the far more appropriate response.

"Captain Carr. Light Company, 105th Foot."

Carr lowered his sword for its point to meet the ground, then placed both hands casually over the pommel. He then tilted his head to one side, whilst retaining his knowing, highly irritating, half grin. Recognition had come to him.

"You may remember, we met in Somerset, back in the year six. We were then called the Fifth Provisionals, you were in some kind of Yeomanry. Seems that things have changed, both for you and I."

Tavender refused to respond.

"Can't you recognise the King's uniform? Is that so hard? You should have cleared the road."

Carr adopted his most languid stance, the thumb of his left hand tucked into his waistband, sword casually sloped onto his right shoulder.

"Ah well, yes, sort of, but yes and no. You see, you know how you cavalry types like to dress up. To make a bit of a show, as it were. Which makes it a bit difficult for the likes of us to tell friend from foe, especially when you're up on a horse!"

He inclined his head insolently to one side.

"But, no matter, now that we know who you are, I think we can have no objection to your riding on through and so we'll allow you pass on."

Tavender's face reddened. "Allow" was indeed the word, because for cavalry, a sunken road was a death trap, if caught by infantry lined along either side. Carr sheathed his sword and stood aside. Tavender spurred his horse into an immediate gallop and all thundered past, but none, including himself, failed to spot the insolent signs and gestures delivered in their direction for the men above them. With them now passed and gone, Miles, out of the view of any of his superiors, was one of those who needed to pull his trousers back up from his knees. As in most armies, there was never any love lost between their cavalry and themselves, the infantry. In the same high spirits as his men, Carr crested the ridge, to see; yet another ridge. He turned to Shakeshaft.

"Get a picket out there, six men. Next, send Ellis down into the town to find the Colonel. We'll be up here for the night and so some rations won't come amiss."

***

Leiria had nothing to distinguish it from any other Portuguese settlement of equal size. It's buildings were solid but displayed no measure of affluence, preferring to demonstrate the virtue of endurance come what may, its buildings seeming to grow out of the ground as a low white wall, terminating in the cheerful bright terracotta of a full tiled, but very irregular, roof. As the British marched in, they were greeted by a population fullsome in their welcome, but doing their difficult best to convey it, as the dense columns of fours forced all out of the narrow street, therefore they had no choice but to hang from the windows and balconies, or crowd the entrances to the alleyways. They were more than relieved to see the anticipated red uniforms and so flowers and oranges were tossed into the passing Regiments or passed over at ground level. Carafes of wine were included and swiftly hidden away in the anonymity of the ranks.

Being one of the first Regiments to enter, the 105th passed completely through to the final buildings on the far side and here they were billeted, 20 to 30 to a house, forcing the occupants into small corners, but none billeted encountered any irritation, nor annoyance from those forced to relinquish space for their accommodation. Rumours of what happened under French occupation had reached this, as yet mostly untroubled, district of Portugal. Rumours were also circulating amongst the soldiers, especially when the order came, that they would spend but one night there only and they would march out on the following day, leaving the baggage, the heavy supplies and the camp followers behind.

All through the rest of the afternoon and the evening the army closed up and the orders were received regarding the strict discarding of any heavy baggage that would slow progress, that any such was to remain in Leiria. For D'Villiers and Carravoy, making the best of their hovel, or more accurately issuing instructions to their servant as to what should be done, this order was something beyond being merely irritating. It was D'Villiers' nasal whine which first registered their reaction, when their servant first spoke the news.

"What? Nothing more than pack and blanket?"

Binns, the servant, cowered back slightly at the growing fury, which was the inevitable response to what he was being required to impart.

"Yes, Sirs. Strict orders, nothing that will impair our mobility, Sir. Those were the exact words."

This was met with stony looks and a stonier silence, but Binns felt able to perhaps put a better gloss on things and spoke up, in his best cheery manner.

"It won't be so bad, Sirs, dry ground and warm and, in the Rifles, Officers is required........."

"Out! Get out!"

Carravoy's explosive bellow stopped him in mid sentence, enough even to turn him around to face the door and so, now facing it, he hurried out from the small, hot room, all the better to escape the growing rage that was building before him. Soon they would be throwing things, so a hurried retreat was now in order. Carravoy looked at the glum D'Villiers.

"Damn this Wellesley. No tent and no camp bed! Barely tolerable."

D'Villiers face matched his companion's anger.

"More like intolerable. And unheard of, to sleep in the open, same as the men. It's not to be countenanced, an offence against good order, is how I see it. What if we wake up face to face with ......"

Carravoy held up his hand, his face suddenly much brighter, he had had an idea.

"Not to slow things up, was the order. Right?"

D'Villiers nodded.

"Right."

"Then we'll hire a Portuguese servant and with him a mule! They can go anywhere. And….."

He paused for extra emphasis.

"……..they can carry our tent, the small one, that is, and perhaps some extra bits and pieces."

Then his face fell.

"But the beds, well, perhaps both a bit too bulky and awkward, I'd say?"

D'Villiers nodded agreement.

"Probably, yes, but, we'll see, we'll see. I'd prefer a bed to a washbasin! Perhaps just one and we can take it in turns, or something. Now, as you say, to hire a Portuguese and his mule."

He turned to the door.

"Binns!"

Binns heard, but was disinclined to react immediately; his soldier's patience was at a very low ebb with his two overly aristocratic charges. He had been watching the camp followers passing his door and felt more inclined, for this was where his sympathies lay, to answer the question of Bridie Mulcahy and Nelly Nicholls, who had spent the last ten minutes asking all and sundry where Number Three Company was, with Mary enquiring the same for the Lights. Knowing that the followers were to remain behind, Frederick Binns gave them full directions before answering his masters' call and soon the two matrons were leading the children where they needed to go, but with Mary tailing on in a state of great disappointment. However, she cheered up when Jed Deakin told her where the Lights were and the route they would take through the town come morning. The cheer increased when all sat talking by the firelight with bellies full of beef stew thickened with biscuit, and peas!

The following dawn saw the same forward brigade leading out from the town. The 105th Lights had passed through and then out, whilst darkness still reigned and Mary had to run to catch up with Joe just as they were dispersing beyond the town into skirmish order. After hugging him as best she could, because his equipment prevented her from getting her arms fully around him, she pressed on him a hollow loaf, filled with boiled beef, this being a useful portable food as taught her by Nelly Nicholls. A Corporal began to rebuke Joe Pike for his tardiness, but his harsh words were soon halted by Tom Miles.

"Leave him be, Corporal, he'll not be late, I'll see to that, and he has more of a goodbye to say than most!"

Both took their leave of Mary, Joe with a kiss and Tom with a smile and a wave. Soon they were out on picket, as before, patrolling inside the cavalry. At Noon the army halted for a welcome rest and a meal. For the afternoon march there were changes, the full battalion of the 60th Royal American, were out on picket, which placed the four Companies of the 95th at the head of the column, then the 105th next, with their Light Company leading their Battalion. Early evening saw their arrival at another town, but very different from Leiria and depressingly familiar to any that had followed the French into several towns across Europe. Every door was smashed in and some buildings still smouldered; the wood, thatch and the furniture that once created peaceful domesticity within the stone walls now continuing to throw dark blue smoke at the clean blue sky. Blackened window shutters hung at drunken angles and many small personal possessions, looted, but quickly discarded, lay in the road. At the head of the 105th column, Davey, Miles and Pike looked knowingly at each other, but it was Miles who spoke.

"Nothing different about these Frenchers!"

They had seen similar, in fact worse, when campaigning in Italy, but they had just reached the centre of the village, when people began appearing, walking back apprehensively, carrying large bundles, leading precious animals and guiding small carts, either propelled by themselves or with donkeys between the shafts. Once more the 105th were to pass fully through to the furthest buildings, there to make their billets in whichever constructions presented themselves and, to the joy of the men in the columns, the locals soon proved to be as welcoming and friendly as those at Leiria and, miraculously, as so they thought, the 105th received gifts of wine, bread, sausage and fruit.

A well-appointed stable was the billet of Chosen Man John Davey, Miles, Pike, Zeke Saunders, John Byford, and Private Leonard Bailey, an "Old Norfolk", taciturn and dour, but a good man in a fight. As the first three settled to boil their salt beef and peas, augmented by gifts of sausage, Byford returned to their billet with his messpot full of very hot, aromatic and, seemingly, very appetising stew. Miles was the first to enquire what it was, suddenly in umbrage over what he was plainly missing out on.

"What's that?"

Byford hurried on to Saunders and Bailey.

"Portuguese stew. Pork, beans, sausage, and something called tomate. That's the red element contained therein."

"How'd you get it?"

Byford set down the cauldron in front of his messmates and looked indulgently at Tom Miles, which did nothing to ease his mood.

"It's amazing what currency is gained in foreign parts by a few gee gaws and some odd bits of coloured ribbon!"

Tom Miles looked daggers at Private Byford, who had clearly worked a move that he, himself, would have been thoroughly proud of.

"How'd you make yourself understood?"

Byford returned a querulous look.

"Before we sailed, I bought a book, "A Treatise on the Portuguese Language", I thought knowing a few phrases might prove useful, and so it has transpired."

Tom Miles evil look turned to contempt, but it was benign, and comradely, containing no deep animosity.

"Book learned bastard!"

The reply was a small laugh as the stew was spooned out, but Tom Miles knew his due.

"Well, give us all a taste, then!"

Some was deliberately left over and carefully apportioned between Davey, Pike and Miles. All agreed that it was "estufado comida" of the first quality; they each had soon discovered from Byford the Portuguese words for "stew food", but Miles had not finished.

"How come this lot got out before the Frenchers showed up? Did you find that out?"

Byford nodded.

"Well, as best I understood, the French massacred a place called Evora, down South. The story has spread, and when they knew the French were coming, they got out, each and every one, taking all their stores and food up into the hills. That's why they're being generous now, that's my guess, for us keeping the French at bay."

Miles took a swig from the wine bottle.

"I'll drink to that."

They were spending the night in Alcobaca, the pronunciation of the name again taught carefully by Byford.

"The last 'c' is said 'th'."

This produced yet more surly comments from Miles.

"Too bloody clever by half. Brain that size is like to get his head blown off!"

Chapter Two

First Encounters

The army woke at the customary dawn but did not march immediately. The order came for all soldiers to be checked, ready for combat, and rumours spread that the French may still be nearby. In response each Section of the Light Company was inspected by their Sergeant, a prefect opportunity for Ellis to annoy Miles by inspecting every inch of him, pulling at straps, and finally thrusting a feather into the touchhole of Miles' Baker Rifle. The fact that it came out clean did nothing for the satisfaction of either; Ellis would have preferred it dirty and Miles was wholly affronted at the very idea that such an important part of his weapon required inspection in the first place. However, both then counted the confrontation as a honourable draw. Davey's inspection was far more cursory; he dwelt thoroughly within Ellis' regard and respect. Joe Pike's was equally brief; Ellis trusted Davey to get him up to the mark!

The 105th were on the road, ready and waiting when Fane rode up and halted before Lacey and O'Hare.

"I've sent out all of the 60th and the 95th as an advance guard. As I'm sure ye're aware, the French were here just before we arrived and reports say that they've withdrawn to a place called Caldas, about one hour ahead. Proceed close enough to give the Rifles immediate support, should it be needed. In close column, ye may need to quickly form line, facing ahead, if they get into difficulties. General Spencer's Brigade is right behind ye."

Lacey and O'Hare both saluted and Fane pulled his horse around and rode off. Lacey looked up the road, then at O'Hare.

"Get them formed up, then march off, quick time. That must be the Rifles, up there."

Both could see the green clad figures, now far ahead and almost out of sight, their uniforms adding to the affect of the distance, them now blending with the olive groves, but Lacey had more concerns.

"Send out Carr's two sections either side. I'll feel better with someone out on either flank."

Soon, urged on by O'Hare, the 105th were setting a good pace down the road, with their Light infantry either side in the flat, dry fields. Carr was in the centre, with the Battalion Bugler, "Bugle Bates", both being on the road and just in front of O'Hare. They saw nothing of the Rifles, not even when they came to Caldas, wrecked and looted as Alcobaca had been, but at least not burning. No civilians could be seen.

They passed quickly through and Lacey, riding back and forth on a good mare, was pleased to see Spencer's Brigade holding the distance between themselves, but still a good way back, they must have started sometime after the 105th. Caldas was left behind and they pushed on through open country. Deakin was not happy, marching swiftly besides Ensign Rushby, but it was to Halfway that he spoke.

"Them bloody Rifles, they must be keen as mustard to meet up with the Johnnies, too keen, they'm stretchin' out too far. It can only end bad, with good lads dead, that's how I sees it."

Halfway's worried face showed that he agreed. Meanwhile at the head of the column, remaining just ahead of O'Hare, Carr was thinking the same, but, for now, keeping it to himself, being content to push forward to catch up with the advanced Rifles. Another small village came into view and the 105th now seemed definitely to be drawing nearer to their advanced guard, who were now closing up their files in front of the village, as if for an assault. Suddenly, they saw the white smoke of gunfire, then the sounds of combat reached them and Carr stopped and whipped out his Dolland glass to see the detail. The Rifles were advancing forward by files, the leading man firing, then the next advancing past to stop, fire and load. Some green shapes were already lying on the grass, some still, some writhing in agony, but the attack was being pressed forward.

O'Hare reigned in his horse besides Carr. He was worried himself and, being unknowing of what was ahead, he was unsure of what to do.

"Carr. I'm assuming that the Rifles will clear that village and stop. That would be prudent. Lead the column through with your Lights and secure the far side alongside them, if they do what's sensible, that is, to halt and hold. I'm riding back to relay that to other Captains and to the Colonel. Your thoughts?"

Carr was surprised that he should be asked, however obliquely, to pass judgment on his superior's orders, but he saw no cause for concern, other than that which they both shared; what would the Rifles do?

"A safe assumption, Sir, I would've thought. So, as you order, Sir."

The two exchanged salutes and O'Hare turned his horse to ride back, leaving Carr to lead the Regiment into the village. From the village came the sounds of heavy fighting, but no soldiers of either side could be seen.

"Bates. Call them in."

Bates licked his lips and sounded the notes of recall. The two sections came rapidly back to the road and formed into a column. Soon they were passing the casualties of the conflict, first their own Riflemen, then, inside the village, more and more French, with a few British casualties, some of both dead, some grievously wounded and probably dying, some lightly wounded and trying to tend themselves. All the unscathed Riflemen had passed on and, judging by the sounds ahead, they were still fighting. It was but a small village, Brilos, Carr assumed it was called, this being written on the most imposing building alongside the Church. He came to the final outskirts and saw what he hoped he would not. The Rifles were out and beyond, in the open fields. He halted the column, but gave no order to take defensive positions, realising that they may well be ordered to march on, in which case they needed to remain closed up, as they were and ready to hurry on. He was grateful when both O'Hare and Lacey came riding up.

"The Rifles have pursued further, Sir."

He felt the need to state the obvious, but he noted the consternation on the face of both his superior Officers. Lacey spoke his concerns.

"They're going to meet more than they can deal with and be sent tumbling back, most likely. Us remaining here may be too far back to give them aid. It's safe for us, but we will not be supporting them. Spencer is some way back, but I feel we must go forward; you'd agree O'Hare?"

"I do, but what's best? What formation?"

Carr, with the advantage of knowing the situation before Lacey and O'Hare, had been giving the problem some consideration of his own.

"Sir, may I suggest dividing us into our two wings, then advance in firing line, one behind the other. If the Rifles are forced back, they can form on the advanced half, and, thus supported, that wing can hold against any French advance for a short while, then fall back upon the second, stood waiting as a final rallying point. The two now together forms a full line. Add on the Rifles who should have rallied and soon after General Spencer should be up. Sir."

Lacey and O'Hare looked at him, then at each other. The judgement was made in an Irish accent, with a resigned intonation

"Sure, that's as good as anything I can think of."

"Right, first five Companies forward and into a double rank firing line, then advance. Final five, the same, but three hundred yards behind. You'll lead the first line."

Both saluted as Lacey rode off to tell all elsewhere. O'Hare looked at Carr.

"Then onward, Captain. You take left of centre, Number Three the centre and the rest as they come. Quick time, I think."

The column trotted forward and the formation was changed quickly and perfectly, this much to O'Hare's satisfaction, viewing all from the height of his horse.

"Ah now boys, fine work, fine work indeed! The Prince himself would be clapping his bejewelled hands together at the very sight. Then raising a glass or ten to the toast of the green of the Wessex!"

Each company, most laughing at O'Hare's humour, swung off the road, in column of fours marching right or left according to O'Hare's directions. This initially placed each Company, as they halted, in a four deep firing line, but the rear two ranks ran further on to extend the line of the front two. Waiting left of centre for the line to form, Joe Pike was stood beside John Davey, Tom Miles behind in the second rank. Scattered casualties, mostly French, littered their field and before them was a dying Frenchman, his cheeks and eyes already sunken. He could barely lift an arm, but he did and also manage a croak.

"L'eau, pour l'amour de Dieu. L'eau."

Tom Miles was in no good mood.

"What's he want?"

John Byford was two files away.

"Water. He says he wants water. For the love of God!"

Joe Pike set the butt of his rifle on the ground, pulled around his own canteen and stepped forward. He had not taken two paces before he was halted by a deep Yorkshire growl.

"Hold tha' place!"

Pike stepped back.

"Leavin' of the ranks; that's a floggin' offence!"

The voice was unmistakably that of Sergeant Major Gibney, who now stood immediately before Pike, looking down into his face, from a distance of mere inches, mouth, eyes, brows and chin all regulation furious.

"Tha' knows!"

But Gibney straightened, turned and took himself to the dying Frenchman, to pull the canteen from under the prone body and give the man a drink.

"Merci. Merci."

Gibney nodded and propped the canteen against the man's side, then marched on to his place at the end of the line.

In two brief minutes O'Hare commanded a closed up, two deep firing line, of just over 400 men, extending 200 yards. He sat his horse before the centre, drew his sword and motioned his men forward. Deakin, beside Ensign Rushby in the centre of Number Three, knew full well what should have happened already.

"Time to break out The Colours, Sir. You too, Mr. Neape, Sir."

Consternation came over the face of both; this would now have to be done as they marched forward, to pull off the long leather cover at the end of the awkward flagstaff. Finally, clumsily, it was done, but not without the aid of both Colour Sergeants. There was little breeze but just enough, just enough to extend both huge squares of colour out above the heads of those who bore them. On the right was the Union Flag with its centre a circle of olive leaves and the roman numerals CV; this carried by Rushby. On the left was the Regimental Colour, a square of bright emerald green, but not entirely, for it had a Union Flag in the top corner, the top quarter nearest the flagstaff. The bright green matched the facings on all cuffs and button stripes, the Regimental Colour of the 105th, chosen by the Princess of Wales. Now, with all as it should be, they dressed their ranks and marched forward, but now with no embarrassing confusion.

"All set then, Sir."

Deakin spoke cheerily, for, across from Rushby, Ensign Neape was not looking at all well, but he was cheered, after a fashion, by Captain Heaviside, marching at his side.

"Be not moved away from the hope of the gospel, Colossians, 1, verse 23."

Neape looked askance at his Company Captain, unsure of the correct response, but at least it distracted his mind from the sights waiting ahead.

Soon, they had advanced forward almost half a mile and the sounds of skirmishing were still coming from ahead, but, without warning, the sound intensified as if from a set piece battle, whole battalions in volley fire. Veterans, including Carr, O'Hare, and Deakin knew what this meant, but it was Deakin who spoke out loud.

"Right. This is it."

O'Hare raised his voice without turning around.

"Halt."

A slight pause.

"Load!"

The neat appearance of their formation broke as each man began the process of loading their weapon, a 20 second operation, maximum, for a competent soldier, the seconds of which O'Hare counted.

"Shoulder arms."

All weapons went to the vertical, supported by the butt cupped in the palm of the right hand.

"Lock on"

The rear rank shifted slightly to the right to now enable them to see between the heads of the front rank. O'Hare dismounted, sent back his horse, then stood and waited while Carr looked over his shoulder and through the heads of the two ranks behind him. Lacey was doing the same for his "wing" of the battalion. Of Spencer's Brigade there was no sign. Carr faced his front and saw what he knew was coming; the Rifles were pouring back and a long rank of blue French uniforms could be seen beyond, seen through the gaps in the Rifles disordered ranks. Another volley from the blue line saw more Riflemen fall, but the survivors were soon ordering themselves into files of three and they began a rapid and orderly retreat, still keeping up an incessant, if light, fire upon the French line. Soon all Riflemen were in retreat and ran quickly back to the 105th, not to run past, but to extend their line. Highly trained soldiers all, they knew what to do and quickly formed two ranks and their numbers doubled the length of O'Hare's 'wing", half the 105th.

Gibney, out on the left, looked with great displeasure and disgust at the scruffy, powder stained Riflemen forming up at his side.

"If I ever sees thee again, make sure thee has th'sen far more gradely in appearance, thee shoddy bag of bobbins!"

The Rifleman took no notice, being far more concerned to load his Baker Rifle. This done he looked at Gibney.

"You a Sergeant?"

Gibney fixed him with a look that would have boiled milk, but he did answer.

"Aye!"

"Then how come you've no sponson, your spear? You've got a musket."

Gibney leaned down until their shakoes touched.

"Dos tha' Sergeants carry sponsons?"

The Rifleman shrank a little.

"No."

Gibney's gaze intensified, to one that would melt rock.

"An neither does we, in the One-Oh-Five. An' that's a battle honour! Tha' knows!"

The Rifleman nodded submissively and looked to his front. Gibney allowed himself a look of further disapproval, then did the same.

O'Hare had more weighty concerns. He could only delay the French; ambition could not run any further, he could not defeat them, only hold them up and gain enough time to make their own orderly retreat back to Lacey. There were no more Riflemen before him, at least not alive and the French were nearing 100 yards, with more coming on behind. Before came their customary Tirailleurs, sharpshooters in skirmish order, whose bullets were zipping past, to the side and above, and he heard a grunt behind him and the clatter of a soldier and his equipment hitting the ground.

"Make ready!"

All weapons came to the vertical, fingers on triggers, musket muzzles now high in the air. O'Hare retreated to the Colour Company.

"Front rank, present."

The first row of muskets came to the horizontal.

"Fire!"

There was a vicious explosion of sound, then all before was dense white smoke.

"Rear rank, present."

The muzzles of two muskets from the rear rank came down either side of his own face. He allowed the smoke to clear a little until a vague blue line could be seen.

"Fire"

The muzzles blasted forth more flame and smoke to add to that before and O'Hare counted ten seconds. By then, he knew that the front rank would have reloaded and so, satisfied, he ordered one repeat volley from both front and rear ranks.

"Fall back. Form on the Colonel."

The endless practice on the hills of Ireland was paying off. The firing line dissolved, taking the Riflemen with them, to reform on the ends of Lacey's line, back waiting 300 yards behind. With the line formed, Lacey took over, but he remained patient, awaiting events. A minute passed, then two. The wind that had extended out their Colours took away the smoke to reveal the blue line still there and still advancing, but O'Hare had held them up just enough. This time the French were faced with a full battalion, greatly extended by the green uniformed Riflemen. Lacey looked along his line, off to the right, then the left, all was in place including The Colours; his men had formed up, muskets vertical at the 'make ready", no orders had been needed. The French were not done, but this time Lacey was content to allow the range to fall, this time the blow would be harder.

"Front rank. Present!"

The order made its progress by example down the long line as he waited to see what the French would do, he knew that the moment they stopped was when they would prepare to loose off a volley of their own, but for the moment he was content to let them do what they were doing; coming on, with their muskets held diagonally across their chests. Then he saw what he had been waiting for, their Officers turned to face their advancing men, with their swords horizontal; the French were forming their own firing line.

"Fire!"

At 70 yards range, the volley was merely destructive, not devastating, but the French line had been dealt a severe blow.

"Rear rank. Present."

A pause as the hammers of almost 800 muskets and rifles were cocked fully back.

"Fire!"

Again the roar of noise and smoke and then, for some minutes, Lacey maintained fire from both ranks, finally allowing his Company Captains to control the fire of their own Companies. Soon there was nothing but the ear splitting noise of continuous firing up and down the line as Companies fired independently, the Rifles gleefully adding their firepower, but then a voice came to him from the rear.

"Sir. We're being reinforced, Sir."

Lacey turned and pushed his way back through the double rank to see General Spencer and his Staff at the head of his own Brigade, and his battalions were immediately running off both left and right to form their own line. The General cantered over and Lacey saluted, but Spencer asked the first question.

"How many are you against?"

"I've seen at least two battalions, Sir, but I think we've now held them up."

Spencer nodded.

"Good. Well done. Pull your men back and we'll take over, but remain in line as a reserve."

Lacey thought that any form of withdrawal was easier said then done, but, by the time he had returned, O'Hare had ordered their own men to break apart into Companies, to allow Spencer's individual Companies to run between the intervals and form their own firing line. It was smoothly done, but, in fact there was little left of the conflict. The French, seeing the reinforcements, had the choice to feed in men piecemeal and begin a battle against an enemy able to do the same, or pull back to perhaps better positions. They chose the latter and soon there was little to mark their presence, bar a long line of dead and wounded where they had been hit by the volleys of the 105th and the Riflemen.

In the Colour Party, the Colours were lowered, the brass ends of the staffs placed on the ground. Rushby leant wearily on his, whilst Neape, emerging from a mental daze accompanied by a loud ringing in his ears, examined each part of himself and found no damage. However, this was not the case when he examined his Colour; the wide expanse of emerald green. Four musket balls had passed through, each evidenced by Neape thrusting through his finger, which immediately rekindled his state of shock. Deakin looked along and found Rushby examining the silk of his for the same.

"Ah, now don't thee worry too much about those, Sirs. There'll be plenty more of that before we're done 'ere, an' my Bridie will take care of those, Mr. Rushby, Sir, and I'm sure we'll find someone for yours, too, Mr. Neape."

***

"The General's come up."

Lacey looked across to his left, across the now ordered lines of dead and wounded, some with light wounds being attended, some awaiting their turn with the surgeon's probe, knife, or saw. Beyond the sorry scene, Wellesley was talking to the Colonels of the two Rifle Regiments, or, more accurately, thoroughly conveying his displeasure, his wagging finger and aggressive stance fully conveying the fact that both Senior Officers were the recipients of a savage dressing down. That done, and with a final nod of his head, Wellesley mounted his horse. Both Lacey and O'Hare remained watching, but it was O'Hare who first predicted what was coming next.

"Oh God and b'Jesus, he's coming over!"

Lacey gave no reply, being engrossed with his own thoughts, until one fear spoke out.

"Will he say that we were too far ahead?"

"We'll know soon enough."

As Wellesley and his Staff rounded the rows of casualties, both Officers came to the attention and saluted, the reply being a touch of a riding crop to the peak of Wellesley's bi-corn.

"Lacey?'

"Yes Sir."

"And your Major?"

"Sir, may I have the honour to present my Senior Major, Padraigh O'Hare?'

O'Hare bowed his head, and Wellesley eased himself in the saddle, both hands now on the pommel.

"You handled your men well, so I hear. Well done."

A very slight pause, but inadequate for any reply.

"Fane has your orders."

With that he pulled his horse's head around, leaving the two both astonished and saluting. With Wellesley now leaving, Lacey looked at O'Hare, both exhaling a sigh of relief.

Fane was nearby; anxious himself to hear Wellesley's words to the final pair of his immediate inferiors, but his first words explained what had just happened.

"The Rifles ran into the whole French army, in front of Obidos, two miles up. They were roughly handled, but at least we now know where the Johnnies are. The General's going to wait, now, for the rest of the army to come up throughout today and tomorrow, then we move. Stand your men down and camp here."

He paused and shifted his feet.

"Blake and Webster of the Rifles send their thanks. The sight of your men stood formed up as they fell back was most welcome. Well done from me, too."

Lacey and O'Hare spoke their thanks in unison, then Fane departed and they were left alone.

"Right. Stood down for a day, or more!"

Lacey looked around, then spoke further.

"Tell the men to make camp and take their rest. A "make and mend", I think the matelots call it."

The order soon circulated and the 105th made camp more or less where they had fought, although many moved on, the sounds of the surgeons at work gave them a chill presage of their own possible future. It was exactly such a group that strategically set themselves up near enough to the road to immediately see the arrival of supplies, but sufficiently out of the way for peace and quiet. This group which habitually cleaved together consisted of; Davey, Pike, Miles, Byford, Bailey and Saunders from the Light Company and, from Number Three; Deakin, Halfway, and "the twins", Alfred Stiles and Samuel Peters. Although wholly unrelated, the last two were practically alike in appearance, speech, stature and their outlook on life. The first thoughts of all were relief at seeing each of their mates still alive, but they soon fell to squabbling about sharing out biscuit, dried salt beef and water, until rations arrived. However, the whole group was cheered up by the arrival of a "happier" Tom Miles from the site of their conflict. He'd found a Light Infantry shako that almost fitted, but just too big, it needed but an extra sweat band sewn around inside, which he'd torn out of another, but then the squabbling began again when he found that he did not have enough black thread and needed to borrow some.

Elsewhere, others settled to post battle activities. Carravoy and D'Villiers sat on their single campbed, both still shaking, but strangely euphoric and this not only stemming the simple reason that they were still alive. This was not their first battle and so the sights and horror had not impacted upon them so powerfully, but they had still been frightened to the point of terror. However, they consoled themselves that their Grenadiers had fought well and they, despite their almost paralyzing fear, had given the correct orders and stood with their men. As they sat in silence, each was now grateful for the mug of coffee prepared them by Binns, himself covered in blood, the result of the death of the filemate who had been stood in the rank before.

Blood was very much on the mind of Chaplain Prudoe. The wounds and injuries of the men surrounding him were enough to turn his stomach, to the extent that he felt unable to approach those whose wounds were clearly fatal and had little time to live. He had been able to bring himself to stand and say the Last Rites, but had been unable to emulate Sedgwicke who knelt besides all, this evidenced by the bloodstained knees of his trousers caused by kneeling on the blood-soaked earth, the better to help them to drink and ease their suffering in whichever way he could, either by a drop of rum, or a drink of water, or a few comforting words. Many greeted him familiarly, "Hello, Old Parson," and, for one or two, those were the last words they ever said. For Prudoe, whilst uncomprehending at his own lack of ability to administer The Sacrament to dying men from a position close enough for them to feel the comfort of it, he reassuringly consoled himself that Sedgwicke was "from amongst them, after all," therefore more insensitive and unconcerned by the close proximity of men choking their last breaths on their own blood. He moved on and around, convincing himself that his simple presence and the enunciation of the correct spiritual formula constituted duty done. Meanwhile, back in Leiria, news of the conflict was reaching the camp followers and all took themselves to the road, to garner news, especially from any in the carts of wounded that had facings of emerald green.

The day wore on. Brilos was too small to accommodate a whole army and so those of the 105th enjoyed the entertainment of watching other Regiments march through and off to one side or the other, there to make their own camp. Come nightfall, the sounds of the surgeons at work ceased and all slept undisturbed, bar Lieutenant D'Villiers when Captain Lord Carravoy woke him to claim his time with the camp bed, this occupying one half of their tiny tent. After breakfast, knowing that a battle was imminent, Lacey ordered another check, this time to be carried out by both NCO's and Officers, the former checking the verdict of the latter. Wellesley rode back from examining the French, accompanied by his Staff and a squadron of the 20th Light Dragoons. Those who could, including Davey's group by the road, examined his face for any sign, but there were as many opinions as there were people making them. With the dying of the day, orders arrived and Lacey and O'Hare read theirs. They were to be withdrawn from Fane's command and placed as the reserve battalion for Brigadier Caitlin Crauford, which was itself in reserve on the left, behind the Brigade of General Nightingall. Fane, at the moment of their reading, came to their small camp, looking none too pleased, a mood added to by the growing dark.

"Ye've read your orders?"

Both nodded and Lacey answered.

"Yes Sir, but it doesn't say why."

Fane took a suck on a pipe that he had been cradling in his right hand. The exhumed smoke added to the gloom between them.

"Ye're in reserve, at my request. That was a good fight ye put up yesterday, ye don't deserve involvement in another, let some others tak' a turn. Besides, my Rifles have been given a skirmishing job out on the left, nae place for a full Line Regiment such as yourselves."

This was the first time the 105th had been recognised as such and both Lacey and O'Hare felt a small swelling of emotion, but Fane had more to say.

"Johnny has set up a defensive line beyond Obidos, defending a village called Rolica. My men are part of a double flanking move on the left, ye see, because out even beyond me, further off in the hills, will be two Brigades under Ferguson and Bowes."

He took another puff, then spoke as he exhaled the smoke, including a repeat of what they already knew.

"Just tae put ye both fuller in the picture, Crauford will be in reserve for the left centre behind Nightingall, with Hill right centre. On the far right, far out, will be Trant, trying the same as my men, but with Portuguese brigands."

He sucked expectantly on the pipe and seemed nonplussed when he exhaled no smoke, but his description continued.

"Wellesley means to gobble them up with a double bite!"

Another puff, but the pipe was dead. Fane fished for his tobacco pouch, but O'Hare handed over his own, and while Fane affected a re-fill, Lacey went for the candle lantern, the better to read the order.

"Sir, there is something here that I do find, er, unusual. This last sentence, "All Regiments are required to advance in their best order and to maintain the highest standard of drill and appearance when within sight of the enemy".

Fane was still filling his pipe, but he answered.

"Wellesley's hoping tae mesmerize the French by what they will see facing them from in front and not notice Ferguson and the Portuguese out on their flanks."

The pipe was filled and, after a burning brand fetched from the fire had been thoroughly applied, Fane gave his opinion.

"And good luck to us all with that one!"

With the pipe lit and smoldering, Fane took his leave, a cloud of aromatic smoke mingling with the darkness. Equally unconvinced, Lacey and O'Hare took some supper and went to bed, which meant being rolled up in their blankets fully clothed, lying amongst their men.

***

The following day proved to be a bright one, perfect for Wellesley's attempt at his "coup d'oeil". To become almost the last battalion in the formation, the 105th watched and waited as the rest of the army marched past to take up their positions. All were immaculately turned out, with perfect dressing of their ranks, Colours brave, prominent, and eye-catching, reaching out above the ranks in the good breeze. The army marched in column for about two miles, through Obidos and then deployed. The 105th had but little to do, other than to march out to their left and position themselves behind the 91st Argyllshire Highlanders. 'Scots girlies' according to Miles.

Caitlin Crauford galloped up with a small Staff. This was the first time they had seen him and he reminded Lacey of an angry gamekeeper, accosting a group of wayward picnickers.

"Lacey?"

"Sir."

"Keep up close."

With that he was gone and, with all arranged and resplendent, the army advanced, down the floor of a "horse-shoe shaped" valley, with hills on both sides. Officers and NCO's were soon screaming about alignment and dressing, as the ranks broke up to negotiate the variety of obstacles; buildings, paddocks, and ponds. As they pressed on, with shouldered arms and maintained step, from time to time, they could see, off to their left, in the hills, the red columns of Ferguson and Bowes, but Fanes Rifles, although closer to them, were blending with the dark green hills of the background. For almost an hour nothing happened, then white smoke burst out on the near hillside, finally revealing the whereabouts of Fane and his men. At that moment, coincident with the sound of musketry, O'Hare, on a burnished horse besides Lacey, noticed a substantial group of houses appearing on their right.

"I do believe that to be Rolica. Shouldn't we have done some fighting, to get up to here?"

"I feel you're right, but look over yonder; you can see Ferguson. He's come in from the hills to spring a trap that didn't work, but, if I'm not mistaken, he'll be sent back out. Johnny's not been taken in, he'll now be pulling back to a better position. No guesses, Padraigh, but I'd say that one, right ahead."

O'Hare looked in both places, the first to see Ferguson marching back to the hills, the second to see a formidable ridge, split, but only far up, by four small gaps or ravines. Through these, blue uniforms could be seen filing up and in, taking up positions in a position wholly formidable for defence. A windmill stood on the lower heights of the far left, presumably hurriedly abandoned by its miller owner, its sails still mournfully turning, waving it's distress at what had come upon it, in all its destructive power.

"This, I don't like! Sure, that's the same as a fortress, only twice as high!"

The order came to halt from a rather distraught aide-de-camp, riding up on a huge horse and, as he departed, Lacey gave the order.

"Stand down, make a meal. There's going to be some changes."

The morning had been wasted, the French had stood for just so long, warily suspecting a trap inside the shallow valley, and then they had expertly pulled themselves out, leaving the British stood as bachelors at a dance with too few partners. The men sat and lounged around their fires and, after eating their stew, they cleaned again what was already clean and drank water flavoured with wine. Officers sat on whatever could be found, most often furniture brought from the houses of Rolica. Somehow appearing both strained and at leisure, they drank tea after consuming their own stew, but, for them, there was the addition of fruit, picked from the trees all round. However, both men and Officers had one thing in common, they looked with foreboding at the French position, from which could be seen the Frenchmen's own campfires. Joe Pike was sat alongside John Davey, each mirroring the other's position, each had their rifle erect between their knees, hands grasping the barrel just far enough up to support their chins, but Davey's outside his right cheek, Pike's outside his left. Neither would leave their precious Baker lying in the dust.

"Will they be sending us up there, John?"

Davey lifted his face from the rest provided by his hands.

"Well, one thing, we'll not be first, boy, not from back here. That's the job of those up front, and, for it, I'd not swap places."

He chased a flea down the back of his neck, then spoke from a strained mouth as the chase proved difficult.

"But we'll be up there, sometime, no fear of that. Those lads out in the hills will winkle 'em out sometime, let's hope that happens before the likes of us have to tackle the thing head on!"

With that, he stood and went off to a nearby stable to relieve himself. Pike remained in gloom, his head sunk lower on his hands, dreaming of his Mary. Meanwhile, from their higher position, sat on chairs, Lacey and O'Hare could see that their rest would now be soon over. A battery had arrived on the only small hillock besides Rolica, a site now shared with another windmill, built on this high ground near the village. They watched the guns being unlimbered, the horses led to the rear and the guns loaded. An Officer checked the elevation and alignment of each, took himself beyond the rightmost gun and then raised his arm. His cry of "fire" was dwarfed by the shattering crash of six field guns firing as one, then the gunners moved to rapidly reload. Lacey got to his feet.

"I think that opens The Ball. Get the men formed up."

The men had seen for themselves the arrival of the battery and all eating equipment had been immediately packed. They had only to stand and re-form their ranks, their assembling being copied throughout the whole army, but few could bring themselves to raise their eyes and study the formidable sight of the ridgeline topped with French uniforms. The guns fired again, but the effect of solid shot against an open hillside was impossible to see, but this time, French guns answered, not at the battery, but seeking targets somewhere on the British right. These guns must have been sited there early in the day, proving that the French had fully planned to make their stand on the ridge, not before as Wellesley had hoped. The aide-de-camp arrived again.

"General Crauford wished to keep you aware, Sir, that he is now the centre, with General Hill on the right. Hill's 5th will attempt the gully on the right, the 29th will attempt the two gullies right centre. The 82nd will try the left hand gully and the 45th will make an attempt up the height with the windmill, the farthest left. You are to support the 45th, Sir, therefore if you could position your men, Sir. Those are the 45th you can see now, Sir, to your immediate front."

The aide-de-camp pointed to the rear of a firing line now being exposed, as the 91st marched off to the right to support the 82nd, these now drawn up before their own gully, the most leftward of the French position.

"Very good, Captain. Now I would be obliged if you could return via the Colonel of the 45th Nottinghamshires and inform him that the 105th Wessex is close behind and he can be assured of our fullest support."

The aide-de-camp fully appreciated the emotion within Lacey's request and saluted smartly, then, as he left them, Lacey looked at O'Hare.

"If we are to go up that narrow slope, our ground will be constricted. The Lights and Grenadiers will lead off first, skirmishing order across our front. Then columns of Companies, six Companies, with two held back, in reserve. Make those Three and Four. The six Companies in formed fours, there'll be room for nothing wider."

O'Hare ran off, calling for the Captains. Soon the 105th were formed up, a main line of six columns, each company forming its own column, merely four men wide. The 45th moved forward as the guns sent round shot over their heads and, 400 yards distant from the ridge, the 105th halted, leaving the 45th to move on. Finally the effect of one British shot could be seen; the windmill was at last put out of its misery, a shot hitting the centre of the arms sending at least two to the ground. This was not lost on Carr.

"That's scotched some poor Devil with a telescope."

He was using his own telescope, from a position almost equally perilous, stood up on the corner fencing of a paddock, leaning for extra balance on a long pole. Drake was impatient for news.

"What can you see? What's happening?"

His answer to the second question came not from Carr, but from the hillside; musketry erupting from the right centre, beginning with the French above, then answered by the British below, but at no other point along the line. However, Carr was forming an answer to the first question.

"Well ...."

He moved the Dolland to his right eye and tried to focus one handed, for his left hand was clinging to the pole.

"......... it looks as if someone's taking his men in early. Got there too soon and didn't stop. They're going up. If they get in, they'll be up there alone and taking on the whole lot that Johnny's got waiting!"

As an endorsement of Carr's appraisal, more firing came from the same point, with redcoats pouring forward, but then firing erupted along the whole line. Soon smoke covered the whole hillside, perpetuated and thickened by a furious exchange from both forces, above and below, soon Carr could see nothing more and stood down. Lacey was standing with O'Hare, watching what could be seen, this being only the backs of the 45th assaulting their portion of the hill. Fane's Rifles emerged from the hillside on the left and began an assault in from the extreme left. It was intense fighting, in which the British left was making little progress. As the 105th stood and watched, they saw small groups of both Rifles and 45th gain a foothold on the summit, only to be thrust back down by counter-attacks. Solid shot sighed overhead, heard even above the din of the incessant firing, but to little effect, at least for the assault troops, bravely mounting attack after attack, but all in vain. They watched the almost hopeless contest for almost an hour, then Lacey turned to O'Hare.

"Pass the word. Weapons, water, and cartridge box only. Leave all else behind. Then form up, as agreed."

Lacey was relieving his men of almost 40lbs of equipment, to better enable them to scramble and climb up the steep hillside. He waited and within minutes his men were formed in their six assault columns, with the Lights and Grenadiers skirmishing out in front. A hare, startled and frightened, ran panicked across the front of the 105th, spurred on by the hoots of the men, some following the terrified animal with their muskets, as it dodged, helter skelter, off to safety. It was a moment of humour that relieved the tension of the moment and all were grateful.

Lacey had received no orders, but he could see for himself that the assault of the 45th was waning. Their efforts had been resisted for over an hour; they must be tired, short of ammunition and more then a little dispirited. O'Hare had taken himself over to the right and so Lacey placed himself before the Colours, held by Number Three, one of the two reserve Companies. Major Simmonds could be seen positioned to the left. Lacey turned to Rushby and Neape.

"When we go up, I do not want to see you two leading the way! When we go forward, you follow the assault, not lead it. Clear?"

Rushby nodded, but Neape was paralysed at the sight of what was before him, a cacophony from Hell, shots, noise, shouts and screaming, made worse by the sights that went with it, men with dreadful wounds, limping back or dragging themselves to relative safety. Lacey turned to Deakin.

"Sergeant, run over to Mr. O'Hare, on the right. He's behind the skirmish line. Tell him to advance."

Deakin saluted and ran off. He found O'Hare standing facing his front behind the Grenadiers and not at all confident about what was about to happen, but he turned when Deakin arrived.

"Sir. The Colonel says advance, Sir."

O'Hare nodded.

"Captain Carravoy! Forward, if you please?"

Carravoy, his head spinning from the mayhem all about, pushed his left foot forward, followed by the right and kept going. His men followed, then O'Hare walked forward himself, the signal for the six Companies to begin their own advance.

The Grenadiers on the skirmishing line's right, with Carravoy in the lead, advanced steadily on. They crossed a small ditch, full of stagnant water, but now stained red. He was grateful for the smoke, but his fear grew as the ground rose in front of him, they were now ascending the slope. The smoke was so thick that he could not see if they were heading for one of the ravines or one of the impossible cliffs that lay between. He could only lead his men on and hope and wait, hope to meet a possible way up and wait for the time when the defenders would see him and his men and choose them as their target. On his right was a Regiment whose name he did not know, but he saw groups of men, several with superficial wounds but still in the fight, being gathered up by whoever was there, Officer or NCO, to again renew the assault. With their Officer or NCO in the lead, he saw them lean into the steep gradient and push their way upward, through the difficult scrub and bush.

That before Carravoy thickened and he had to use both arms to push the branches and trailing brambles aside. Finally it became too thick and he was brought to a halt. He found that he had trodden on something soft, to look down to see that it was a red-coated body, so he shifted his feet to only stand on another. Stuck fast in the bush he was unable to move, until he felt a huge shove against his back so that he was pushed through. Two of his Grenadiers had put their shoulders against him and levered him onward. He turned in anger, but was met by two expectant faces, both as wide-eyed, dirty and scratched as his own.

"Right with you, Sir. As you order!"

The rebuke died in his throat. Carravoy gripped his sword and swallowed hard, which helped with both his fear and his anger at being manhandled by his own men, both of D'Villiers' Company. He scrambled further up, brambles and gorse clinging to his legs now already labouring with the slope impossible to walk upright upon. The smoke cleared, just enough for him to see that they were almost at the top, but the crest was thick with blue shapes. The Grenadiers behind him immediately levelled their muskets either side of him and fired, the smoke obscuring all. What must have been a musket ball stung his jaw in its passing, before dislodging an epaulette, which remained hanging by a few threads. He crouched a little lower as one of his followers collapsed in a heap.

On the left, Carr had led his men right up to the final rock face, an advance giving both dismay and revulsion as they passed over the many wounded of the 45th. However, as standing orders dictated, his Officers and NCO's forbade any man to stop to answer the pitiful pleading of those still able to speak, from amongst the bodies now thick over the final yards. He was nominally leading his men, but this in theory only, for they were going no-where, stuck fast merely yards below the crest, but a distance too far. The French seemed to making good use of some kind of stone bulwark, which was serving as a kind of parapet, but at least they had to lean over to find a target from amongst his men, making them targets themselves for his accurate rifles. He had linked with Fane's Riflemen who were attacking from the left and some 45th Lights and others from their Line Companies. All mingled amongst his own men, who were working well in their files, maintaining the sequence of the lead man firing, then being replaced by one from behind; but there could be no advance. The leading man could only step back and down to give his replacement the room he needed. Thus his men were circling to climb, fire, and then fall back. Carr looked around for ideas but found none. He sheathed his sword, picked up a discarded musket, pillaged a handful of cartridges from the box of a dead 45th and began firing. For the moment they could hold, but either the French fell back, or soon they would be forced to do so themselves and it would not be too long.

Carr did not know it, but he had lost one Officer. Shakeshaft was unconscious, sprawled on the ground, feet uppermost. He had gained the bulwark only to be thrust back by a shove from a musket butt, to tumble back and hit his head. Ellis had checked for life and found a pulse, but could do no more. He was now in command of the section; he could not see Captain Carr and so he decided to join with the nearest file, this being Davey, Pike and Miles, the four then sending accurate shots into the blue ranks above them, but they never seemed to be thinning, always there was another to take their place.

Tom Miles was growing angrier; firstly at the "God-awful" situation they were in and, secondly, at having to "file in" and co-operate with Sergeant Ellis, him endlessly making some needless comment.

"I've bloody well had enough of this!"

When his turn came, he fired, but did not come down to be replaced. Instead he scrambled up, using the butt of his musket as a lever. He reached the stone bulwark, knocked aside a bayonet, smashed the butt of his rifle upwards into the moustache of the Frenchman who was wielding it, then swung his weapon sideways into the arm of another. He then flung his rifle at the nearest Frenchman, drew his sword bayonet and stabbed another, but he was then kicked to the ground himself. Struggling to get up, he found that he could not, a bayonet had gone through his shako, pinning it to the ground, also some Redcoat was standing on top of him. He ripped away the chinstrap to look up and see the distorted face of the Frenchman who had made the bayonet thrust at his head whilst he was on the ground, the face in agony and shock with Ellis' bayonet sideways through his neck. Ellis' thrust had altered the French bayonet's aim just enough, however, just for good measure Miles punched the dying face, before finding his own Baker and reloading.

68

Many British were coming up from below into what now proved to be a shallow trench and they gained several yards of it but could go no further, neither forward nor right nor left, The French were fighting desperately to hold back the British assault at this end of their line. The Redcoats in the trench began an intense rate of fire into the French before them, their muzzles but yards apart, yet the French gave no ground. Carr had seen the small gain and hauled himself up, using a gorse bush growing from the edge. His hand came away covered in blood, but this claimed his attention for barely a second. The situation of the men in the trench was utterly desperate, already there was no room to stand, but on dead bodies only, both red and blue. He seized the bush again and swung back out over the edge to see more Redcoats now gathering there. He shouted down to the powder blackened, sweat stained, faces.

"Re-load, then fix bayonets."

Looking down at his men, he waited unbearable seconds until all stood there were at the make ready, an agonising time, for he could hear the conflict behind him continuing with appalling ferocity, French curses mingling with English insults, both barely heard between the continuous blasts of musketry. He threw his own musket into the trench and drew his sword.

"I want you up, stood on the edge and ready for a volley. Then charge, with me."

He looked along the upturned faces.

"Now, up!"

They all scrambled up, some being pushed by others, some using projecting rocks and bushes, some pulling up those behind them by seizing the bayonet socket around the muzzle of the proferred musket, but all got there and immediately came to the present.

"Fire!"

All fired as one. The effect was lost in the smoke, but Carr didn't wait for it clear.

"Charge, boys, charge! Hurrah for 45th, hurrah for the One-Oh-Five!"

Carr's men and those already in the trench all sprang forward, screaming at the top of their lungs. The French gave way and those staying to fight were quickly despatched with butt, boot, or bayonet and soon they were through the smoke, but the sight that awaited them gave Carr no choice.

"Form to receive cavalry."

Forming a square where they now stood was obviously ridiculous, on the broken ground, but his men understood immediately what was needed. They instinctively grouped closer together, those at the front extending their bayonets. They were being charged by French cavalry.

Carrovoy was equally stalled, not so much by the French as by his own state of indecision. Himself and D'Villiers' section of Grenadiers, were at the mouth of a small ravine and, from what he could see through the smoke, it was thoroughly occupied by French infantry, all loading and firing from a solid line. Should he advance or be content to hold what they had? He was one side of the ravine, D'Villiers on the other, and they were both supervising pairs of Grenadiers who emerged from cover to fire and retire, to be replaced by the next pair. Carravoy was content that they were causing casualties, but their task was to capture the top and the dead from his own command were building because the French above had learned to wait for the two pairs to appear.

Both Carravoy and D'Villiers were concentrating ahead, up the ravine, and so did not see what now arrived behind them, their first realisation of new events came to their ears through the smoke.

"Come on, lads! We're Grenadiers! We belong to the One-Oh-Five!"

Lieutenant Ameshurst, on the left of the Grenadiers, when confronted by a sheer cliff, had led his men off to the right, and found the ravine. Not noticing Carravoy, nor D'Villiers, he had seen Grenadiers of his own Regiment grouped around apparently leaderless and so now he was encouraging those of D'Villiers to join them. Stood before his men, sword in hand, he formed them into a solid line across the opening.

"Present! Show them who's at the door, boys."

The muskets were all raised to aim up into the ravine.

"Fire!"

The volley crashed out and all sprang forward. Without any orders from D'Villiers, all his Grenadiers followed upwar and Carravoy and D'Villiers had little option but to follow themselves. Ameshurst and his men were fighting their way up, he leading a knot of men that seemed to be acting as his personal bodyguard. Any Frenchman who aimed at him was shot by a Grenadier ready loaded and prepared for such and anyone who attacked him with sword, butt, or bayonet, found themselves assaulted and despatched by at least two. With his Sergeant at his shoulder, Ameshurst reached the top to find himself in a shallow valley, with a track that led onward from the ravine, deeper into the hill. Serious resistance seemed to have ended, for the French were now streaming in disorder right to left across his front in their tens and hundreds. The British assault on the right must have broken through and, on that flank, the French were in full retreat, but too many were moving for one section of Grenadiers to stop and capture.

"Firing line! At your best rate. Bring them down, lads, but watch out for our own, they'll be following, certain sure!"

As the Grenadiers loaded and fired at the best rate they could manage, Carravoy came up and found Ameshurst. He had picked up a French musket and, standing beside his men, was loading French cartridges and encouraging them with promises of a guinea for anyone who could match his rate of fire. It took some time before he noticed his Company Captain.

"Hello Sir. Did you get stuck somewhere? I had a bit of luck, found this ravine."

Back on their start line, Lacey was studying the assault along his own front, him standing with the two reserve Companies formed up and waiting. When he saw Carr's men scramble up and deliver their volley, he drew his own sword and began walking forward.

"Advance!"

Besides the Colour Party, Captain Heaviside clenched his Bible for the final time and thrust it into his pocket.

"Be strong, and quit yourselves like men. First Samuel, 4, verse 9."

The sight of the rocklike Heaviside, quoting text, stern faced and now much in need of a shave, did much to steady the shaking Neape. He gripped the flagstaff and fell into step with Lacey ahead, but that was soon lost on the steep, rocky ground. Lacey, not a young man, was soon labouring, but he fronted his men for as long as he could, until, inevitably, they caught up, first in the shape of Heaviside, who appeared beside him. Lacey had led them up to the point where he had seen Carr going over the edge and, with laboured breath, he gave his orders.

"Captain Heaviside. Captain Carr is above you. I believe he has made the summit. Take your men up and support him."

In answer Heaviside brought his sword up to his nose and pushed on. Just below the crest he turned to his men, but nothing as mundane as "Come on, boys" emerged from him.

"Let no man's heart fail because of him; thy servant will go and fight with this Philistine. First Samuel, 17, verse 32."

With that, he was over the top, followed closely by his men.

Carr was stood amongst a growing company of men, his own, some 45th and some Rifles. They could go no further; they were blocked by cavalry, who were being handled superbly by their Officers, making controlled charges to discourage any advance and perhaps to pick off any redcoat not solid amongst his comrades. Carr was grateful that the term "charge" was misplaced. The ground on top the ridge was a jumble of rocks, hillocks, gullies and trees, but the French horsemen were making good use of the spaces available in between. Carr's men maintained a good rate of fire and the French were suffering but Carr was in a state of intense frustration. The whole French position had now been carried and the French were in full retreat into a valley over to the British left which would provide an escape route, but the cavalry before him with an infantry rearguard were holding the door open to enable their comrades to withdraw from their previous positions on the ridge over to the right and now escape.

Heaviside appeared at his side and there commenced an exchange, absurdly formal, as though on the boundary of a cricket field, ridiculous amongst the mayhem all around them.

"Henry."

"Joshua."

"What would you propose?"

"Take your men off to the right, see if you can link with ours, some must be over there, pushing the Johnnies our way."

As Heaviside examined the ground he would be crossing, Carr continued.

"Where's the Colonel?"

"Below, or right behind. Whatever, not far."

Whilst Carr nodded acknowledgment, Heaviside pointed.

"Seems I'll not have far to go….."

Carr followed the direction of his arm.

"……seems those are our side, coming over now."

With that, Heaviside saluted with his sword and in return, Carr grounded his musket and saluted. Although, as characters, they were so very different, between the two there was the deepest respect. Heaviside waved his sword for his men to follow and was gone. What Heaviside had seen were dense crowds of redcoats coming over the farthest hillocks to their right that made up the back of the ridge and he knew that the task was now to kill or capture as many French as possible, before they escaped. Carr looked again at his front; the cavalry were riding away, leaving the infantry rearguard standing steady, whilst a dense crowd of French infantry streamed past behind them. His men were still formed against cavalry, a line three or four deep, but now almost 50 yards long and growing. He yelled above the continued din.

"Reload. Prepare for volley fire, then forward."

He waited until all were at the "make ready". He could only hope that his men could decide for themselves how best to follow his orders.

"Front two. Present."

A pause.

"Fire!"

A loud crash amidst the clouds of smoke, heard above the din of the ongoing conflict.

"Rear ranks. Present."

The muskets came down besides the heads of the front ranks, these all now waiting with bayonets "en garde".

"Fire."

He did not wait for the rear ranks to recover.

"On! Come on! Charge! Follow me."

They plunged into the smoke and soon were amongst dead and dying Frenchmen, but none were standing in their path, the line had withdrawn; an experienced Officer must have seen the British prepare, but Carr did not hesitate.

"Forward. On!"

All that remained of the French army were stragglers running over the hills at the back of the ridge, chased on by a thickening cloud of redcoats. Carr led his men down and to the left, to the bottom of the valley and onto a track, which the French were using to escape upwards, but once there, again he held up his sword and called halt. To his left, in the valley bottom, the French cavalry had reformed, leaving enough of a space for their fleeing infantry, but ready to charge into the flank of any pursuing British. Carr could do no other thing.

"Firing line!"

His men, now an even bigger variety of Regiments, obeyed and held their ground. Carr now looked up the valley and felt even more justified at halting his men. Some French battalions had rallied and reformed, blocking any advance, certainly against so small and polyglot a force as his. The Officers of the Redcoats who had chased the French across from the right were of the same opinion, as they were also coming disorganised off the ridge and onto the track leading up the valley. They had no choice, same as he, but to attempt to get their men grouped into some kind of fighting formation. However, at least Carr no longer felt alone and also there were trees on the far side of the valley which they could get into and, from there, give the cavalry an unpleasant time from their musketry, especially from his men and the Riflemen, both with their Bakers. Running to the end of his line, he waved his sword and lead them over.

Lacey, with the Colour Party, had climbed the ridge and had also now descended onto the valley track, where he suddenly found himself surrounded by horses, but one look up found him staring into the face of General Wellesley. It was Lacey alone who saluted, Wellesley had more urgent thoughts.

"Lacey. Are your men in some sort of order?"

Although uncertain of the accuracy of his reply, Lacey gave it anyway.

"Yes Sir."

Wellesley looked over at Carr's position.

"Are those yours, over there?"

Lacey looked and recognised Carr, him giving directions to place his men in their firing positions.

"Yes Sir. Those are my Lights, with some add-ons, of course."

"Right. Get yours over and reinforce what's there now. Push up that side of the valley. Expect Ferguson to come in on your left."

Wellesley sat stock still, waiting for his order to be obeyed. Lacey anxiously looked around and was grateful to see so many uniforms with the bright green facings and some Officers he knew. His men had kept together, so he took a deep breath.

"Prince of Wales Wessex! Follow me. Form column as we go."

From all around, Lacey's men ran forward, Officers and NCO shouting instructions for their men to form column on the run, the final achievement of which did not go unnoticed by Wellesley and his Staff. Lacey came up to Carr and, from some place, O'Hare and Simmonds also arrived. Carr's men were already causing casualties amongst the cavalry as Lacey spoke.

"Carr. Take your men forward. Firing line but open order, then wait for another Company to file on through you. Push them out of there, they won't stand."

As Carr led his men forward, Lacey turned to his two Majors.

"Get the men into three Companies and we'll take one each. We'll push up behind Carr and file through. Wellesley wants us advancing up this side and we can expect reinforcements from the left. Ferguson!"

Three minutes saw some form of order established and they advanced on, but against increasingly disintegrating resistance. The affair was practically done. At first the rearguard battalions and the cavalry held back their pursuers, but, as the valley narrowed, the fugitives and the rearguard became increasingly desperate, especially when it narrowed to a defile merely yards wide. Many French were cut off on the right, by the British still coming over from the back of the ridge, but when all the French were through that were going to get through, Wellesley sent an aide-de-camp around his pursuing and now victorious army. The 105th, still on the far side, were the last to receive the message to halt, but Lacey had already halted his men. Ferguson did arrive, his men all fresh and eager, and they had vigorously joined in, at which point Lacey concluded that his men had done enough; there were now more than enough British pushing up to the defile.

The 105th sat calmly on the warm, dry, grass of the hillside, drinking water and eating biscuit from their pockets. Their main rations were in their packs, left behind in front of the ridge, but for a while this was ignored as they watched three captured French guns being manhandled back down the valley and a large column of prisoners. However, soon hunger asserted itself and Lacey, feeling his own appetite, ordered them to form up and retire, this time in proper Company order; he was determined that they would return in the proper style. The order was to return to their start line and eat, but, as they walked over to fall in, Miles, as hungry as anyone, as usual made his feelings known.

"Food! Yes, if some bloody peasant hasn't made off with our kit!"

It was Byford who replied.

"It must be your optimism that keeps you going!"

75

Many laughed, but Miles simply looked puzzled and annoyed, pondering the meaning of "optimism". He obtained no answer from within his own vocabulary, but the moment for a reply had passed. Then Ellis caught up with him

"Miles! No headgear. Get that put right; come rollcall."

Miles, unsurprisingly not yet calm from the fever of the battle, would have hit him there and then, but John Davey saw all the danger signs and grasped one of Miles' crossbelts and pulled him away and then on down, further down the valley to the others of the Light Company, now forming up.

"Come on, Tom, I 'spect it's still up where that Frencher stuck his bayonet through it. Let's go see."

***

Chapter Three

"A Field of French Ruin"

For greater or for lesser, after a battle, all survivors are casualties. From the controlling General, him perhaps regretting the mistakes of the battle, right down to the lowliest private, now a casualty and barely conscious in a wagon of similarly wounded. Him still in shock from his time with the Surgeon, unable to flick away flies from the stump of a limb, or a weeping wound, but hoping that he will be amongst the lucky that avoid gangrene, and will still remain alive to join the beggars of London, Bristol and a dozen other cities where a maimed soldier may earn an existence rattling his tin cup. However, the severest casualties, most of all, were the surviving wounded, neither yet discovered nor moved to safety, being pillaged, killed or further maimed by marauding peasants, cloaked by the night for their pitiless deeds.

Exactly the same for the British army after Rolica. Wellesley sat regretting the misfire of his plan to outflank the French, and also chafed over the number of their army who escaped. Beneath him, the survivors of his Battalions now stood in line, still whole and hale to a degree sufficient to answer their roll call, but mourning lost comrades; then down to those who were filling up the casualty carts, awaiting the start of a hideous journey back to the coast. Then finally, the dead, now being brought down from the ridge for formal burial. The battle had begun for real in the afternoon and now the sun was setting, mercifully reducing the fierce heat which so many of the wounded and dying were forced to endure. After roll call, all Colonels allowed their men to merely sit at their campfires, except those whose men were sent out to guard against a French return, these being justifiably the Regiments that had taken little or no part in the fighting. There they cooked their rations and talked in hushed tones about the day and the fate of those now gone from their Mess. Just so the Officers of the 105th; Carr and Drake, both now relieved at the recovery of Shakeshaft, thanks to Morrison's ancient remedies for concussion Similarly Carravoy and D'Villiers, each alone with their unspoken Devils called up by the memory of their own indecision and being so evidently outshone by Ameshurst.

Jed Deakin and Joe Pike sat by the evening campfire more cheery than most, for word had spread that the camp followers were coming up and would be with them, the day after tomorrow, or even late tomorrow; yet even they could be more accurately described as being quiet and subdued. Also, after a while, whilst the evening rations were still cooking, Joe Pike started shaking. Jed Deakin gave him a drink of hoarded rum, this within the permanent keeping of Zeke Saunders, and then lay him down, to cover him with a blanket, despite the warm August night.

"Shock, is what the Medicals calls it. Shock, from what he saw, up there, in front of that ridge."

He took a sip of the rum himself and then passed it on. None spoke; all knew that Jed Deakin had more to say.

"That were a bad go. I've seen the like, but not often, and Joe here, not at all. I've heard tell that our Generals is pleased with the low casualties in takin' such a place, but they can tell that to the folks of Nottingham and Worcester. Seems the Colonel of the 29th took his lads forward too far and led 'em straight up. Got half of 'em killed or captured, but we'll never know 'is story, 'cos 'ee's dead too."

He took off his shako, pulled off the bandana knotted beneath and wiped his face; all sat waiting for more words, but none came. The pork and peas were eaten and then the veteran Deakin eased himself down off the log until it supported his neck and he went to sleep, despite the still bright sun. At this, the rest turned themselves to their own affairs, including Tom Miles, who was assessing the need for final stitches in the two bayonet holes in his shako. Concluding that more were needed, he set to and, being an acknowledged expert wood carver, he was equally adept with needle and thread. He took a final look and mumbled out loud, more to himself than others.

"There! That'll serve, even for that fault findin' bastard Ellis."

The night came and wore on, but even with the emergence of dawn, wounded were still being discovered and carried back down from the ridge, for some had fallen into the numerous holes and crevices. Also, some had been severely wounded in the fight amongst the thick gorse and brush, to also lie undiscovered until daylight. Joe Pike woke refreshed, the rum and the sleep had completed its work, but his hand went immediately to the side of his face, it felt as though there was a deep burn. He turned to Tom Miles, showing him the painful side of his head.

"Tom. What's that? There?"

He pointed with his finger, but didn't need to. Tom could see enough. He chuckled his reply.

"Seems like you've been grazed by a bullet, boy. It's made a nice plough furrow down the side of your head. That's the end of your good looks, I d'reckon!"

He chuckled some more as he ladled out the breakfast porridge to the discomfited Pike.

"We shall have to see what your Mary thinks when she comes up. Perhaps she might take exception to takin' up with a cove as've got a slot full down the side of 'is head!"

He chuckled some more, until silenced by Jed Deakin.

"One day, Tom, that tongue of your'n will see you six foot under and some poor sod swingin' off the crosspiece for it!"

A genuine reprimand from Jedediah Deakin gave them all pause and Tom was no exception. He fell silent and consumed his own food, whilst Joe, after his porridge was eaten, set gingerly about the business of shaving and not aggravating his wound, preparatory to the arrival of Mary. This happened in the late afternoon, when the camp followers of the whole army arrived to do their best to find their men, by wending their way through an army allowed to rest and re-supply itself. Bridie saw the group first, but it was Mary who arrived first, running forward to fling herself into Joe's arms and kiss him fiercely. She took one look at his now livid wound, kissed him again, then she looked up into his eyes, worried, almost terrified.

"Joe, I've got some news. Can we go off aways?"

Joe now looked equally as worried as she and allowed her to take his hand and lead him off to a group of trees, currently occupied by tethered horses. Jed and Bridie kissed affectionately, then Bridie automatically began the process of cooking the evening meal, carefully assembling all the necessary utensils and ingredients from the children, who were then, each in turn, kissed by "Uncle Jed".

"Uncle Jed" had seen Mary and Joe walking off and detected the signs of an event of great import.

"What's all that about?"

Bridie looked at him and there was no joy in her face.

"Mary's expectin'. She's pregnant."

Jed repeated the word stupidly.

"Pregnant?"

Bridie looked at him impatiently.

"That's what I said! Not just a little bit, nor somethin' that'll go away. She's havin' a baby!"

Deakin's head sank down to look at the dusty earth beneath his feet.

"When did that happen?"

Bridie hit him with a spoon.

"What are you talkin' about, you old fool? How do you think babies comes about and who knows which time?"

She allowed the scolding to sink in.

"She's havin' a baby, sure enough, an' that's what it's all about!"

Deakin's mind returned to a measure of sensibility, his face full of worry, but Bridie had changed the subject.

"I've some news of my own."

Now Jed's face did register deep concern.

"You're not pregnant!"

She hit him again with the spoon.

"Jasus, no! How often do we sleep under the same blanket for that to happen? No, I've got a little position, you might say. Mrs Prudoe the Chaplain's wife, she's asked me to be her servant, sort of like. It'll not be much, and will not take me away too much, but it'll mean a bit of coin and some extras, I daresay. She does her own cookin', but she needs me, us, for washin' and mendin', sort of thing. So….."

But Jed broke in.

"Us?"

"Yes. Me and the children. 'Specially Eirin. She could practically do the whole thing herself, but it's myself that's best at the mendin', although she's comin' on."

She looked into the pot, now steaming.

"When this lot comes to the simmer, someone'll have to take over the stirrin'. I've got to be off and see what needs doin' with her. Beatrice she's called, an' she's a very nice lady."

She held the spoon vertical in the pot, poised there by one finger, looking querulously at Jed. The intimation was clear; Jed should take over. He exhaled a long sigh, took the spoon and eased the uncooked ingredients around the iron pot.

"Right, I'll be over there, by that wagon. Not far."

Deakin looked over to see Sedgwicke unloading the Chaplain's few items of furniture. He nodded his last words to Bridie.

"Then we'll see you later?"

"You will. I'll be back to finish the stew and warm up some bread. You'll see."

The look on her face and the smile in her eyes took away the last shreds of any ill temper as he watched her gather the two eldest, Eirin and Patrick and take them over to the Chaplain's camp. His attention was brought away by the return of Joe Pike and Mary and Deakin was struck by the competing emotions that passed across the young man's face, a contest between deepest anxiety and the fullest elation. Mary sat beside him, seemingly consumed mostly by the former.

***

The day after the rest day, that being the 19th, orders arrived, read out by Ellis to the assembled company, in the early morning.

"Army to be ready to march and engage the enemy on the morrow. All men to be in the fullest state of readiness and preparation."

He looked at the blank faces.

"Dismiss and see to it! I'll be checkin', be certain."

Nothing was to escape examination, from the stitching of every pack strap to the greasing of the artillery wheels, and thus, all through the day, every soldier was inspected and issued with all that would be required for another battle. Rumours abounded that the main French army was marching up from Lisbon. As usual Ellis inspected Miles himself, but it was neither as detailed nor as irksome to Miles as the previous, but Ellis did pull off Miles shako and examine it carefully. After examining the stitching he nodded and grunted.

"There'll be a damn sight more damage like this on all of us, afore this lot's done!"

With that, he thrust the shako back against Miles' chest and walked on, leaving Miles feeling more than slightly appeased. There had been nothing extra, nor insulting, about his examination. Ellis had done the job he was ordered to do and that was that.

The following day, as they had when leaving Mondego, Fane's Brigade led off, with the Light Company of the 105th together with the four Companies of the 95th, alternating picket with the full 60th, these now being first out, spread before the advancing army. The followers of each Brigade were allowed to immediately join the end of their own Battalion column and they did, the other Brigades then joining on the road in their own allotted place. The day was hot and all soon began to suffer, especially those with wounds that were painful, yet not so bad as to give them excuse for falling out from the column. Veterans helped where they could and luckily, there were several streams on their route where sympathetic NCOs allowed men to fall out and wet neck cloths and bandanas to help with the dressings and ease the incessant heat upon necks and shoulders, thus many limped on, holding their place as best they could.

Beatrice Prudoe had asked, or perhaps more accurately, allowed Bridie's children to ride in the wagon when they grew tired, on the pretext that she needed their help within it's comfortable confines when arranging the personal affects of herself and her husband. It was her first meeting with the likes of the Mulcahy family and she had been uncertain as to what to expect, but she had been pleased, almost delighted. Eirin was what she was, a girl grown up quick to meet the needs of a soldier's family, but she was pleasant and respectful and not without initiative. Patrick and Kevin were two boys, full of energy that emerged when allowed, but most often terrified in the company of such as her husband, the stern and lugubrious Chaplain Prudoe. However, they were polite enough to her, but, unbeknown to Beatrice, both were under the strictest instructions from Bridie regarding their behaviour. Thus charmed, Beatrice decided to undertake some education of the four. The youngest, Sinead, was barely able to carry water, but she seemed the brightest of the four, and so it began, in the back of the jolting wagon, on a slate with a piece of chalk. Eirin well knew her role - to please her mistress and so she absorbed all as quickly as she could; it was important to show willing, which she well realised. The boys, unsurprisingly, struggled, but Sinead made progress almost as quickly as Eirin, to the delight of Mrs. Prudoe, but to the strong disapproval of her husband who saw little point in her schooling endeavours. When the children were gone, he, sat at the front with Sedgwicke, had inevitably overheard the lesson.

"I fail to see, my Dear, why you give your concern to the whole business. I am of the firm opinion that, as long as they can commit their prayers to memory, that will be fully sufficient. Book learning and the ability to read can only disturb their inner equilibrium, which I would regard as a cruelty. Besides, I very much doubt that they have the capacity to achieve such. I do wish that you would desist!"

Beatrice smiled and pretended not to hear, whilst examining the words each had formed on the slate. Sedgwicke, holding the reigns, had heard all, and felt inclined and capable, even justified, to give his own opinion, born from his own experience.

"If I may, Sir, I myself taught the daughter of one of the men, Chosen Man Davey of the Light Company, and I found her to be a gifted and astute pupil. A pleasure to teach! She remains back in England and I do hope that someone is now helping her to continue with her studies. She could practically read, write, and cipher as well as any full time scholar at most schools!"

He was referring back to the time when he taught Tilly, John Davey's adopted daughter, but Chaplain Prudoe, having given his opinion, was not to be gainsaid.

"I'm sure you're wrong over that, Private. You were overcome by the minor success that you did have. The very idea of achieving a standard that will enable independent study is quite beyond them, I feel sure."

The downturn in Prudoe's voice at the end of the sentence showed that Prudoe was set and firm in a verdict pontifically delivered. Sedgwicke, for his part, felt annoyed that his own experience should be discarded in so offhand a manner and he straightened his back and lifted his head to make reply, but Prudoe raised his own hand in rebuff.

"No! I'll not hear another word. My wife can indulge herself if she chooses, but I find myself in total disagreement with your judgment. And, what's more, you are not to include yourself in this foolish exercise. I do hope that is plain, Private Sedgwicke."

The stern use of his formal title told Sedgwicke all he needed to know regarding the requirements of his superior Officer. He flicked the reigns and held his peace, whilst seething inside at the cursory dismissal of his own opinion that was, what is more, supported by evidence, as he saw it. The wagon rumbled on, jerking all from side to side.

Meanwhile, there was discussion elsewhere amongst the 105th. Davey, Pike, Miles and Byford were marching in their column, just reformed from their turn out on picket. It was Davey who opened the discussion.

"So what's the verdict about the Johnnies at Rolica, just passed?

They gave it a better go than at Maida, I'd say. What from you two veterans?

He was referring to Miles and Byford and the former looked across to Byford and received back a blank look, this being wholly typical of him, content to remain in the background. Therefore Miles formed his own opinion.

"I'd say true. They took a damn sight more shiftin' than that tassle swingin' crew what took off after no more'n three volleys!"

He looked at Byford.

"What were they called, Byfe?"

"Voltiguers. Light Infantry, like us."

Davey joined in.

"Only better dressed!"

Miles paused to allow himself a nod or two.

"But, yes, I'd say this lot of Frenchers is a cut above what we saw the backs of in Sicily."

Davey turned to look at Miles, changing the subject.

"You goin' to get another Frencher backpack, like last time, back then,Tom?"

"You're damn right I am! Couldn't get one from the last go, 'cos I 'ad to worry about findin' an' stitchin' up this damn shako. Yes, I do fancy havin' a good cowhide job that'll stand up, what won't come apart over the next few months. We'n here for some time, I fancy, in no easy set up."

He paused and looked at all three.

"If I were you, I'd do the same. King George's don't last much more than one soakin', and then where are you? Havin' to carry your kit in a cloth bundle."

He paused, his face written over with further productive thoughts.

"An' a pair of Frencher boots an all, if I can find some as fits!"

He nodded and smiled, satisfied at the unassailable wisdom of his experienced words. Joe Pike said nothing, but, instead, fingered the wound on the side of his head. It was healing, but still disconcertingly deep and he winced as his finger found the deepest part.

***

Evening found them in the rounded hills around Vimeiro, a place practically on the coast and the reason for Wellesley's choice became clear on the following day, as the 105th sat in their camp beside the road that ran between the sea and the village itself. Unknown to them, a disembarkation was taking place, similar to their own at Mondego; Wellesley was being reinforced and the first the men of the 105th knew of it was when the first column came marching up through the gap in the hills which held the road that led up from the beach. Lounging at their ease and unoccupied, the three, Miles, Davey and Pike, decided to take themselves down there and watch proceedings. Perhaps some of these newcomers would be spilled into the water as they had been themselves, here was perhaps some entertainment. However, they had not gone 20 yards before they were observed by Sergeant Major Gibney.

"Thee three! Hold fast! Wheers tha' think tha's goin'? Hold there, no farther."

Miles was the first to think of anything like a plausible reply and began waving his canteen."

"Off to fill our canteens, Sar' Major. We've all run dry."

Gibney produced his best fume, his face reddening."

"Down stream! Full of all the filth from camp! Get th'sens back, and sharp, and not let me see thee down this way again. Back! "

But Miles had not surrendered the fight.

"There's a spring as comes out the rock, Sar' Major. That's pure, an' we'n all hankerin' after some good water! "

Gibney was now near and towering over all.

"Water from the butt's gradely enough for thee! That's pure, ah knows that because ah sees over the fetchin' of it m'sen. Back t'camp at thy …….."

The order tailed off. Gibney could see the road, whilst the three had their backs to it. He paused, silent, his eyes growing wide in surprise, then anger. He raised his arm to point, then produced his best bellow.

"Thee! Ont' road. Halt"

The order, delivered at such volume, would have halted a Regiment, but the five it was intended for stopped immediately. Gibney strode onto the road, the eyes of Davey and the others, now fascinated, following his progress, then they turned bodily to see the object of his attention and surprise grew on their own faces. They had recognised the fifth member of the party, for he was a prisoner, the other four an escort. Gibney placed himself before them on the road and addressed the leader of the party, a Corporal.

"And wheer's thee takin' this 'un?"

The Corporal reached into his pocket and produced his orders. He couldn't read them, he could only assume that Gibney could, but it would only be to confirm what he now said.

"This is a prisoner, Sar' Major, a deserter, sent out here, rather than hung, is my guess. I'd say his size've saved him from the rope. Seth Tiley's 'is name, I'm to get him to the 105th Wessex, 5th Provisionals as was. Can you help, Sar' Major?"

Gibney looked at the prisoner, a level look, for both were of equal stature. Seth Tiley was a huge man, large in every component bar his eyes that were small, set back and mean. Hatred and loathing burned within both, but Cyrus Gibney had seen all of such before and was well capable of matching any level of malevolence with interest. He stood and grinned, imitating the perverse pleasure of a gaoler or even a hangman, someone who thoroughly enjoyed their line of work, which he was about to embark on.

"Seth Tiley! An' what evil doin's has thee be aboot, eh? 'Oo's tha' robbed, even murdered, since tha' absconded? Eh?"

Tiley's response was to spit on the ground, to immediately be clubbed in the back by the musket butts of the two escorts behind. They had no liking for their huge, menacing prisoner, a known criminal and deserter. Tiley staggered forward, but soon regained his balance, but Gibney remained as he was, grinning with evil pleasure.

"Well, now tha's back amongst us, we'd best get thee on t'Colonel, an' we'll see what fate he's a mind to bestow upon thee."

He turned to the Corporal.

"I'll take 'im. Does thee need anythin' signin'?"

The Corporal was as intimidated by Gibney as he was at the thought of Tiley breaking free, so passing him on equalled the relief of escaping a punishment of some sort on himself. He thrust the document forward and fished a pencil from his jacket pocket. Still grinning maliciously at Tiley, Gibney took the pencil and accurately signed his name, precisely in the required place. The Corporal stuffed the paper into his pocket and held out the rope that led from Tiley's hands for Gibney to take, but Gibney made no move, merely to continue to return Tiley's look of hatred with equal intensity. Gibney took one pace forward, seized Tiley's shirt and yanked it open to reveal a burned in brand, just healing.

"D! Deserter! What else?"

There was no change in Tiley, not even when Gibney pushed him back as he closed his shirt. After a second or two, Gibney motioned to the three, still stood watching. Naming no-one, he motioned them forward.

"Take the rope. Bring 'im wi' me."

It was Miles who moved the quickest. He had his own score to settle with Tiley and he placed himself squarely before the giant felon as the escort scuttled off. Although Tom Miles barely came to Tiley's breastbone, size meant little to him. He looked up into Tiley's eyes, his own expression conveying all the malice he could combine together.

"Tiley. I've not forgot what you did to Joe Farley. Never got over it, he didn't. Now he's back with his family, an' not much use for nuthin!"

Miles checked that Gibney was now out of earshot, then he closed right up to Tiley's face, pulling the rope down in the hope of lowering Tiley's head. It did not but it made no difference to what came next.

"You piece of shite! A strong word of warning, don't you go puttin' yourself in front the muzzle of my musket. Front or side, I'll blow the backbone out of you, and piss down the 'ole they stuffs you in!"

With that he gave the rope a vicious tug and Tiley obediently followed, with Davey and Pike either side. All three had their own memories of Tiley; Miles, when he was escort to the group of prisoners from whom Tiley had attempted an escape, which had included Sedgwicke and Davey, and Pike, who had been near to fainting when Tiley was flogged for that very offence. With Gibney now some way ahead, Miles pulled the rope hard for them to catch up and, when there was no response, Davey did not hesitate to give Tiley a shove in the back. He felt no affinity with him, even though he had been sentenced at the same Assize Court as himself, because, to Davey, Tiley was a thief and a murderer, who preyed on everyone, rich or poor, weak or able. Given the chance, had he not been caught again, he could have attacked Davey's own family, pounding any who resisted with his huge fists and escaping with whatever he chose and could carry.

They progressed through the camp and many came forward to watch, whilst those who did not, looked up from their chores to see the return of Seth Tiley, the unmistakable figure from their time at Taunton Barracks. Colonel Lacey had his headquarters under a large oak tree, there for shade from the sun, and the disturbance caused him to look up from his papers and drop the quill into the inkpot. He then sat with his fingers poised together, watching the procession behind Gibney. He was as astonished as anyone when he recognised Tiley, the man he had caused to be flogged and who had escaped during their march to Weymouth, en route to Sicily. Some yards from the Colonel's table, Gibney took the rope from Miles and spoke words to all three that they were staggered to hear.

"Off, and get tha' canteens filled. From the spring, if tha's a mind."

All three nodded vigorously and turned to leave Gibney to pull Tiley up to the Colonel's table. Gibney saluted.

"Beg pardon, Sir. Private Seth Tiley, Sir, has been returned to us. Just off the transports, Sir."

Lacey studied Tiley for a good few seconds, a gaze that Tiley could not meet, but instead cast his eyes to the trodden grass.

"Tiley!"

Tiley looked up to hear Lacey's words, spoken as though cast in iron, wasting no time with any niceties.

"One thing, Tiley, for you to remember, the first thing you do against army law and you I will hang! Whatever it is, the slightest or the smallest. We are in the presence of the enemy, Tiley, it requires but one word from me, but one word, and we leave you swinging from a tree!"

He continued looking at Tiley who had again averted his gaze, then Lacey turned to Gibney.

"Take him to the Grenadiers. Tell Captain Carravoy that he is to be tethered and guarded each night and, when on duty, put him with a Corporal with orders to shoot him at the slightest suspicion."

Gibney saluted and made to leave, taking Tiley with him, but Lacey held up his hand.

"Tiley."

The silky threat in the softly spoken word once more gained Tiley's attention.

"Tomorrow or the day after there will be a battle. Serve your place in the line, as your comrades undoubtedly will and no one will shoot you. And perhaps the French won't either. Perhaps!"

He paused to allow his words to be absorbed, then nodded for Gibney to leave, but not before acknowledging a blistering salute. Meanwhile Davey, Miles and Pike, were off down the road, hoping to get to the beach and to actually do what they had used as a poor excuse to Sergeant Major Gibney, to fill their canteens from the pure spring. However, they were now sure that all hope of entertainment had passed and also they found their progress blocked by a column of infantry advancing up the road, so they were forced to stand aside and wait for them to pass. With this gone, they progressed on down, whilst the infantry continued marching up, to soon approach the camp of the 105th, which sight caused Jed Deakin to look up. He studied the column, then recognition grew within him. They had broken out their Colours and this confirmed his thoughts, these including the pale cream yellow of their Regimental Colour.

"Hold up. Toby, look who's arrived."

Halfway looked up, as did "The Twins" and Deakin continued.

"That's the 20th. Them good lads from Maida. We should go down, say a few words."

With that he rose, as did the three he had spoken to and then several others, who also recognised the Regiment who had supported them at a desperate moment during that battle. Deakin and his companions went to the road and stood waiting, but before any discourse, as between rankers, they must allow the Colour Party and the Senior Officers to pass, but, with them passed and gone, it began.

"All right lads? Done well since Sicily? Been anywhere's special, somewhere where you was welcomed? Anywhere?"

Several heads turned, but only one spoke.

"Who're you?"

"The Hundred and Fifth. The Fifth Provisionals as was. We marched off from Maida together."

Recognition came within the ranks and all was smiles, but the banter soon began, first from within the 20th.

"It's the "Rag and Bone Boys", now done up in smart green. You now some kind of "half-arsed" Paddy mob? What's it like, yer, in this Portugal?"

Laughs and smiles.

"Just as hot, but the locals is just as friendly. Plenty of fruit, wine and good water, but you might need summat to trade with, these is canny farmers."

"Who's yer Brigadier?"

"Fane."

"Never 'eard of him. Ours is Acland."

"Never 'eard of 'im! But the Frogs is more lively this time, it'll take your best."

The conversation, with one participant moving quickly, was necessarily short, but it was cheerful and warming for both sides, cheering to be re-united with men who had fought alongside them, men who had stood up, as they had themselves, to desperate peril. Each had supported the other and together, they had emerged victorious and, as they marched away, they had shared precious water on a baking field.

The men of the 105th stood to watch and exchange more good-natured insults with the last files and, with their passing, they returned to their own camp. Cheered by the event they returned to their fires and, as the day progressed, their situation became yet more comfortable, even salubrious. The valley was sheltered with numerous trees and contained clean water, but, best for them in their situation, was the arrival of locals with carts covered in all varieties of the provender so often denied them on the march: fruit of all kinds; meat and fish, smoked or salted; fresh bread, cheese and vegetables; and, of course, wine, a rough red. Soon all messes were enjoying the best meal they had eaten since arriving in the country, all diners praising not just the quality, but also the generosity of the local Portuguese. Few coins had been required to obtain the food which soon found its way into every mess pot in the camp, including those of the Officers and, even these soon became used to the taste of the wine and found their teeth less set on edge. For the rest of the day, all was peace and tranquil in the camp of the 105th, all enjoying the bucolic, even holiday, atmosphere.

Carr, Drake, and the ashen faced Shakeshaft all sat on various sections of tree trunk, hauled into position by Morrison, to make a kind of circle around the now dying cooking fire. Their third beaker of the local "red" was, at last, not now tasting quite so acid, and even seemed to be perking up Shakeshaft, the alcohol dulling the pain in his head. Carr took another drink and this time his mouth did not contort quite so much as last time, although his teeth still felt the affects. The idyll of their surroundings was turning his thoughts to home.

"Morrison! Did you get that letter away which I gave you the other day?"

Morrison did not pause from his domestic duties.

"Yes Sir. It went with despatches before Rolica. That was when you give it me. Sir."

Carr nodded and looked at Drake.

"It was a good letter, was it not?"

Drake pursed his lips and nodded gravely.

"I feel so, yes. In fact I'm certain. I would certainly think so, the number of times you drafted and re-drafted. But the kernel of the matter was made plain, to whit, your commitment, undying affection and the fact that only she, her, can make you a happy man. That made the mark, plain and clear, the rest is now up to her."

An unpleasant thought suddenly entered Drake's head, one that he would not put past the likes of Henry Carr.

"You did include that sort of thing, did you not?"

Carr nodded vigorously whilst setting down his beaker for another re-fill.

"I did, yes, I did. Those were your specific instructions, were they not?"

"They were, and of the highest import."

He looked up to see the three Grenadier Officers passing by; Carravoy, D'Villiers and Ameshurst. The wine was working up Drake's natural bonhomie and he raised up their carafe.

"Gentlemen, would you care for a glass. What's at the bottom is slightly more palatable than what's at the top!"

Only Ameshurst grinned at the ridiculous comment, whilst the remaining two remained stony faced. It was Carravoy who answered.

"I thank you, but no. Wine somewhat closer to vinegar than anything remotely drinkable, does not lie within my taste. Perhaps some other time."

D'Villiers had not even looked in their direction to acknowledge the invitation, but Ameshurst felt able to accept and did so, joining Drake on the log, and happily accepting a beaker from Carr, which Drake filled. The ensuing conversation between them was cheery, non-military, inane and not a little drunken.

However, if there was a discourse within the whole camp on any level of serious content it was taking place around the wagon of Chaplain Prudoe. His wife, Beatrice, had the Mulcahy clan assembled around her, plus Nelly Nicholls, and it was she that was merrily describing the life of a camp follower, about which Beatrice Prudoe knew nothing, from which grew her deep fascination. Nelly Nicholls soon pronounced her judgement.

"Sure, it's not so bad, to follow the army. You're with your man, to keep an eye on him, and 'tis better than losing the draw and staying in barracks. I say "stay", 'cos sometimes they throws you out, into some God Forsaken camp, under tents, without too much thought on the weather."

Beatrice Prudoe's brows knitted.

"The draw?"

"Yes. You draws lots to follow the Company. About six to ten families comes out, and only them as has childers as can help goes into the hat, in the first place. All mine can, as with Bridie here. We've been lucky and come out twice, but there's not that many families in the 105th that goes into the hat. Most has babbies and such, or little more than."

Bridie nodded vigorously.

"She's got the right of it. 'Tis better to follow, an' put yourself forward, despite the hardships. Here in the sunshine, 'tis more healthy, is what I'm thinkin', to bein' stuck in a stinkin' barrack. I've heard tell of men returnin' home to find their loved ones dead, died of gaol fever or somesuch, when they themselves've come out whole from a battle!"

She sat back to arrange her arms under her bosom.

"Out here, there's fresh food from the locals, like now, and army rations when there's nothin' else."

Mrs. Prudoe remained spellbound.

"But you march with the army. You must keep up, in all weathers."

Nelly Nicholls fashioned the reply.

"Ah sure, but you get used to that, and, as often as not, we gets left behind, if there's fightin' comin' up. That's a rest, you understand."

She had a further thought.

"An' sometimes they marches back to us. We don't do the same distance as themselves, if you understands me."

Beatrice nodded, but continued with a new subject.

"But the danger to your husbands. You could be widowed. Easily."

Both Bridie and Nelly grinned broadly, and knowingly, but it was Nelly who spoke.

"Ah sure, you get used to that, too! Bridie's on her second and I'm on my fourth! If your man gets kilt, 'tis not long before another makes an offer, sometimes two or three of the pantin' sods!"

Nelly slapped her hand against Bridie's arm, inviting her to share the mirth, but it was already there. Beatrice Prudoe smiled at the gales of laughter.

"I must say that you are extremely resilient. Whatever comes, you cope. You have my admiration."

"Ah well, there's not too much of a downside, but there is some. We are under military orders, you understand, just the same as our men. We have to do as we're told, and, whilst they can be flogged, so can we. Not yet, in my case, but close, a couple of times. Same for you, Bridie?"

Bridie nodded.

"I've heard and seen, but as for meself, no, not yet, and I hope not ever."

The gentle wife of the Chaplain sat back shocked, her hands falling into her lap. Bridie prepared some more bread and butter, whilst the Chaplain himself, returning from his rounds and seeing his wife now in the society of camp followers, abandoned his return to the wagon and strode off, hands pointedly clasped behind his back, for another tour of the camp. The children played quietly, practising their letters in the brown dust and Mrs. Prudoe, pleased at the sight, but still shaken by the revelations of their Mothers, sliced some oranges and lemons for them to eat. When the youngest found the lemons too bitter, she corrected that with the help of some sugar.

***

The order came at 6.00 in the evening, shattering the surreal atmosphere of peace and tranquillity to quickly bring all back to reality. Everyone was to be in position before dawn the next day and so the army immediately prepared for bed. Carr, as a Captain, had been called by Lacey to hear the written orders now arrived and he was immediately questioned by Drake on his return.

"Where are we?"

"On the ridge above Vimeiro. Far left."

"Not in reserve?"

"No. Front line and right in the way!"

Drake nodded, knowing what that meant, and rolled himself into his blanket, besides Shakeshaft and Morrison. After but a few hours and very pre-dawn, Sergeants and Corporals went around the camp to wake their charges. Bridie woke with Jed Deakin to "bring im up" with a "drop of tea", whilst Jed himself, threw water on his face and donned his full kit. Nearby, Miles, Davey and Pike were doing the same, Joe Pike the last, for he spent much time kneeling by Mary. Now aware of his attention, she fully opened her bleary eyes

"I should get up, make you some breakfast."

"No, you stay. Tom's doing that. I'll be fine."

He smiled at her with all the tenderness he could muster.

"That's fine for me. His cooking's coming on, you know."

This was said loud enough to produce a scowl and an evil look from Miles, but a full smile from Mary. Joe pulled the blanket up further.

"You stay there, get all the rest you can. We may have to move soon and keep going. Bridie will take care."

He stood up and took the mug of coffee from Miles, to be followed by a thick slice of bread covered in last night's stew. Soon, and almost too soon for them to consume their food, the bugles sounded "fall in". With their dying notes, Mary did rise to hug Joe fiercely, whilst Bridie and Jed shared a gentle embrace, which was soon transferred to the children. The 105th fell in and marched off, towards the faintest blue beyond the hills to the East.

They passed through Vimeiro, this a well appointed and prosperous village, made up of solid whitewashed houses and a tall church on a mound off to the left, but throughout the village all was quiet and deserted, save the odd hungry and curious dog. The inhabitants had left; an army nearby meant strife and conflict, right on their doorsteps, so no lights showed behind the shuttered windows set deep within white walls. Their ancestors had chosen a spot to build on that stood sheltered from the Atlantic storms and winds from the North, but now bitter winds blew from the East in the shape of conflict borne out of French ambitions throughout Europe. When this wind blew, there was no kind of shelter.

Soon, after the village, they were climbing a steep hill, but on the ridge, above the village and in position held just back from the crest, the 105th stood to in the lessening gloom. With the full dawning of the day the order came to stand down and wait, but to remain in position on the rear slope, therefore the men, in their two deep line, sat, lounged and talked, whilst the sun rose and the heat grew.

Within the full light of morning, Carr, Drake and Shakeshaft, took themselves to the brow of the hill and examined their surroundings. The 105th were on the far left of the British line that was holding Vimeiro ridge, making the Light Company, on the left of the 105th, at the very end of that line. Before them was a shallow valley that became deeper as it ran down to their right, but on their left the valley began, at a track that used the extra height to run out of Vimeiro that was behind them, to then curve around the head of the valley and lose itself amongst the distant olive groves and vineyards that girded the ridge opposite. This ridge was similar to their own and, presumably, stood waiting to be occupied by the French, whilst behind them, if they turned to look, could just be seen the bell tower of Vimeiro church, marking the position of the village itself. The three looked across the valley but made no comment, until the questions came, inevitably from Drake.

"What does he know, d'you think?"

It was Carr who answered, Shakeshaft had not a clue as to the answer and, besides, he felt out of sorts, but more from the wine of the previous day than from his experience at Rolica.

"I take it you mean our esteemed General. Well, as far as I can ascertain, we are stood here because he knows that they are coming, to find us. When they are coming he seems not to know, he seems to have got that wrong. After all, they are not here!"

He took out a white kerchief and took off his shako to wipe his brow, fully exposing the two scars, now livid with the heat. He let out a long sigh.

"Find Morrison. Get him to make some coffee, or tea or something. The same for the men. Noon's not far, and they've had nothing since breakfast. Whatever's in their haversacks, let them eat it! If we're still here come evening, we'll get some more rations brought up."

He placed his shako on the ground, swung his sword away behind him and sat on the ground to contemplate the slope in front. Drake and Shakeshaft also sat near to look at the same, but it was Carr who expressed their thoughts, as much to raise confidence within them all, as to reassure the novice Shakeshaft.

"They're coming from the South, so they'll have to come up that slope."

He paused.

"Very open."

He paused again, as if carefully finding the next words.

"That's a long way, for any army, in any formation."

Again he paused, looked at his companions and smiled.

"Especially against the likes of us!"

Just then the coffee arrived with slices of fresh bread. Morrison had anticipated Carr's order to take food before Noon and he had brought extra, this being three local fruit, of a type unknown in England. Drake was the first to bite and his face expanded in pleasure.

"I say, Morrison. These are rather good. What are they called?"

"It sounded like "pessaygo", Sir. So I'll continue with that."

"Continue indeed. Whenever you see some, buy some. You'd agree?"

He was looking at Carr but he was too busy dealing with the juice running down his chin. Shakeshaft was on his third bite.

Meanwhile, one hundred yards along the line, refreshment was also being taken but by others of a higher ranking order. Brigadier Fane was stood with Lacey and O'Hare, alone on the ridge top, studying the Rifles of Fane's Brigade positioned in skirmish order down in the valley. The 105th sat in their line behind where the three stood, every man now eating or drinking. Fane took a drink from his spirit flask and passed it to Lacey.

"Colonel, I hear that in the 105th ye've been practicing some kind of fancy firing drill? Wellesley expects the French to attack in column and I'd say he's correct. It's their usual way of going about their business, so do we hear."

Lacey cleared his throat and searched for words to best explain what could prove difficult.

"Not too fancy, Sir, we still have half Company volleys, the front ranks of the centre two companies firing together to start us off, Three and Five in our case, and then the volleys commencing outwards along the line. What's different is that we don't wait until the end Companies of the front rank give their volleys before the second rank commences. The second ranks of Three and Five give theirs as soon as the smoke clears. By then any dead or disabled at the front of the column have fallen over and those marching behind should, er, not be

96

left to walk forward unmolested, for however short a time. That's how we see it, Sir."

He looked at Fane.

"I hope that's clear, Sir, and I hope you approve?"

Fane took another serious nip from the flask.

"Oh, aye, Colonel, I approve. Anything that sends bullets into a French column at a rapid rate has my approval. Let's hope we send in enough! Eh?"

With that he toasted the pair with his flask, took another nip and walked to his horse, the mare held steady and waiting patiently in the charge of Fane's own mounted aide-de-camp. Lacey and O'Hare contented themselves by staring out at the shimmering hills, baking in the heat, and viewing the distant road, dancing snakelike in the thermals, up which the French must come.

Further right still, with the Grenadiers at the far right end of the 105th, Carravoy and D'Villiers were sitting down on the reverse slope, their backs to the same distant hills. They looked with some distaste, Carravoy more so, at Ameshurst walking amongst his men, sharing jokes and comments, relieving the tension. Whatever he was asked by them Ameshurst tried to answer and this his men appreciated, always thanking him and parting with a smile. However, in truth Carravoy and D'Villiers paid little attention, the minds of each were far too preoccupied with the forthcoming battle, one that evidently was going to be set piece and serious. The societal rules that both had been brought up to adhere to, required some neutral conversation, about one's tailor, or hunter, or whether to chose a phaeton or a cabriolet as one's next sporting carriage, but no such topic occurred to either, bar the one that was barred, the one whose very mention would betray the fear and anxiety that dwelt within each. They were not alone; all along the ridge Officers gathered to act out the required blasé insouciance, yet all, frequently, sent glances in the supposed direction of the oncoming enemy. Finally, D'Villiers did find something which was sufficiently neutral.

"Shall we take a bit of a wander, up over there? See what those gunners are up to?"

It was Carravoy raising himself from the ground that gave D'Villiers all the affirmative answer he needed and the two strolled over to the battery of six guns positioned between themselves and the next battalion along the ridge. The battery Captain saw them coming and walked forward to greet them and shake hands.

"Morning. Ellis Broughton, Royal Artillery."

Neither Carravoy nor D'Villiers were certain if that was a double-barrelled name or not, but they introduced themselves and began to study the six guns, now at close proximity, which was new to them both. Carravoy took the initiative.

"Mind if we take a look?"

The Captain became even more cheerful, plainly eager to accommodate anyone showing any level of interest in his six charges.

"Why no, of course, allow me."

They walked behind the line of guns, each with its blank, threatening muzzle, and saw the preparations of the gunners, if preparation it genuinely was, for all had been thoroughly prepared for a long time. They had been given the task of carefully examining the bags of powder stored in the ammunition caissons, six "boxes on wheels" placed down the back slope, each behind its own gun. Each bag was taken out and examined for leaks, then carefully replaced. The expressions on the gunners' faces showed that they knew this to be an utterly pointless task, which they had been set merely to keep them occupied. Then Ellis Broughton seemed to have a good idea.

"Have you seen these?"

He pointed to a box of what looked like normal cannonballs. Carravoy and D'Villiers looked querulously at him, then turned to follow the direction of his arm.

"They are called shrapnel, invented by a chap of the same name. It's filled with musket balls! We call it a shell, because it has a hollow inside, like the shells you find at the seaside."

He raised one up and both could see, albeit circumspectly, the difference between it and common solid shot. Fastened to the side that had been facing down in the box, presumably for safety, was a wooden disc with a hole in the centre. Ellis Broughton pointed to it, accompanied by a very enthusiastic description.

"This is a fuse hole. Before we fire one of these, we insert a timing fuse. Very short, so that it explodes somewhere along its trajectory, but before it reaches its target, that's very important. Then the musket balls, a whole cloud of them, carry on to the target with the same velocity that the shell had. Very nasty for them but good for us;

with these we can do them a lot of no-good way before they get anywhere near."

He paused and screwed up his face, conveying an element of doubt.

"At least, that's the theory and today, if we do fight today; is the first time we'll have used it in anger."

He shifted his jaw to one side, closed one eye and looked at the sky with the other.

"Well not quite. We fired it at some Dutch a while back and they quickly surrendered! We have great hopes for it and the trials have been good."

D'Villiers and Carravoy leaned forward to take a good look, but not handle. D'Villiers continued the questioning.

"You mean this thing goes, "bang", in the middle of them?"

"Well, that would be sort of good, or above their heads, but just before would be better, giving the musket balls a chance to spread, you see."

Carravoy took up the point, equally curious.

"So, who sets the fuse?"

"Me, that's my job, to get the length right."

"And you light the fuse, before putting this "shell thing", in the barrel?"

Broughton looked shocked.

"Oh no! Oh my word no! When we fire the gun, the explosion inside the barrel ignites the fuse, or it should, as it goes on its way off to Johnny. The wooden disc faces the gunpowder in the barrel, so's the fuse gets the explosion."

He twirled the shell above his head.

"All quite simple, really."

It was D'Villiers mind that had moved to technical considerations.

"What's the shortest? I mean, when can you no longer use it, because the range is now too short?"

Broughton nodded, at the posing of a sensible question at last.

"Well, the shortest we've managed is 300 yards. After that, well, it's canister. Or case, as some say."

He pointed to another box, just back from the gun, containing shiny tubes of metal, each about seven inches long. Both had heard of canister, but its precise design was unknown and D'Villiers proved to be the more ignorant of the two.

"What's in those?"

"More musket balls, packed in sawdust. It's like a shotgun cartridge, only much bigger. The case splits on firing and the balls spread out. But that's only much use under 300 yards."

He then pointed to another box.

"Between the two we could use that."

He allowed time for his audience to shift their attention.

"Grapeshot. Same kind of thing as canister, but with larger balls."

With that Broughton adopted a look both cheerful and quizzical, his mouth smiling, but his eyebrows raised, as if to enquire if there were any more questions. None came, for both Carravoy and D'Villiers, as infantrymen, had realised that French gunners could apply the same to themselves, so Ellis Broughton continued himself.

"So, there you are! We'll be doing the Frogs a lot a damage, way before they get to you!"

He allowed the encouraging words to sink in.

"Then it's your turn to take a crack."

Both Officers nodded, encouraged but not convinced. Carravoy was the most sceptical.

"Well, here's hoping, for all our sakes."

They shook hands and returned to their men, to find Ameshurst sat alone, looking for the French. They walked past him, saying nothing, then on, to the centre of their Company. It was D'Villiers, him holding inside the most anxiety concerning the future, that the conversation with Ellis Broughton had had the most impact on.

"Well, seems we have a trick up our sleeve! Perhaps this job won't be so hard, after all!"

Carravoy nodded, but said nothing, which in turn, did nothing to reassure the worried warrior D'Villiers.

The afternoon wore on into evening. Wellesley and his Staff rode along the front and a Major detached himself from the group and approached Lacey, him stood with O'Hare before the Colour Party.

"Sir. The General says to tell you that the French will arrive tomorrow. So, for now, allow the men their meal, but you are to remain in position throughout the night. Stand to before dawn. Sir. The Rifles will stand picket."

Lacey nodded to the Aide de Camp, then to O'Hare. Within two minutes, what had been said, and completely overheard, had been passed on and out to both ends of the line. The rations were sent for and the men allowed to gather wood. Soon could be heard the pop of gunpowder being fired in the flash pan of many muskets for the burning powder to then be tipped over the dry kindling, which all soldiers kept a supply of in their packs. Soon after that, could be seen many cheery fires, all across the back slope, along the full length of the ridge. As the three, Pike, Miles and Davey, sat finishing their meal beside Byford, Saunders and Fearnley, some speculation was voiced about the following day, but the reply was mostly, "No way of telling!" Joe Pike turned to Tom Miles.

"What will they do with us, Tom? We're Lights, could be anything!"

Tom Miles smiled in the dying light of the cooking fire.

"You just answered your own question, boy. Could be anything!"

***

Most were awake before dawn and the NCO's had little work to do in rousing the army. The fires, by now already lit and kindled, gave light to those who passed through and around, needing to do but little to persuade the men to prepare for the battle that all now accepted as inevitable. Breakfast was done before the sun departed the horizon, after which Sergeant Fearnley moved amongst Shakeshaft's section checking that all was in progress. A "horse of a different colour" to Ellis, Fearnley was content to trust those whom he knew he could, and to spend more time with those more suspect. He had few enemies, thus to Miles and Davey there came but a brief word.

"All set, Tom? John?'

A brief raising of a hand was the only reply he needed.

"You'll look over Joe?"

This time came a nod and George Fearnley moved on.

Some of Fane's Rifles from the 60th and 95th, held in reserve on Vimeiro Ridge for those that had remained far out in the woods and valley floor to the South, were the first to rise and walk over the ridge top to disappear down the far side. Keeping themselves apart, they had made their campfires at the closest point to their picket position and so had but a little distance to walk to resume their watch places of the previous day. Binns followed a group through the line of the 105th Grenadiers, carrying bread and a cup of tea for his two Grenadier Officers. He brought ill tidings.

"Benito has absconded, Sirs."

Both looked up, startled, and waited for more, so Binns continued.

"Gone, Sirs, with the mule and everything on it."

The reply came in unison.

"Everything?'

"Yes. Sirs, tent, bed and spare blankets. He kept on saying "Oh franchez nah oh poday ser bateedo" I know it exactly, Sirs, he said it so often. I think it means the French cannot be beat."

Both were on their feet, clenching and unclenching fists, both stamping the ground in their anger, but it was Carravoy who spoke.

"Where did you hire him?"

"Leiria, Sir."

Carravoy now pounded his right fist into his left hand.

"Damn him! If we ever return there, I will seek him out and personally shoot the damnable villain myself. Damn him! Damn him!"

Binns nodded.

"Yes Sir. Lucky you carry your personal effects yourself, Sir, and before this day's out, I fancy there will be more than a few spare blankets. Hopefully more French that ours, Sir."

Binns did not wait for a reply, he knew that anything that came would not be pleasant, so he beetled off to see to his own affairs and found all intact. He had left nothing in the charge of the local peasant Benito. His trust in such did not rise above his bootstrap, an opinion now thoroughly confirmed, but he soon mollified himself. Would he stay if his own armies had been annihilated time and time again?

Orders came that all battalions were to remain behind the ridge, but Officers could please themselves if they took themselves forward to examine the distance for the approaching French, and several did. However, they were disappointed, just as they had been on the previous day, staring again through the telescopes at the empty hills and the emptier road. The heat grew and many of the men arranged their blankets on their bayonets and sat within that meagre shade, sipping their precious water and not saying much, save Joe Pike, growing more animated by the minute.

"Will they come, Tom? We all heard the Officer, but perhaps they won't."

Miles was chewing on a biscuit, so Davey answered, whilst chewing a straw.

"The rumour's that Johnny's on his way. Best get your mind around that."

"Why're we kept back here?"

Davey looked at Miles.

"One for you, Mr. Experience!"

Miles looked angrily at both.

"How do I know? Seems to me that soon enough, when they comes, if they comes, we'll be over that ridge and wavin' halloo to our Frog friends, and kickin' off for a real set piece."

He paused and waved a dirty finger for emphasis.

"But I'll say one thing! I do hope that us walkin' over that hill and standin' out on that slope is delayed until the very, very last minute. I do not fancy standin' out there in full view of them Frogs, like them with the tassels on their hats, them backed up by that Frog artillery."

He terminated his tirade with a ferocious nod of his head, then returned to his biscuit. Pike remained silent and Davey re-checked the lock on his Baker rifle. Meanwhile, Fane had again found Lacey and O'Hare.

"They're on their way. Aboot midnight, the Dragoons heard them crossing a bridge. That means wheels and that means artillery, lots of it."

Lacey nodded and O'Hare raised his glass again to his eye. He studied the view for the briefest moment, lowered it, and then spoke.

"I do believe they're here, Sir."

He handed the glass to Fane and pointed.

"There, Sir. In line with the road, a cloud of dust.

Fane took the glass and adjusted it for his own eyesight. He spoke whilst he moved the glass slightly from side to side.

"Aye, that's them, right enough. Seems M'sieu Junot has arrived with his full Corps, or near enough as makes no difference."

He studied some more.

"And not in blue! Some kind of white, a summer uniform would be my guess."

A pause.

"And ah'd say that's the man. Junot!"

He handed back the telescope and this time he pointed.

"There, on that knoll to your right front. Arrived just this instant!"

He fished for his own watch and failed.

"What time do you have, Lacey?"

Lacey's huge hunter came out from one pull, hauled up by its chain.

"9.22. Sir."

O'Hare had taken the glass and then trained it in the direction indicated. What swam into view through the heat haze was a collection of Officers, very obviously French Staff from the gorgeous design of their uniforms. One was advanced just slightly and in the centre. O'Hare handed the glass to Lacey, who was doing his best to see with only shaded eyes. Lacey took the glass and focused it quickly. He held it for but a minute, before lowering it.

"Gone!"

Fane looked puzzled.

"So measly a glance! Ah well. What's the difference?"

His face grew serious.

"Now. Strict orders from Wellesley, your men are to remain on the reverse slope, until called forward. By me. You can stand where you like, whilst ye wait, but your men stay back, out of harms way! If they see ye, their artillery and skirmishers will be on ye in two shakes. Like flies on a dung pile! Ye're tae stay back, until the time. Ye follow?"

He nodded to both for emphasis and received two in reply, plus Lacey's reassurance.

"Yes, Sir. That has been made very clear."

Fane nodded agreeably.

"Right. Not long now, so let's get the men some rum."

The order was given to form their firing line and, within the hour, the rum arrived, but almost all, because of the heat, watered it down, doubly so for Joe Pike; Davey knowing his incapacity for strong drink. Miles, however, did drink his neat and made a fierce examination of the whereabouts of the thumb of the orderly dolling out the measure. The digit was well back on the handle and all received full measure. The good wishes, spoken by the Orderly, were acknowledged by all.

"Good luck, boys. Come the finish, eh?"

"Come the finish, mate. Right enough."

Whilst the men sipped their strong spirit, all Officers were on the ridge top, telescopes extended out to study the oncoming French, these in three large divisions, spread across the wide river valley that led up from the South, a river being sometimes fed by the parched ditch at the foot of the valley before them. Now bayonets and shako plates sparkled through the dust confirming that this was indeed the enemy, in all his puissant strength. Time passed and the French came on, their Divisions easing over to their own right, revealing to all that they were attacking the British left, ignoring the troops that Wellesley had placed over on the steep hills of his right. The same telescopes watched as the British Regiments Wellesley had placed there, now quit their positions as ordered by him and hurried back to ascend the second, much higher, ridge behind Vimeiro. Wellesley was reacting to the French move, but the telescopes soon returned to the French deploying before them. Little was spoken, the only answer to the central question, 'who were they going to attack', could only come from the French, shown by where they halted and deployed. The answer came, when what appeared to be the main body of the French could be seen arranging itself on the ridgeline directly opposite that of Vimeiro ridge. The telescopes were snapped shut and Lacey spoke what was expressed in various versions all along the ridge.

"He's going to try us first!"

This was confirmed by the sight of the Riflemen filing back from the wooded hills of the ridge that was now occupied by the French. However, little musket fire was exchanged, for they were not being pursued. The green clad Riflemen had withdrawn out of range, then reformed and waited, unmolested. Plainly Junot was arranging his men and the telescopes on the ridge were employed again to observe the next significant development, that being two substantial bodies of troops and cavalry moving off, in Division strength, to try the British left on the high, steep, ridge behind Vimeiro. However, it was plain that the largest French force had been placed before Vimeiro ridge to make their main assault. Within the Colour Party, there was the greatest contrast in the level of agitation, or lack of it, amongst its members. Deakin leant casually on the muzzle of his musket, hands draped one over the other, but in marked contrast, Rushby and Neape craned their necks as if that would help them see what was happening over the ridgeline, twenty yards before them. Rushby had only Deakin that he could consult to any effect.

"Sergeant. What do you think is happening?"

Deakin took a deep breath, ready to indulge, as he knew he must.

"Well, Sir, I cannot give you the detail, but I'd say the French have arrived and our Officers is 'olding us back here, out of sight, so's to keep them guessin', them bein' the French. Sir."

Deakin studied the face of his Ensign and saw that the deep concern remained.

"Always a good idea, Sir, I'd say, to keep your enemy in the dark, as it were."

Deakin shifted his eyes to The Colour that Rushby was leaning on.

"But I'd say that it was time to get the covers off, Sirs."

Rushby looked even more concerned, as did Neape, this simple act would confirm that soon things would become very serious. Both nodded and pulled off the leather cover and Rushby's was stuffed into Deakin's haversack, beside the bag of hard biscuit and dried fish.

"I'd just keep 'em rolled; for now, Sirs."

Both Ensigns nodded and did no more, allowing both Colours to remain wound around their flagstaffs, for their minds were elsewhere, occupied by the growing sounds of conflict coming from beyond the maddeningly blank, brown, top of the ridge, but six yards before them. These were the sounds of skirmishing, at first distant sounds, but as the minutes passed, growing louder. Carr, Drake and Shakeshaft thought it right to remain below the ridge with their men, but Carravoy and D'Villiers joined many other Officers atop the ridge. From there they saw a sight that did nothing to boost their already meagre confidence, nor in any way diminish the nagging fear growing within each. The two looked in awe at a full Napoleonic attack, in fact two, one on the far right, but the other, disturbingly, aiming straight for them. It was a column, 30 men broad; Carravoy counted them through his telescope, and its depth appeared half as many again, about 45 men. They were already ascending the slope, sharpshooter tirailleurs skirmishing in front, three horse artillery guns each side, and cavalry behind. It wasn't hard to discern their intentions; that the artillery would unlimber at effective range, and blow a hole through the British line, through which the column would drive. But first, to open the assault, the tirailleurs would soften up that part of the British line, picking off Officers and individual men. Both Grenadier Officers could look at nothing but the

106

column, for even at long range the front rank had their bayonets "en guard" with the black of their high spats below the all white uniform emphasising their onward march. Their height was emphasised by the over large French shako.

Carr, meanwhile, had changed his mind and was now atop the ridge. He saw the column and gauged its threat. He had seen one before and knew what they must do to defeat it, but his attention and admiration was fixed on the Riflemen below in the valley. The French tirailleurs were making no progress against them at all. The Greenjackets were holding the range at 150 yards, a very difficult range for a French musket, but perfect for a Baker rifle. They were working in threes, one man always loaded, whilst the other two were either beginning a reload or completing one. Any Frenchman who was seen taking aim at the three was shot, and any that attempted to close the range was similarly dealt with. It was only when the front of the column closed up to the stalled tirailleurs that the Riflemen withdrew, to open the range back to 150. In fact, now, with the column advanced yet closer, the horse artillery had come within rifle range and both men and horses were being brought down to wholly hamper the progress of the supporting field guns. Carr saw three, so mauled by the fire from the Bakers that they were forced to stop and those remaining were now suffering severely. However, the main threat, the column, independently and of its own momentum, came on. He judged the time right to now return to his men.

Carravoy and D'Villiers had eyes only for the column and details emerged to be seen as the distance lessened, the Regimental Eagle above a crossbar with horsehair tails suspended from the ends. And as the distance lessened, so grew the sounds of the drums, for inside the column were drummers, beating the rhythm of the advance, driving the column on. Ameshurst arrived beside them, took one look and gave his judgment, not on the sight before them, rather on what to do next.

"Right. I think it's time for a word with my lads."

With that he left, leaving the two either exchanging glances between them that questioned their long term futures, or staring down the slope in mesmerised examination of the column. Ellis Broughton's six guns opening fire to their right jolted them back to reality and they left the ridge to stand before their men, not waiting to see the result of the shrapnel, that now forgotten. The explosion of the six guns gave

Deakin warning that soon they would be in action, time for The Colours.

"Best unfurl now, Sirs. 'Twon't be long now."

Both Rushby and Neape obediently shook out the folds, but retained the poles resting on the ground, giving cause for Deakin to say more.

"Best to get them up high, now, Sirs. Give the lads a sight, as it were, let's 'em know what they'm part of, so to speak. Gives a lift to the spirits when they sees the Old Bits of Rag go up."

Both Ensigns looked at Deakin, more to comprehend his attempt at stirring words, but then both poles were raised to sit in their holders and the cloth of each gently blew out, just enough to catch the eye of the Companies closest, and Deakin was right. Jaws tightened and hands took a better grip on the warm wood of their muskets. A bugle blew, which Deakin recognised as "recall". Recall for whom was quickly revealed as the Riflemen came back over the ridge to form up on the ends of the 105th. Lacey and O'Hare were the only Officers remaining in view atop the ridge and Fane joined them, his horse restless from the cannon fire.

"They're all yours, now, Lacey. You've got my Rifles on your flanks, but 'tis you as is needed to stop them. This'll need good timing, so wait for my word."

He studied the column himself for a moment.

"Seems our guns are doin' them a bit of no good, that at least."

The three saw shrapnel shells exploding before and above the oncoming column, causing a trail of white clad fallen bodies to mark its passage, but the oncoming march, through the smoke of the shrapnel, only added to the drama and the impression of an inexorable advance. The column had begun to shout, loudly, loud enough to be heard between the cannonfire. Lacey and O'Hare listened, long enough to make out the words, before they would have to descend the slope to their men, who could now hear well enough themselves.

"Vive l'Empereur! Vive l'Empereur! Vive l'Empereur."

The drums, if anything grew louder.

"How many would you estimate, Sir."

Fane took another look.

"Fifteen hundred would be ma guess, Lacey."

Fane looked again.

"Two minutes, Lacey, and I'll be sendin' ya down."

Lacey nodded and returned to his men to stand and face the Colour Party. Formed in their long, two deep line, barely half his men could hear, the guns to their right were firing for all they were worth, but he said it anyway.

"Men! There's a French column coming our way. They think that they are going to go through us, like we were no more than a bunch of pitchfork peasants. They're shouting and beating drums like they expect that's all that's needed; to make us cut and run."

He paused to take a deep breath, then to continue. All that could hear were listening, all that could not hear were watching their Colonel, with sword drawn, stood before his men.

"I know that's not true and you know that's not true. I think we should teach these cheeky beggars, now coming up, that we don't run at the beat of a few drums and a bit of bawling; that we are the One Oh Five and we've seen off the likes of them way before now!"

There were cheers in response, which began in the centre than ran down the line. Even though many did not hear, a good shout and cheer eased their nerves, but plenty who were near and had heard, shouted agreement.

"Right with you, Sir."

"Take us on then, Sir."

Lacey responded.

"Load!"

The whole battalion rapidly, but carefully, went through the process of loading, carefully, to ensure no misfires in the first volley. Within 20 seconds, the whole line was stood with muskets against their right shoulders."

"Lock on!"

Lacey saw the faces of the second rank appear between those of the first, as the rear shifted slightly to the right, all as set and grim as his own. He looked around to Fane, still on the ridge, who nodded, then Lacey turned to stand with his back to his men and took another deep breath.

"The One Hundred and Fifth will advance."

As the first foot was lifted and fell to earth, all along the line friends and comrades wished each other well and good luck, bar Captain Heaviside, who spoke to all.

"I am as strong this day as I was in the day that Moses sent me. Joshua, 14, verse 11."

Within the ranks canteens were let fall after a last drink to wet mouths suddenly very dry. A deep breath was sucked in to steady a beating pulse and palms now sticky with sweat took a better grip on the polished wood of their weapon. The 105th, holding their line, took the few steps up to the top of the ridge to take their first sight of the oncoming column. French Officers could be seen cavorting at the front, waving their swords, in confidence supreme, because, until now, they had seen nothing of any concern that stood in their path. To them it appeared as though they had but to march on, through and over.

Neape nearly collapsed, suddenly his knees would not work and he sagged back against Sergeant Hill, close behind him. He could not stop himself from mumbling.

"They're coming straight for us, here, in the centre."

Hill placed his left hand between Neape's shoulder blades and straightened him.

"That's their trick, Sir, to make us lads in the centre turn and run."

Hill paused, thinking of more comforting words,

"Just take a look down the line, Sir, both ways. Steady as a wall! The lads off to the flanks will see this lot away. Johnny don't know what he's in for."

Nervously, Neape obeyed, to look both ways and see the line extending away, long and unbroken, and already the furthest companies were edging forward to close with the flanks of the oncoming column, but Hill had more to say.

"Just keep The Colours up high and steady, Sir. There's nothin' those Crapauds wants to see more than The Colours waverin'. The steadier you are, the more worried they gets to be! So don't 'ee let 'em see it, Sir. Not even think it."

Neape slid his right hand higher up the pole.

The Officers of the 105th, Carr and Heaviside in particular, looked first at their French equivalents, advancing opposite, and immediately saw a marked change. Their antics immediately ceased, plainly brought about by the sudden appearance over the ridge of the long line of red coated British, muskets at "shoulder arms", as though they were about to parade down some high street. The line was extended by the green uniformed Riflemen, a line that reached beyond their column by a hundred yards each way. All in the line was parade steady, and when it had halted just over the ridge summit, waiting, unmoved, and unintimidated, something more than apprehension arose in every Frenchman who could see what lay ahead. At that point, at a range now down to little more than 70 yards, the British field guns exploded with one murderous volley, using what most veterans identified as grapeshot, the way the heavy balls passed through the whole column. This brought it shuddering to a halt. Lacey recognised the moment and came back to stand just before The Colours.

"Make ready!"

Every musket was raised high in the air, the movement spreading outward like a wave.

"Front rank. Present!"

The muskets behind him came to the horizontal, to be quickly joined by those further along the line.

"Fire!"

The front ranks of the two Companies either side of him fired in unison, to be quickly taken up, in their turn, by the front ranks of the Companies further out along the line. In under ten seconds the whole of the front rank had sent their bullets into the reeling column. Heaviside, Captain of Number Three in the centre, looked along his Company. The front rank was furiously reloading, the rear rank remaining at the "make ready". He peered through the thinning smoke and discerned for himself the vague line of black shakoes that marked the front of the column. He shouted above the din, drawing out the words and syllables.

"Rear rank. Present"

The muskets of the rear rank came down.

"Fire!"

Again the appalling noise and all was again obscured by smoke, but the noise now became incessant, from the half Company volleys running down the firing line, out from the centre.

The French column had marched up to confront the Colour Company, leaving the Light Company and the Grenadiers looking far over towards the centre to see any kind of target. On the left, Carr, unhappy with his ineffective position, approached the Captain commanding the Riflemen next to him, these even further from the French, more even than his own men. The exchange was absurdly formal, but requirements dictated that it be so.

"Carr. Hundred and Fifth."

"Runiack. Ninety Fifth Rifles. How do you do?"

"I feel that we should wheel in and close with the side of that column. I'm taking my men forward, will you support?"

Runiack grinned and nodded.

"Happy to oblige!"

However, on the far right, with their Grenadiers, Carravoy and D'Villiers stood in a quandary regarding what they should do. His men were firing obliquely at the column and at long range. To swing across to confront their side of the column would mean breaking the link with the field guns, leaving a hole on their right, a potentially dangerous development, but there were no French in view to exploit it, as yet. He stood in a state of uncertainty, looking at D'Villiers but gaining nothing, for he was transfixed in a state of near shock from the noise and carnage around and before him. Finally, the decision was taken for him. The Riflemen on his side were using their own initiative and running across to close the range, accompanied and without higher orders, by Ameshurst's Company, them closing up at the trot into two perfect ranks. Ameshurst was out in front and leading his men on. D'Villier's men also began running behind to catch up, led by their Sergeants. The two Company Officers, Carravoy and D'Villiers had little choice but to add themselves on to the rear of the advance and hope, from there, to exert some influence.

The British line, with the 105th in the centre, was now curling around the head of the French column. The Riflemen, still skirmishing down the slope, ensured that not one of the French guns supporting the column fired a single shot, then, with all the gunners dead or in retreat, the Riflemen, too, turned their firepower onto the column. For the British within the firing line, it was a place from inside the Gates of Hell; one of incessant noise, explosions inches from their faces from muzzles and flint locks, whilst all the while they hastily reloaded, biting open cartridges that filled their mouths with gunpowder. From the cartridge they had to carefully tip some powder into firing pans, close the pan, then the remainder into the half-inch muzzles of their weapons. Finally, to ram down the paper of the cartridge which included the ball, whilst ignoring, but registering, the grunt and scream of a comrade hit by a French musket ball. When reloaded they raised their musket to the vertical "make ready", the signal to their Officer that they were waiting to add their weight to yet another volley. This madness persisted for mere minutes, but it seemed endless, the cacophony of noise, screaming and the automatic motions of loading, awaiting the order, firing, then re-loading again.

Fane, still mounted, could see more than most and knew the moment had come, but such was the noise that he could shout but one word.

"Lacey!"

Lacey knew what he meant.

"Cease fire!"

A pause.

"Fix bayonets!"

The firing ceased from the centre as the order was relayed out to the flanks by the Company Officers. The smoke cleared to reveal the scene before them, but this time it was that of a charnel house. The front rank of the column was now several yards further back, not because the French had retreated, but because the front ranks of the column had all been downed by the continuous, half company volleys. There were so few ranks now remaining at the front that the Drummerboys in the centre and the Farriers guarding the Eagle could now be viewed through the one or two ranks still standing, all now shocked and badly shaken. Between them, the ground was covered with dead, dying, and writhing wounded, the whole made more sanguine by the contrast between the white uniforms and the blood of the many wounds.

Lacey waited but three seconds.

"Level bayonets."

All came down to the horizontal, a menacing and ominous line of bright steel. Lacey took himself five paces forward to stand beyond The Colours, but Redcoats were still falling; the French column was not entirely beaten, some were firing still. Lacey drew his sword and held it up high.

"The 105th will advance."

He set the pace, slow, purposeful and determined. His men followed, stepping over their own casualties to close the short space between them and the remains of the column, soon reaching the French casualties caused by their musketry. The long, majestic overlapping line, the leveled bayonets, the stern advance, was too much for the badly mauled French. They broke and ran, long before any bayonets were used, but then Lacey held up his sword, forbidding his men to follow. However, not the Riflemen, they, all trained for individual action, ran forward with their Officers, first, to continue picking off individual Frenchmen and, second and most important, to capture the six guns. With the guns quickly secured, their own pursuit ended, then the Riflemen made use of any horses remaining and triumphantly pulled the guns back and through the British line, pulling on the harness themselves to speed their progress.

Tom Miles was less than pleased at the order that came next. It took him away, from what was for him, the main point of the whole affair.

"About face! Retire!"

The whole line, turned around and marched back over the ridge top, leaving the camp followers, who had come up behind them, to tend the wounded and help those who could be moved back to the Surgeons.

"No bloody chance of a bit of plunder. Not stuck back yer again!"

Ellis had overheard.

"Looks like your pack and boots will have to wait a while longer, Miles."

Miles was incensed enough to ignore the difference in their ranks.

"An' don't tell me you won't be aimin' for a bit of French kit. Nor a drop of brandy, neither.

Ellis for once took no umbrage, he was, as were the others, elated at the ease of their victory; a full French column defeated and dispatched in no more than five minutes. Relief and exhilaration ran through the thoughts of all. So also for Lacey and O'Hare, stood atop the ridge amongst their own dead and wounded, but very satisfied to see that the other column that had also assaulted the ridge further over to their right, had been similarly dealt with and was now but a rabble, streaming back to the French side of the valley. They were enjoying the view when no less than General Wellesley himself rode up, accompanied this time by merely three Staff only. Lacey and O'Hare sprang to the attention, but Wellesley was as jubilant as themselves.

"Lacey! O'Hare! Damn fine job! Damn fine! They couldn't stick it, not here against you, nor over in front of Anstruther. Your one battalion saw off two of theirs! Damn fine work! My word, yes, deuced fine work!"

He paused but to draw breath, as elated as anyone.

"That's the answer, you see. Us to hold back on a reverse slope, then come at them, at just the right moment. It makes for a very nasty surprise, you see. For them, but not for us!"

He looked across the valley at the disappearing masses.

"Ha!"

Then his eyes were drawn to the movement of Fane joining them, mounted with more Staff than Wellesley.

"Fane! Well done! Well done indeed. Now. Hold your men in place and ready."

He looked again across the valley.

"M'sieu isn't done yet. He has plenty left in him, back over there, and he'll try again, be certain."

He wheeled his horse away, this action corresponding with his last words.

"Hold here. This isn't yet done."

Back amongst the 105th the men were drinking water and receiving fresh cartridges, at the same time making gaps in their ranks to allow the dead and wounded to be dragged back off the ridge to clear the way for any required advance. Anyone not needed in the firing line was now attending to this task, orderlies and transport waggoners having joined the camp followers, including Bridie, Mary and Nelly Nicholls, but Mary went straight over to the Light Company, where she could check on Joe. Much to the amusement of his fellow Lights, she clamped him tight around the neck for almost a minute, before moving on to help with the wounded, some now mercifully unconscious, others too much awake, screaming at the pain from their wounds. However, few of those remaining in the firing line, paid much attention, even to those screaming from hideous wounds. They were too busy cleaning muskets and checking flints.

Out before the Colour Company, Chaplain's Assistant Sedgwicke was stood with Prudoe, both saying the Last Rites over those allowed to die before being dragged back over the ridge. Still remaining in their ranks, Halfway and Deakin greeted Sedgwicke like a long lost comrade, a familiarity that still oddly rankled with the ex-Cleric, but even more so with the more elevated Battalion Chaplain Prudoe.

"Look Toby, 'tis Parson. Hello old Parson!"

Deakin had noticed something missing with Sedgwicke, who, never being that efficient a soldier, had forgotten his canteen, but Deakin was ready to put that right.

"Parson, thee'ce need a drop of water? Here, take a swallow."

Sedgwicke suddenly realized that he did have a severe thirst, but refused.

"I thank you, but you'll need it. I'll take a drink from the canteen of one of the dead. I'm sure there will be no objection. But I do thank you again."

Deakin nodded and smiled. All in the ranks had some level of affection for the hopeless soldier Sedgwicke, but they respected his learning and his attention to his duties, this now being fully occupied to comfort the dying and wounded. As Sedgwicke went back up to the ridge, Deakin looked at his Ensigns, both as elated as schoolboys and still stood holding The Colours aloft and erect, hands just beneath the cloth. Deakin smiled at their trancelike happiness and attended to cleaning his own musket. Then the order came to sit and rest.

Accompanied by Brigadier Fane, Lacey and O'Hare stood on the ridgetop, it now cleared of casualties, and all three were watching the French position beyond. The sounds of battle had now died away completely, to be replaced by the groans of the wounded and dying from the French column, whom no-one attended to, neither did they pay any attention to the odd French soldier who managed to rise and begin his painful journey, to hobble or stagger back to his own side. All three were waiting for Wellesley's prediction to come true and between them little was said, in fact, Fane's eventual observation was amongst the first words spoken, after some fifteen minutes.

"Here it comes."

He used the singular, when, in fact, there were two, but the column far on the left was already disappearing down the slope to its right, to another part of the battlefield, this being the valley much beyond the 105th's left which contained the main road into Vimeiro. This was, in addition, much beyond and below the track that could be seen at the head of their own valley spread in front of them. The three telescopes came up as one to examine their new opponents, but it was Fane who pronounced judgment.

"Grenadiers!"

Lacey and O'Hare lowered their instruments to look at him, waiting for more, but when none came they raised their telescopes again. The uniforms were certainly different from those of the first columns; it was now genuine French blue, with red epaulettes, and large shakoes also much decorated with red. This uniform was evidently being worn by tall men, therefore Lacey and O'Hare had no cause to disagree with Fane's pronouncement. Junot was sending forward his elite assault troops and the size of the two columns indicated that he was using all the Grenadiers he had. They were supported, some way back, by what seemed to the British to be probably the rallied wrecks of the two previous attacks, evidenced by their white uniforms.

Telescopes were not required to answer the next question which each asked, silently, of themselves. "Who would they be opposite?" The specifics were unknown, but it would not be the 105th. The Grenadier column was aiming for the centre of the British line on the ridge, a point somewhere beyond the battery of guns to their right. Lacey spoke aloud the thoughts of each.

"That's murder, damn murder. He's sending one column against a line that's just beaten two! Even supported, they are out front and exposed. Murder!"

Fane nodded sagely.

"Aye, ye may be right, but dinne' forget yon that's just gone off to the left, down to the main road, I'd say. He's holding us here with yon column, knowing it'll get beat, but hoping t'other will get in behind. Keep it in your minds, ye may need to re-deploy!"

The three Officers looked at the oncoming column, a variety of emotions competing within them. The magnificent sight of the lone column, the best soldiers of the best army in Europe, coming on against hopeless odds. Sorrow, admiration and relief ran across the minds of all, relief that a victory, at least on their ground, was all but assured.

Even as they watched, the column was being mauled, not by any Riflemen, but by shrapnel. At least half of the explosions from the new device were exploding perfectly above and before the column. The British artillery, with none of their own skirmishers out front to hinder their field of fire was firing at the best rate they could manage. The French column continued advancing, leaving a trail of dead and wounded on the trodden grass behind. The one point in their favour was that they were bringing forward eight guns, four each side, these with progress unhindered, for there were no Riflemen out to oppose them.

Fane stood evidently pondering.

"This has to be done quick. If nae man has any thoughts to this I have. There'll be doings to our left and behind, more serious than anythin' we'll be getting' from these poor beggars."

He spoke as though thinking out loud, then turned to Lacey.

"Get yours up and ready to move against their flank. I'll send down my Rifles to deal wi' the guns on this side. You close up to the column and see them awa'. It'll no take long, then we'll see what comes next. As I say, be mindful that ye may need to face left. Or even behind ye!"

With those final dire words, Fane mounted his horse and rode to the guns adjacent to the 105th, still assailing the column with bursts of shrapnel. Lacey looked at O'Hare and they walked back to their men, who stood up at the sight of both coming into view from over the ridge. Lacey sent a runner each side to gather his Captains and when they arrived he spoke differently to those from each wing.

"Right wing Companies, there is another French column approaching, alone, heading for the centre of the ridge, over to our right. Use your judgment, you may need to swing down against them, as our left wing will be, but I see you stood almost in their path and opposed to their right front."

He looked at Carravoy.

"You're leading our far right, I expect your Grenadiers to close right up to their front. Muzzle to muzzle! This needs doing quickly!"

He fixed Carravoy with a direct and unequivocal stare to add emphasis to what he had ordered.

"Now go! See to your men."

As the five Captains of the right wing ran off, Lacey turned to his Captains of the left.

"As you heard, you swing down and round to come upon their flank. Make the whole column feel your fire. Their guns will be taken care of. Now go!"

To add emphasis to his words, Riflemen were already disappearing over the crest to tackle to French field guns on their side. Lacey then looked to his centre.

"Set The Colours."

Without looking to ensure that his order was being carried out, Lacey turned his back, and drew his sword.

"Load"

He counted the twenty seconds.

"Shoulder arms".

He felt as, much as heard, the heavy muskets moving to right shoulders, then walked forward, his movement giving the order for his men to follow, again a long line, two deep. As before, men took a last swallow, but not with the deep trepidation of the time previous; the ease with which the French had been dismissed had not been lost on them. Another such triumph was eagerly anticipated and perhaps, as lay uppermost in the thoughts of such as Tom Miles, this time some plunder. The line crested the ridge and advanced on, over the casualties of the first French column, some endeavouring to crawl away, expecting a bayonet as a "coup de grace", but few, if any, of the 105th looked down at the desperate faces of the wounded, nor the blank faces of the dead. All looked to examine their next opponents.

Carravoy, on the right, was best placed to see a sight which greatly reassured him. The whole British line, as far as he could see, was advancing down, conforming with his own battalion. Despite all four of their supporting field guns having been brought to a halt by the Riflemen, the French Grenadiers were, with the utmost courage, still advancing up, but it was not long before they were also brought to a halt. The Battalion immediately against their front, now at a range below 100 yards, opened fire with a two rank volley; front then rear. The front ranks of the Grenadiers fell as if cut by a scythe, but the following ranks came on, their shouts undiminished, their drums defiant as ever. Whoever the British battalion was immediately in the French path, this battalion advanced further down the slope for another volley at murderous range, but the effect was lost in the smoke of the Grenadiers themselves returning fire, sparse as it may be.

Carravoy's Grenadiers, acting as a pivot for the whole 105th to swing right, were the first to come into effective range. He held out his sword horizontal.

"Halt! Make ready."

A pause. Carravoy was thoroughly controlling his men. He felt fully assured.

"Lock on."

If any French Grenadier was able to look across at the new arrivals, the precision of the arms drill displayed would have done little to lift their spirits.

"Front rank. Present."

The muskets came down.

"Fire!"

The muskets roared out, but Carravoy did not wait for the smoke to clear. The range was too low.

"Rear rank. Present."

As one the muskets were levelled.

"Fire!"

Again the crash and the smoke thickened further. Carravoy, now very calm, a feeling born of confidence in the outcome, continued to work his men, front ranks and rear rank. To his left the other Companies formed on the end of his Company and added their weight of fire to that of his Grenadiers hitting the French column. Thus, it was not long before, through the smoke, they could see the large black and red shakoes falling back down the slope. They had stood up to the dreadful musket assault for less than three minute, and they had had enough. Down on the 105th left, Carr's Light Company barely had time for one volley, before the French before them began to fall back, not quite in panic, but in no kind of controlled retreat. The French infantry in support, following on behind, had not even got up to effective range, before they fell back again at the sight of the elite Grenadier's own failure before them and the appalling sound of the continuous volleys .

However, Carr did not order ceasefire. As long as there were French within musket shot, a fleeing mob though they may be, he would order no ceasefire. The second volley had just been delivered when he heard a horse gallop up behind him and he turned to see his Brigade Commander, Brigadier Fane. Carr had no chance to salute before Fane was addressing him.

"Name?"

"Carr, Sir. Light Company."

"Carr, get your men back and over to the road. Back there. Beyond yon track."

He gestured with his thumb, back over his shoulder.

"Ye know where I mean?"

"Yes Sir. The main road in the valley."

"Aye, the same. The other column has got into the village, but are being held at the Church. Get over with your Lights and get onto their flank. My Rifles are on their way. Quick aboot it now!"

Carr took one look behind him, which confirmed what Fane had said. The Riflemen, having dealt with the Grenadier's field guns as effectively as previously, were scurrying up the slope to the head of the valley, the best runners already crossing the track there. He called for his "Commanders".

"Drake, Shakeshaft, Ellis, Fearnley. Get the men turned around and following those Riflemen back there."

A quick wave of his hand indicated the disappearing green uniforms. As the four ran off to their respective sections, Carr took charge of those nearest using the top of his voice, above the continuous musketry.

"About turn!"

When he saw their faces, he gave his final order, before running himself to obey those of Fane.

"Follow me! Load as you go."

Miles was at first distraught, then incensed, when he heard the order from Ellis. In front of him was a pile of the most flamboyant soldiers in any army, Grenadiers, always to be relied on to have gold rings and earrings, besides the usual wine, brandy, and other useful provender. And Officers with much silver on their uniforms!

"What's we got to do now? Ain't we done enough, done our share?"

Ellis was in no mood to argue. He had fully understood the urgency in Carr's voice and had seen the running Riflemen for himself.

"You get yourself running, Miles, with the rest. See you do, and your best speed, or I'll see you on the triangle! See if I don't."

Miles was enough of an old soldier to recognise the threat of a flogging and possessed of enough good sense to realise that Ellis was deadly serious and the business urgent. With not another word he turned and sprinted to catch up with Pike and Davey. They ran unhesitatingly across the wreckage of the first French column and their boots soon crunched over the gravel of the track, then they were on the grass slope above the road and there they halted. Below them was a scene of intense conflict. The Grenadiers were halted on the road along the valley floor, still in their column, but fighting furiously against all those that assailed them. The loudest sound of conflict came from further into the village, where the French must be held up by some opposing force, probably at the church, whose tower could be seen, above the thick, clinging smoke. Just below them, the Rifles had already formed a thick skirmish line and were sending their fire into the side of the column, whilst on the far side of the French column more

Riflemen were adding their weight to the assault. They must have run down from the high ridge behind Vimeiro.

Carr looked around and saw that his men had kept well together. He held up his sword.

"To me. Wessex Lights. To me!"

Those who saw passed on the order and soon Carr had virtually all his Company gathered around him. He could see little point in adding his men to the already thick Rifle skirmish line, because they were already very effective and any more would provide a target for the French Grenadiers impossible to miss at 50 yards range. He quickly reasoned that it would be better to threaten to cut off the French by closing with the road behind them; the fear of being cut off would, at least, cause the rear of the Grenadier column to retreat, hopefully dragging the rest with them. Doing that would do more to cause the French to retreat than simply adding 70 odd weapons to that of the Rifles already in place and his small Company would just be enough to at least create a level of concern. He raised his voice above the din of the conflict, restricting his words to the very minimum.

"Skirmish line as you go. Form on me. This way."

He led his men forward, down the slope and to the right, beyond the end of the line of Rifles engaged with the embattled column. Looking at the enemy all the while, even whilst running, he saw a Grenadier Officer waving his own sword and leading some men back to meet them; he had evidently identified the threat that Carr hoped to create. Carr knew that his one Company could not hold back a column two battalions strong, but now, at least, one French Officer was concerned over what was happening and that was all that was needed. The Grenadier Officer led his men back, many others joining, all matching Carr's Lights for pace. Carr stood and held out his sword for a line above the road, 100 yards up, where his Rifles would be accurate, whilst a French musket not quite so. His own muskets couldn't miss so dense a target. His Light Company formed a line and immediately opened fire, the first casualty from their fire being the French Officer; many had picked him out. However, they were answered furiously by the French and Carr's men began to take casualties, but the French Grenadiers, the whole column, was now falling back the way they had come. Whether from Carr's threat, or another from a different direction, or they had just had enough, their Commander had judged the task as hopeless and they were being assailed from every quarter. The reason was impossible to tell. What could be seen, even through the shrouding smoke, was the mass of Grenadiers now streaming back, still hundreds of men, for they had advanced as a column of over one thousand.

Carr looked left. Halfway down the slope as they were, they would be fully in the way of such a mass and, with it retreating quickly, they soon would be. He grabbed his Bugler.

"Bates. Sound Retire."

As the last note died, Carr gave the order.

"Fall back. To the top! Form again there."

Few could hear him, but all conformed to the bugle call and their firing line was again formed at the top of the slope. The mass of Grenadiers poured past, sullenly, many obeying their Officers to stop and form a firing line to answer and to halt their pursuers. Their defeat was obviously not as catastrophic as that of the other French attacks, but, nevertheless, the French were pulling back as quickly as could be, albeit in good order. Carr's Lights and the Rifles to his left, continued to fire, but soon all the French were gone, save the pathetic wounded, staggering back, using what support they could, and many falling, not from continued British fire, but from the severity of their own wounds.

Carr looked at the wreck of the Grenadier column, the road into Vimeiro almost invisible under the covering of blue uniforms, most unmoving, some still with life inside them. He looked once more, then pondered his own next move and decided; they had no further business there, their own Regiment was back where they had come from, still formed on the ridge. He found his Bugler again.

"Sound Recall."

As the notes sounded, Carr set out himself for the ridge where he knew the 105th had last been. He felt very tired, three conflicts in but a little over one hour had taken its toll. He glanced back and saw that his men were following, in no particular order, but at least more a column than a line. Drake and Shakeshaft joined him. He felt light headed.

"Hello you two. All hale and hearty?"

Both gave wan smiles from weary and dirty faces, then both heads nodded, but Shakeshaft pulled up his sheathed sword to show a significant dent in the bellguard. Carr recognised the shape; it had been hit by a musket ball.

"Where was the sword when it was hit."

"Held aloft."

Carr grinned at the thought.

"Very Lionheart."

He took the sword himself, to give a better examination.

"You are not to have it beaten out. That…."

He spoke the word loudly to give it better emphasis, and pointed, to give extra potency to his next words.

"……… is something you will show to your Grandchildren. I carried that sword the first time we beat the French in a set-piece battle! The battle of Vimeiro."

He smiled in Fatherly fashion at his youngest Lieutenant, then looked back at his men, all looked tired and grimy, but he felt pleased, even proud, his men had fought well, there could be no complaints. A sudden bugle call called him back from his self-congratulatory reverie, this further broken by an observation from Drake.

"Perhaps this is not all yet done!"

He need say no more, nor make any indication, because across their front, a cavalry charge, no less, was in progress. The strident notes of the bugle had sounded "the charge", and now that was exactly what they were witnessing, a headlong charge by the only cavalry the army possessed, the 20th Light Dragoons. However, it was not the fast canter, in a controlled and ordered line that would take them up to the enemy, to then accelerate at the last minute to maximize impact. This was a headlong gallop, almost out of control, and already beginning to lose formation. Their Colonel was way out front, with a few Officers in close attendance, which was drawing the Regiment out into a loose diamond shape.

Carr drew in a deep breath and sighed.

"I'm no detailed student of cavalry tactics, but that can only end badly."

He looked at his two Officers.

"Get the men in skirmish line. I suspect we won't be alone. They are going to need some cover before long."

And so it transpired. The Dragoons full galloped across the valley, smashed through a light cavalry screen placed there to protect the French retreat and plunged into the disordered and fleeing French infantry. The sight, albeit at a distance, of the raised sabres showed that they were now dealing out some measure of death and destruction, but back on the British ridge, Lacey's thoughts were matching Carr's concerns. He spoke aloud, caused by his deep disquiet, speaking of what he wanted to hear, that which should come from an proficient cavalry Commander.

"Sound retire. Come back, you've done enough. Retire."

But the conflict continued and could be seen right up on the crest of the French ridge. Then came what Lacey had been dreading, a counter attack by French cavalry and soon French sabres were being raised amongst those of the British. This lasted but minutes before odd and disordered groups of British Dragoons began to appear out of the scene of confused conflict and gallop back, pushing their horses to the limit, with most pursued by vengeful clusters of French cavalry. Lacey turned to whatever Officers and Sergeants he could see.

"Get the men forward! Out on the slope. Open order."

Suddenly, all along the line of the 105th, orders were being screamed out and the men were running back over their ridge, to descend down the slope of their recent conflict. They ran into the formation ordered and stood in their ranks, but they were not alone. The ridge slope was covered in Redcoats, all in open order, leaving enough space for any distraught Trooper to ride through to safety. Soon this was the case. The Light Dragoons, from all parts of the French line, made their escape and found refuge amongst their own infantry and the French cavalry did not pursue too far. Although they stood in open order, not a formal firing line, the French horsemen had concluded, from events not long previous, that this British infantry was of the highest quality and not to be trifled with.

Carr stood and watched. He knew that the French cavalry would not even come within range, therefore he had not even drawn his sword. He watched as the fugitives, on blown horses, their faces registering either shock and terror, rode through his own ranks. One came and halted before them, his horse utterly spent, its mouth agape to suck in air. Its rider was slumped over the saddle, his sword dangling from its wrist cord down besides his horse's right foreleg. Neither horse nor rider could go further, but Carr recognised an Officer's uniform and concluded that he should intervene himself. He walked forward and quickly saw that blood was dripping from the right sword arm of the cavalryman. Also, he had no helmet and blood was matted in his hair. Carr increased the pace of his approach, afraid that the rider would soon fall.

"Can I, or any of my men, be of any assistance?"

The head lifted to reveal a face streaked in blood and utterly weary, but Carr immediately recognised Captain Lucius Tavender. Carr was himself shocked by the depths of Tavender's evident pain and exhaustion.

"Tavender! You must let us help. My men will take care. Let us help you off your horse."

Some life seemed to come back into Tavender.

"No!"

His head slumped back down.

"I thank you, Carr, but no."

A ghost of a smile seemed to pass across his face.

"It is not that I reject your help."

He gulped some air.

"I am very grateful."

He swallowed hard.

"It's just that, …. that….. I don't think I can walk."

The ironic smile returned, matched by the same from Carr.

"I have a wound you see, a lance, in my leg, the right one. Best leave me up here, and get me in. If you'd be so good."

Carr could not help but notice the extreme change from the arrogant Tavender that he had known before, but he dwelt on that for but a second. He turned to his men.

"Riley, Evans, Thomson. Get Captain Tavender back into our own lines."

He lifted his own water canteen from his side.

"Here, take some water."

Tavender lifted himself just enough to make an angle for the water to pour into his mouth, steadying the canteen with his good left arm. Some dribbled from his mouth as he lowered the canteen; he could drink no more, he was so exhausted. He spoke as an exhaled breath.

"I thank you."

The three Lights were now stood near and Carr instructed them.

"Find him a Surgeon, any, I doubt his own will be easily to hand. And be careful, he is too liable to fall off. Keep him up there."

He turned to Tavender and had to stoop to see his face.

"My men will take care of you now, Captain. You are in good hands."

The only reply was a weak raising of Tavender's left hand as his horse was led forward. Carr watched him go for a brief moment, before looking for more returning Troopers, and there were many, mostly carrying a wound of some kind.

The defeat of the cavalry, although inevitable, left a sour taste in some of the army, "some" being mostly the Officer Corps. Amongst the men, however, now was the time to see what could be gathered from amongst the fallen of the French. Few Officers stopped what was almost traditional and, besides, it made sense. A French army lived off the land and every knapsack on every fallen Frenchman would contain food of some kind, meaning less food being required from their own Commissariat. In addition, why let it go to waste, although, all knew that the local peasantry would be out, come the night, and it was arguable that, with their crops now wrecked under a battlefield, they had a better claim.

Miles, Pike, and Davey were out amongst the French dead, those of the first column, because to go further over to the site of the Grenadier defeat would require too far a "wander away". The experienced Miles was directing all.

"Once you've got a backpack, look for boots and buttons. 'Specially on an Officer, an' provender too, don't forget that. 'Specially brandy!"

Chosen Man Davey rose up and gave Miles a "Go teach your Granny to suck eggs" look, but Miles was too busy to notice, for he had looked around and saw what he especially desired. Soon the buttons and buckles were removed as were the coins in the purse, but Officers carried no food, but this one did have a flask of brandy. With this now safely stored in his knapsack, he looked at the feet and shouted, still examining.

"John! You got a good pair of boots?"

The reply came from behind him.

"I have."

Miles looked for Joe Pike.

"Joe. There's some good boots yer, Officer boots, an' he's about your size."

As he helped Joe pull off the boots, suddenly firing erupted again, but distant and far to their left. Most stopped, whilst Miles gave it but a glance, but soon all were looking at the high ridge behind Vimeiro. They knew it to be the site of Ventosa farm and there they saw what had become of at least one of the two French Divisions that had marched off to their left, prior to the attacks upon themselves. One

Division was at the top of the long slope up to the farm and under intense pressure. Having made the long, direct climb up the hill, the observers down on Vimeiro ridge could see that it been formed into four columns, but these were now opposed by a long British firing line, which had, itself, split into four, the better to wrap around the heads of the French columns. It was not long before they saw the French break and run, just as those had which they had fought. The firing died away and they saw the French, as a white shape, streaming away up the valley, pushed by what looked like, from the distance, to be two British battalions. Tom Miles returned to rifling through another knapsack.

"That's that, then!"

But within five minutes the firing began again. This time little could be seen of the conflict, but it was fierce, possibly more so that of merely minutes previous. All looked anxiously, whilst nothing could be seen but the smoke of the conflict and the rear of a red line atop the ridge. Next, what looked like a whole battalion of their men was doubling around behind the British line and it swung up right to join the British left and disappeared over the ridge. Anxiety still held sway on Vimeiro ridge far below and they waited, listening only to the sounds of the fighting, which seemed ferocious, but soon this also died away and the red of British uniforms could still be seen holding the ridge top above Ventosa farm. Many asked themselves the question, "Surely that's the end?" and it was. The last act of the battle may well have occurred between Lacey and O'Hare.

"What time do you have, Padraigh?"

O'Hare's watch was already in his hand, he had asked himself the same question.

"10.32."

"Hmmm. A veteran French army, seen off between a late breakfast and an early lunch. That hasn't happened anywhere in Europe for a very long while!"

The battle of Vimeiro had, indeed, now ended. A hitherto undefeated French army had been completely shattered, a large part of it left lying on the fields around Vimeiro, the remainder, in shock from a defeat hitherto unknown, was streaming back South on the road to Lisbon. Veterans of a dozen victories could not, when they closed their eyes, avoid the image of a parade steady, long line of Redcoats, before there came the memory of a horrific experience of smoke, volleys and the sound of their comrades falling dead and wounded both left and right. Junot himself knew that his four attacks on the British line, perfectly launched according to Napoleon's own regulations, had each been dismissed, almost swatted away, in no more than ten minutes. He was wholly at a loss as to what to do next, other than to fall back on Lisbon, when some reinforcements may be obtained.

***

They had but two hours to recover and prepare from the battle before Gibney and all other Sergeants were coursing about the ridge bellowing to form column and be ready to march. All was swiftly prepared, transport, camp followers, Colours and all, to stand, then sit, all afternoon in order of march. As the sun Westered, all in the army were stood fretting, and many spoke.

"If we're to march, let's be on, if not, then make camp, prepare for night."

Lacey and O'Hare sat on the slope, privileged on chairs brought from a nearby house. Both were anxious for their men and for the camp followers, all being required to remain in position ready to obey the simple order to join the column of march. Genuine dusk was falling when Fane rode up, alone and plainly in some distemper.

"Stand down, Lacey. Make camp."

He did not dismount, clearly he had others to inform, but he paid them both the compliment of adding more..

"Sir Harry Burrard has arrived and is now in command. They've been arguing all afternoon, him and Wellesley."

He took a deep breath, which he released slowly in exasperation.

"Wellesley wants to pursue the French. They have nae undefeated troops left in Portugal, bar a garrison in Lisbon, I'd reckon. Wellesley wants to pursue them, but Burrard thinks enough has been done and caution should prevail. Tomorrow Sir Hugh Dalrymple arrives to take command from Burrard, and, from memory, they're both horses of the same colour. Stand down Lacey. We're here for some time."

His face matching the oncoming gloom, he pulled his horse's head to one side and rode off, to convey the news to his Rifles.

At Lacey's order, the column dissolved, with much ill grace and even more bad language, and the 105th re-made their camp of before.

With the night, darkness descended over the battlefield, but not the peace of night, nor the end of activity, some compassionate but much that was malevolent. Good souls of the likes of Percy Sedgwicke, carrying a lantern, walked the line that marked the high tidemark of the French attacks, looking for wounded of both sides, giving water, and, in almost every case, the last rites to those fated to die slowly, often using a Catholic crucifix that was pure anathema to him, but, nevertheless, it gave comfort when pressed to the lips of a dying man. Good souls like Joe Pike, hearing the continuing groans and cries for pity, spoke out, that perhaps they should go and help also. He was silenced by Tom Miles, who made it clear that if he wanted to look after wounded, he should go and help with their own, there was no shortage of them.

But no such concerns dwelt within Seth Tiley. He had passed out through the sentry line, bribing his guard, who had proven to be as likeminded as himself, using brandy and half the booty harvested during daylight. Tiley took himself amongst the French Grenadiers, frightening off any local peasant that may have desires to pillage any likely corpse that he had ambitions on himself. A pistol butt crushed fingers, making them easier to sever to gain rings, and his clasp knife also removed any buttons of possible value; earrings were simply torn out. Where a body showed signs of life, his knife across their throat soon removed the problem. Here were pickings richer than any English hovel or even substantial dwelling, and he'd not even had the chance for any measure of looting as yet. That was still to come, so, perhaps he'd stick with this army after all. Perhaps he could count on riches more substantial than any to be gained from his previous existence as a thief, footpad and thug back home.

***

Chapter Four

A Triumph Discarded

The dawn came with the grating argument of hundreds of rooks, ravens and crows, all eager to peck at thousands of open wounds. With no orders to move against the retreating French and the extra uncertainty caused by yet another transfer of overall command, Brigadiers concluded that the ridge may well be their place of rest for some days yet. They, therefore, ordered their Regiments to create burial parties to work alongside the Portuguese locals who had been hired for the task and rid their immediate locality of the noisome remains of the previous days conflict. However, with this in train, the inhabitants of Vimeiro returned and this was doubly welcome, not only to share the burden of burial, but also for them to offer for sale food, drink and other wares, and at affordable prices for the booty rich British army. Yet the dead were the first priority, the British dead were laid side by side in long graves, one separate for each Battalion. For that of the 105th there was no shortage of mourners for the 38 killed, and Chaplain Prudoe gave a good sermon, for all to then sing the 32nd Psalm, their own tuneful words intermingled with those heard from other nearby ceremonies, and the grave was closed.

Not so for the French dead. Many had been stripped of useful clothing and these, in a deeper grave, were piled in layers, four or five deep. None stood to mourn them with solemn hymns nor prayers, save those who carried the spiritual burden of the whole army, the likes of Chaplain Prudoe, Mrs Prudoe, and Percy Sedgwicke. A prayer was said, crosses described in the air, and the grave closed. Perhaps more ghoulish than the events of the night before, at the insistence of veterans like Jed Deakin, any dead French drummer boy had been stripped of useful clothing for the children of their own army. French boots, coats, breeches and shirts were far better quality than any that covered the thin bodies of the Regiment's young ones and so these were stored, to be carried, not worn. Deakin, and many others, had some foreboding about the months ahead and insisted that, in the warm weather of late summer, what they already wore should be used up. The heavy French drummer boy uniforms would serve better in the winter to come.

The army remained in place, throughout the day, better making the acquaintance of the locals, all of whom offered genuine friendship and gratitude to the occupying British. Their village had not suffered what was usual from a battle raging all around, the usual being that it was raized to the ground. Bar the odd building around the church the damage was negligible and the Vimeiro residents' own places of safety for valuables and precious items were now close to full from their own searching of the battlefield and the additional transactions with their soldier guests. On top, these British paid for what they required, rather than brutally maltreat, so that food be "donated" to Napoleon's cause.

Another night was spent in the same place as before, but this time in peace, save for the unnecessary challenges of the sentries, because the retreating French army was long gone. For the dawn of the following day all Regiments had been ordered to form up for the march and the 105th again formed up their column preparatory to receiving orders to get onto the road and move away. There were none of the precautions taken as for the march from Mondego and, once on the road, only the remains of the 20th Light Dragoons coursed around the head of the column, this again a seemingly endless procession now pounding the road in spirited fashion on South, to Torres Vedras. However, they had gone not five miles when the order came to halt and make camp, full camp, meaning all could assume that they would be there for some while. None complained, the heavy baggage had joined them and soon tents were springing up, substantial cooking facilities built or assembled, even areas for washing, both skin and clothing, were designated at nearby streams. The army settled to its ease and to eat their midday meal, with little to trouble it, bar the growing ranks of crosses on a nearby hillside as some of the wounded finally succumbed to their injuries. Fane's Brigade had been again at the head of the column, as it contained the Rifles, and so they were, if the description were not a misleading, "closest to the French".

Suddenly, the halcyon, post meal atmosphere was shattered by a begrimed and distraught Portuguese Cavalry Officer riding into the camp, accompanied by a half dozen Troopers, all equally alarmed and ill at ease. They rode straight up to the camp sentries, these being from the 105th, to whit, Sergeant Obediah Hill, Corporal Tobias Halfway and Privates Alfred Stiles and Samuel Peters. The four barred the way, bringing the Officer to a halt on a blown and panting horse. The horse may have been grateful but the Officer was not. He was possessed of just enough English, just equal to the task of blurting out three vital sentences.

"The French come. Make the alarm. See me the General."

Hill looked at Halfway, saying nothing, but easily conveying his thought, 'What choice do we have?' Finally he did speak, but to Stiles and Peters.

"Find our Bugler and tell 'im to sound "stand to". Toby, you take this cove to Headquarters, I'll get some lads to form some kind of forward picket."

Saying nothing more, he ran off to attend to his own task, leaving the other three to attend to theirs, this being Halfway to motion the Portuguese Officer forward and Stiles and Peters to run to their own lines. Within a minute "stand to" was being sounded, repeatedly, to rouse the whole camp and minutes after, whole Battalions were falling into their Companies. Hill found some Rifles and some of his own 105th and led them forward to a hedge line that topped a rise. To meet any oncoming French some 20th Light Dragoons galloped past beyond them, down the road, and disappeared through the trees, but soon Sergeant Hill had reached his own vantage point, and could look far down the road. There, indeed, were some French, these being two squadrons of cavalry, his experienced eye told him, all led by a group of Officers, but these were preceded by a single Trooper carrying a white flag. The Light Dragoons of the 20th that had not long ridden out, were on either side, placing themselves at a respectful distance. Hill turned to those he commanded.

"Block the road."

Hill stood out before his small command and watched the approach of the French, the dust being raised in clouds from the road by their careful progress, but drifting sufficiently wide so as not to mire the gorgeous uniforms of the group of Officers leading. However, one stood out, partly because he was in the lead, but mostly because his uniform was markedly plainer than those of his aides, bar the amazing embroidery around his collars. The Trooper bearing the white flag reigned in before Sergeant Hill and the prominent Officer rode past to confront Hill, him stood with musket across his ample stomach, bayonet fixed. Hill looked immovable, as was his attitude, and it was he that began the conversation, bluntly and to the point.

"Halt!"

The Officer, whom Hill rightly took to be a French General, halted his already walking horse, then Hill swung his bayonet to remain inches from the horses muzzle.

"State your business."

The General rested both delicately gloved hands on the pommel of his saddle and looked down at this typical image of a British soldier; surly, solid, uncompromising, and be-whiskered, but he did begin, and in polite fashion.

"I am General Francois-Ettienne Kellerman. I wish to speak to your General Wellesley."

Hill looked up at him, then past him and beyond, to examine his dusty escort. His face showed no compliance, but he did manage a halfway constructive reply.

"B'aint Wellesley no more, 'tis General Dalrymple."

Kellerman nodded.

"Tres bien! Then I wish to see General Dalrymple."

He struggled with each syllable within "Dalrymple", but then remained silent, his expression not angry, but open and businesslike. Hill looked at the members of the 20th , sat their horses nearby, but the highest rank amongst them was a Sergeant, as he was himself. However, relief was at hand, Fane himself was cantering up the road, with his own escort of aides. Hill and his men sprang to attention at the side of the road, allowing Fane the space to approach. Fane reigned in to regard Kellerman with something between Scottish annoyance and English disdain for a foe of so proven slight a mettle. However, Kellerman did salute, which Fane responded to with but a brief nod of his head, then he spoke further.

"As I have a said to your "Sergent" here, I wish to speak to your General, I believe, called Dalrymple."

He managed a better fist of the name this time, but Fane's brows furrowed further as he gave the situation some thought. Then he gave his judgement.

"You and two others can come further. The rest must stay here, with this Sergeant and his men."

He rose up in his saddle, plainly not in the best of moods.

"So! If ye'll care to follow me."

Kellerman looked behind him and spoke rapidly in French so that when he rode forward, but two came forward to follow. Hill, his men, and those of the 20th, stood looking at the French Troopers, but within five minutes they were all exchanging wine, brandy and tobacco and, soon after that, all were smoking or drinking contentedly.

Within an hour the army had returned to its state of domestic tranquillity, meeting its own concerns, both spiritual and temporal. Prudoe and Sedgwicke needed to discuss the best way to conduct a burial service for a Quaker who had just died of his wounds, whilst most attended to the maintenance of their kit, or the welfare of their families. Bridie and Nelly Nicholls were attending to the extra bullet holes in The Colours, Joe and Mary just sitting holding hands.

Come the end of the evening meal, Ellis, Davey, Miles, Pike, Saunders, Byford and Bailey were now on guard, out with the French escort, these by now devoid of any alcoholic drink; all having been exchanged for tobacco. The light was failing, but not by so much as to hide movement back at the British lines, which caught the eye of Davey, who stood up to see more, which gave the signal for the others to stand. A French Sergent de Cheval was dismounted and lounging with them in the midst of a line of smokers, so Byford decided that he had better give him a nudge.

"Je pense que votre Général est retourné."

The Sergent levered himself away from the wall that had been his support and called to his comrades, whilst emptying his pipe. Within a minute all were in the saddle, formed up and waiting. Ellis also got his men into line, to await Kellerman and his two escorts, cantering up alone. They rode straight past Ellis and his men and gave not a glance to their own men, who immediately turned their horses to follow. Davey and Byford had studied the General as he rode past and Davey spoke first.

"'Tain't often you sees an enemy General this close up."

Byford nodded.

"Nor one that has just been so thoroughly defeated, but is now looking so thoroughly pleased with himself!"

***

The British army was moving South, very much in good spirits. Through whichever town, village or hamlet they passed, the inhabitants came out to greet them, effusively, these red coated "soldados" that had thrown the French out of their country. All knapsacks bulged with bread, cheese, fruit, wine, and dried meats of all kinds. Few could remember the time that they had eaten so much and most had taken to eating on the march, for, in the next village, they would be able to fill their knapsacks again. All was good humoured in the ranks. Many had heard that a Treaty had been signed and enough of the details had emerged to form the topic of conversation for almost all social groups that the army contained, including one particular Mess of the Light Company, now well down the road to Lisbon. As usual it was Miles who began the argument, in his usual belligerent tone.

"There's some as don't agree with this Treaty job……. what's it called, Byfe?"

He had been failed by both his memory and his pronunciation, but Byford spoke indulgently and slowly from the rear.

"The Convention of Cintra".

"Just so. What you say. Well, I say that anyone who don't see the sense of avoidin' a battle, perhaps more than one, must need 'is bumps felt!"

Unusually, it was Byford who answered.

"I take it from that, that you hold the opinion that anyone disagreeing with said Treaty should be examined for levels of insanity?"

Miles did not turn around, he merely smiled, the "levels of insanity" part he understood.

"'Sright!"

Byford nodded indulgently, but none spoke further. Miles permitted himself a smirk of satisfaction at having begun and ended the argument himself. Meanwhile, back amongst the column, the Convention was the most common topic of conversation; all knew that it was to be signed that very day. Typically in Number Three Company, amongst the Colour Party, Ensign Rushby felt the need to consult someone as a means of expressing his own concerns, but, intimidated by his own deeply devout Company Commander, he spoke instead to the only oracle he had regarding the machinations of an extensive campaign, this being Colour Sergeant Deakin. He looked across at the veteran on his right hand side, automatically matching his own footsteps.

"This Treaty, Sergeant, or Convention. You've heard?"

Deakin did not change his expression, nor change any part of his marching gait.

"Yes Sir. I have, Sir."

Rushby hoped for more but none came.

"What is your view? Many are for, many against."

Deakin eased the sling of his musket over his shoulder.

"Well Sir, from what I've heard, we'n allowing the Johnnies to march out of the country, with their arms and their booty, and, on top, we'n takin' them away in our own ships, to let 'em off at a port convenient to them. There's plenty as would be none too pleased with that! Sir."

His tone showed that there was more to come and so Rushby remained silent.

"On the other hand, it cleared them out of the whole country with no need for to fire but one more shot. There's them as says we could've rounded 'em up an' shipped 'em back home, but who's to say they might've then not holed up in some fort or somesuch, leavin' us to lay siege or even storm the place. An' most sieges involves disease of some sort, both inside an' outside the walls. If you asks the lads, most'll say that clearin' out the Johnnies by the signin' of no more'n a bit of paper beats clearin' 'em away by defeatin' 'em on a battlefield. Sir."

He re-hitched the straps of his backpack, then his chinstrap, then strode on. Rushby pondered his words for some small time, then offered his own thoughts.

"But don't you think there's an issue of honour here? I mean allowing them, as you say, to walk out, under arms, with all that they have stolen, and then be carried, free of charge, in our own ships. Many would say we have acted cravenly, that we are reluctant to meet the French again."

He paused, finding the right words.

"That we would prefer to sign a dishonourable agreement!"

Deakin smiled humourlessly.

"It seems to my mind, Sir, that honour as you describes it, is a concern only for them as carries the King's Commission. If you was to ask the lads who stands in a firin' line about honour, the nearest thing you'd get to it, is some words about not lettin' your mates down. We knows that winnin' keeps the most of us alive. Battles b'ain't funny and doin' the job by makin' a bargain suits the most of us as carries a musket, beggin' your pardon, Sir."

Deakin's argument from the point of view of the ranks seemed to silence Rushby, but as usual, marching close to The Colours, was Sergeant Major Gibney. Rushby looked over to him.

"Sergeant Major Gibney, would that be your view?"

Gibney's jaw clamped together and his eyebrows closed together before he answered, very carefully.

"Ah tends not to question the decisions of my superiors, Sir, beggin' your pardon, but ah'd say that Deakin here has just about the balance of it!"

Later that evening the argument did become more balanced and more heated. As they neared Lisbon, the 105th were encamped near a large mansion and all Officers were invited by the incumbent Grandee to avail themselves of his numerous rooms for the night, prior to marching on further and, on top, they were to make full use of his extensive dining room and his serving staff. The food and wine also was to be at his expense. Lacey could barely understand a word of the invitation, but he understood the continuous pointing into the dining room and very much understood the emotion behind it when the good Portuguese said the word "Franchez" and simultaneously spat on the floor!

The room was most pleasant, low and cool, with plain white walls, decorated simply by small alcoves halfway up, each containing a silver double candlestick. Remarkably the ceiling was the same as the floor, polished red cedar, whilst the table was unadorned, of polished black oak and, very obviously, extremely old. Lacey, seeing such a salubrious venue and anticipating no small level of luxury, had identified a good opportunity to invite Brigadier Fane and he duly arrived, dressed in full Highland fig; Clan McNeil, he explained when asked; the Fanes were a sub-clan. The dinner was most convivial, the food justified Lacey's faith, and Brigadier Fane, so lugubrious in the field, proved himself to be very genial company at table.

However, come the nuts and port, the topic turned to recent events and it was Fane himself who introduced the topic for debate.

"So, Gentlemen. What's our opinion of the gentlemen we're opposed to?"

Most looked at each other, perhaps seeking inspiration, but it was Carr who formed an opinion first.

"They're very confident, Sir. Too confident. I've now met them twice, and both times they seem to think it quite sufficient to simply march up to the muzzles of our guns for their job to be done, and they are then surprised to find themselves blown back again. That's their undoing. Bold and capable they may be, but the way they go about their business gives them no opportunity to show it. In my opinion, they behaved best at Rolica, manoeuvring well, holding us off, even though greatly outnumbered, and then conducting a controlled retreat."

Several nodding heads around the table showed that few disagreed, but Carravoy, now somewhat "in his cups" from the good wine and with his self regard restored by his conduct against the Grenadier column, felt secure enough to broach the topic that dwelt prominently within most minds.

"Which makes this Convention even more shameful. The French are very beatable, we should beat them again and make an end that way."

He looked up and down the table for support, and got it; more nodding heads. Thus encouraged he continued.

"The terms are an absolute disgrace! The Portuguese are absolutely livid! Why even the French settlers who stole land and farms can remain for a year, then sell the property and keep the money! And that's besides all the booty they are allowed to depart with, under the title 'personal wealth and possessions'. Why, even Portuguese traitors are to be protected by us."

He looked around, feeling that perhaps he had said too much, too vociferously, but heads were still nodding. Nevertheless, he thought it best to bring his polemic to a close. He then spoke much more sotto voce than before.

"I personally feel sullied by the whole affair."

He sat back in his chair and took a drink, but it was Carr who replied.

"I hear you, Charles, and am in some inclination to agree. But, if it came to another battle to clear them out and finally defeat them, I can't see them making the same mistake again, to walk up a slope and give us the perfect opportunity to come over the crest and blow them to Kingdom Come. They now know, that their tactics which worked all across Europe, won't work with us. The next time will be a very bloody affair. We've cleared them out of Portugal, a whole country, with two battles and a Treaty. That has merits which I for one can see."

Carravoy lowered his glass, now empty, but his confidence had returned.

"And I know who's grinning all the way to our boats. Johnny Frog, that's who, him with his arms and his guns and his ill-gotten gains!"

Fane, sat at the table just up from the pair, sagely nodded his head, then puffed out some smoke and removed his pipe.

"You have a good point young fellah, you too, Carr. It is Carr?'

Carr nodded.

"Yes Sir. You sent me across to fight that other Grenadier column, the one on the road."

Fane nodded.

"I remember, but my reply to you both is, that while we have but 14,000 men here, our French opponents still have 25,000, still in Portugal. On the counter, our job is to fight our country's enemies and see them off. One thing is for certain sure, we'll be meeting the likes of M'siers Junot, Loison, Dellaborde and Kellerman again, and not before long, but at least not here."

Lacey took that as a very opportune time to end the meal before the argument became any more entrenched. He knew too well the potential for acrimony between Carr and Carravoy and so he stood up.

"Gentlemen, with those wise words, I think we are done. We must prepare for tomorrow's march, for tomorrow, we enter Lisbon!"

He grinned at his men, then turned to General Fane.

"I will now convey our thanks to Brigadier Fane for giving us his company at our table."

Fane waved his pipe dismissively.

"Och awa' Lacey. It's been ma pleasure, I've grown a soft spot for ye. And ye're crew of villains! Ye're guid lads, all."

This brought much appreciative applause from down the table and voluble reciprocations, accompanied by raised glasses, their owners using the good excuse to refill and drain to the bottom.

**

The following day saw the whole army out on the road when the sun was barely edging the horizon, this in obedience to Wellesley's orders, to enable them to cover a good number of miles before the heat became so oppressive that it could actually cause casualties from heat-stroke. With the Convention signed there could be no threat from the French and so Fane's Brigade were sent to the rear, to "rest and recuperate", but this meant walking through dusty air, along a road already broken into powder by thousands of heavy boots and feet. Nevertheless, the simple business of marching on, anticipating the next rest, lifted all their spirits.

They were following the army though a valley, shallow, but steep enough to make all feel enclosed, when there came a sudden shout of alarm from the rear, to be turned into screams and cries of dire alarm from the camp followers.

"The French! The French are upon us!"

Carr looked back and saw, that indeed, the crest of the valley slope was now crowded with French uniforms. He reacted instantly.

"Lights! On me! Skirmish line as you go."

He ran back down the column and out from the road, to lead his men between the helpless camp followers and these newly appeared enemy. His men rapidly formed a skirmish line behind him and he waved them forward, grateful to hear the shouted orders for the rest of the Regiment to form a firing line behind. They walked on and Drake joined his Captain.

"No guns, nor cavalry."

There was no reply, Carr was assessing the military situation and he did not like it, they were still coming over the valley edge in numbers, with the ground in their favour, but Drake's next words gave him pause.

"Correct me if I'm wrong, but if my eyesight has it correct, they're coming down with arms reversed."

Carr looked questioningly at Drake, then found his telescope and extended it in one movement. A quick study confirmed what Drake had said, for the "French", if indeed French they were, were coming down the valley side, with the butts of their muskets out first, muzzles facing back.

"You're right. They're surrendering, or some such."

He turned to his men, mostly out to his left, all stood with weapons at the final readiness to open fire.

"Close flintlocks. Shoulder arms!"

As his men stood down, Carr halted and watched the long line of "enemy" continue to approach, several men deep. Lacey and O'Hare soon joined them.

"What do you think, Carr?"

"They are coming down with arms reversed, Sir. I don't think they mean us harm."

He looked again.

"Now I'm sure."

Lacey had seen it as well. A leading Officer had a white handkerchief fixed to the point of his sword and held aloft. Lacey studied the scene.

"I think I should speak to them."

He then turned to Drake.

"You speak French, after a fashion, do you not Drake?"

Nathaniel Drake swallowed hard.

"After a fashion, Sir, usually enough to get by."

"Then accompany me, if you please."

He took one pace forward.

"O'Hare."

Drake and O'Hare followed and soon they were within easy talking distance and both sides halted.

"Ask him what he wants."

"Ce que vous fait veut?"

The Officer lowered his sword with the attached white kerchief.

"Wir sind Deutsch und Schweizer. Wir wünschen, den Dienst des Französisches zu verlassen."

Drake turned to Lacey, clearly unsure, but Lacey grinned.

"It's all right, Drake. I think I can take over from here."

He turned to face the Officer.

"Wir wünschen, den Dienst des Französisches zu verlassen?"

"Ja."

He turned to face O'Hare, Carr was beyond him.

"They're deserting the French and they want to join us. They're Swiss and German."

He paused.

"Carr. Send a runner, we should get one of our Generals; Wellesley, Burrard or even Dalrymple back here. They'll want to question these immediately"

Carr nodded to Drake, who trotted back, carrying his sword at a safe angle.

"Miles!"

Miles grounded his rifle beside his left boot and came to the attention.

"Miles. Catch up with the column and find any Officer on a horse. Tell him that about 500 Swiss and Germans want to desert to us and are at the rear of the column. Say that your Colonel is of the view that one of our General Officers should come back to supervise. Do you have that clear?"

Miles eyes narrowed with impatience, but this was unknown to Drake, the angry slits being hidden under the peak of Miles shako, in deep shadow.

"Yes Sir. 500 Swiss and Germans want part of us, Sir, and a General should come back to sort things out, Sir."

Drake nodded, just about satisfied with Miles' synopsis and so he motioned him on his way, thus sending him to make his way back through the ranks of his grinning comrades Many took the opportunity to take advantage of his discomfiture, but Miles' belligerent response was immediate and contained words rarely heard in a Sunday Sermon. His exposure to antipathy continued as he passed the camp followers, now progressing on, following the departed army. Nelly Nicholls spotted the lone soldier as he ran past.

"Ah now, Tom Miles. Sure now, didn't you look a fine sight, just now, runnin' out to put your bothersome self between us and them murderin' French Heathens!"

The look she received in reply would have curdled milk, but Nelly Nicholls was thoroughly enjoying the moment and grinned insolently back. Miles ran on and it was some minutes before he found a mounted Officer, who listened and then galloped off, leaving Miles with no coin. To add to Miles distemper, Wellesley himself galloped past with a squadron of Light Dragoons, covering Miles in dust as he returned to his Battalion. The unpleasant experience of Miles was complete when he was required to endure yet more ribaldry as he set himself back within the ranks of his Company, still at the rear.

***

Their supervision of the deserting Germans and Swiss caused the 105th to fall way behind the rest of the army, causing them to be the last Battalion to enter Lisbon. Their camp followers had halted to wait, and then tailed on at the rear, as they passed by, but, as they marched through the thickening numbers of houses on the outskirts, much of the celebration of independence had died away and, what was worse, none of the inhabitants had much left to give in greeting. Clapping and cheering were all very pleasant, but something to add to the contents of their knapsacks would have been a good deal more substantial. Soon, they found themselves marching along the bank of the Tagus and, slowly, on a hill above the city, a large fort came into view to dominate the horizon. An Aide de Camp rode up to Lacey and all at the front saw the Aide point to the very same citadel and soon they found themselves leaving the river bank, gratefully, for it was growing more noisome by the yard, then for them to ascend and approach the walls. These grew more impressive with each foot climbed, being Moorish in parts, but, newly incorporated, were all the modern defensive features of the age.

The climb ascended through a maze of twists and turns, some wide, some narrow, until finally, they came before the walls that glowered down intimidatingly at all stood before them. They wended their way through a series of intimidating gates and frightening tunnels which led through a whole series of walls that brooded above them, clearly designed to kill as many attackers within the walls as were out, trapped within a "killing zone". Finally, they entered a large courtyard where Lacey found Colonel Webster of the 95th waiting for him. They shook hands, then Webster spoke, somewhat apprehensively.

"You're in here with us and Blake's Royal Americans, and that's not all."

Lacey's brows narrowed in question.

"Half the place is filled up with French! Nigh on 1000, most left behind as garrison when Junot marched out to meet us. They're terrified of going out into the city without an escort, and that won't happen until our ships arrive to take 'em off. Being here some time, they've spread themselves out, to take their ease at maximum comfort. Blake and myself crowded our portion of the barracks to the maximum, but there's not much left for you, you being a whole Battalion."

He paused and shifted his feet. The implication was obvious.

"If you need our help to move them out, you need only ask."

Lacey looked around, French uniforms were to be seen in many places, their wearers stood looking curious or suspicious or both, at the new arrivals.

"Any wounded?"

"There's an Infirmary close to the gate. Their worst are in there. What's in the barracks are mostly hale and hearty, bar a few walking sick, with ailments of one sort or another."

Lacey nodded.

"Other than barracks, where are there? Gun galleries?"

Webster nodded.

"Yes. Two tiers, besides the battlements. And empty stores and cellars, down deep."

"Access to water"

"By the kitchen, same as for the rest of us."

Lacey nodded again and turned to see the last of his men enter the inner gate. He offered his hand to Webster who took it.

"I thank you for your offer of help, Webster, but we'll take care of this ourselves."

Both saluted simultaneously and Webster walked away, as Lacey looked around for Sergeant Major Gibney. Conspicuous by his bulk, he soon saw him.

"Sar' Major!"

Gibney came trotting over, halted two yards away and circled a blistering salute.

"Sir!"

"Sar' Major. There are hundreds of French in the barrack rooms, rooms meant for us. We will be remaining here for some time, whilst they will not. Get the Grenadiers and Lieutenant Drake, who can speak a bit of the necessary. Also, involve the Grenadier Officers, they will wish to be anyway, I feel sure. Move the French down to the cellars and gun galleries and any other empty place they care to go, they know this place far better than ourselves, I feel sure, but the barrack-rooms are for us."

Lacey paused to give the next phrase emphasis.

"There is to be no violence! None. Lieutenant Drake tells them what they are to do, and they are to be given the chance to do it. If any resist or complain, well that is obviously a different matter."

Gibney took one pace back and saluted, before turning on his heel and marching off, straight over to Lieutenant Drake, then both went onto Captain Carravoy. It did nothing for Lacey's apprehension about violence when he saw the Grenadiers fixing bayonets and then see four half sections of them disappearing into any likely doorway. Some of the rooms contained Riflemen and so they immediately re-emerged to move on and scour the other barrack-rooms. Then, soon, accompanied by bellows and shouts that contained much English bad language, came irate, sulky, and dishevelled French emerging into the sunlight, carrying packs and other items in their arms, whilst weapons and other equipment dangled from necks, shoulders and elbows. It wasn't long before other French possessions appeared in the courtyard also, but this time expelled through the doors and windows. The French disappeared through similar doors, but ones that led to the stairways penetrating down into the citadel and some even went out through the gate, so that, after 15 minutes, Gibney came running back over, to halt and salute in the same manner as before.

"Barracks now cleared. Sir."

Lacey permitted himself the slightest smile.

"Very good, Sergeant Major. Report to Major O'Hare who will undertake the allocation."

***

Within Fort St. George, the name they soon discovered, barrack life was soon established, the veterans of the 105th and their equally veteran camp followers immediately turning the barracks into a home from home; family and mess cribs being quickly established by suspended blankets and soon all were sat at tables, or taking their ease on the straw palliasses along the side. Then the gamblers of all Companies soon added the sound of their gaming and rolling dice to the general hubbub. Many stood and compared this barracks with their own, back in Taunton, and the comparison was favourable, the reason perhaps identified by Byford who had noticed the Royal Coat of Arms on the end wall of each.

"I would not be surprised to learn that this was once the barracks of the Guard to the Portuguese Household."

Those around nodded even more approvingly at their new home, as buckets of water arrived and they set to cleaning and washing, both the barracks and themselves, whilst the cooks, both male and female, lit the cooking fires in the several, well appointed fire-grates. That evening was one of contentment and cheerfulness, born from the end of marching and the good billet they were in. Fiddles and squeeze-boxes appeared in many rooms to start off the singsongs and dancing; so all was good humour with the 105th, likewise for the Officers. Byford was correct, this had been the accommodation of the Imperial Guard, therefore the provision for the Officers was also of the highest order, enabling all the Officers of all three Regiments to dine together, in uproarious fashion, for their first night there. For these, fresh food of good quality had been purchased down in the town, whilst the men dined on common rations, but no one minded. For them all, albeit far from home, life could not be much better. Neither, it would seem for the good citizens of Lisbon, for all through that night, even up to the battlements of Fort St. George, came the sounds of singing, dancing, bonfires, and even fireworks, as they celebrated the arrival of their British allies, to finally confirm the liberation of their city from the hated French.

Breakfast the next day, for all, made full use of supplies purchased down below at the harbour and it was composed of fish stew, brought to them from the fort's own cookhouse and prepared by the Portuguese cooks. To be enticed to table by the spicey aroma, mingled with that of fresh bread, rekindled the good cheer amongst all, especially when Tom Miles dug in his spoon, to find himself staring at a fish eye.

"It couldn't happen to a nicer person," was the verdict of John Byford. At this Miles immediately took extreme umbrage and attempted to tip the staring organ out onto the table, banging his spoon onto the surface when it showed reluctance to fall away. John Davey told him to keep looking, "You might find an ear."

A whole British army was billeted around the city and so patrols from their barracks were not required to go far to cover their allocated area. In fours and sixes, the men walked their nearby streets at leisure, bayonets fixed, but rifles slung over their shoulders, where they generally remained. They could see scant military reason in incessantly wandering up and down the steep alleyways that surrounded Fort St. George, but there was the compensation that it brought them into close proximity with the local girls. This happened especially when their patrol, by pure and natural chance, brought them to the local fountain, where the washing of clothes took place for the families from some way around. It was also a good place to sit, mid patrol, and drink some water, which soon became a beaker of wine, brought out for them from one of the local houses.

Tom Miles, out with Pike, Byford, Saunders and Bailey, with Chosen Man Davey in charge, were all imbibing in just such fashion and Miles had singled out one particular female, a young, dark haired beauty, not too much adorned, but just enough, with beads, and rings, both in her hair and in her ears. It was not long, as though she were possessed of some sixth sense, before she noticed Miles fixed stare, which produced a knowing, if slightly embarrassed smile. Miles nudged Davey.

"John. That one there, the one with purple in her skirt, more off to the right."

Davey looked and nodded, but his face showed no small amount of worry.

"Tom! She b'aint no Taunton doxey. You mess with these girls an' Father will be after thee with a hatchet, or a blunderbuss or somesuch."

Miles was in no way deterred.

"Do 'ee think she'll do my washin', if I were to give 'er the coin?"

Davey looked askance at his awkward messmate.

"Theece'll have to mend 'em first! Thee can't ask no fair maid such as that there, to wash what's not much more than a set of holes!"

Chuckles came from all round, but Miles was set on his plan.

"My kit be sound enough. 'Sides, now, most 'tis French. I've a mind to run back and get my spare shirt an' drawers. B'ain't no 'oles in they."

Davey looked at him. They could be remaining there for some time; there was no set timetable for their nominal patrols.

"All right. But bring back coin, so's we can say we was buyin' supplies. An' five minutes only, the fort be just up there aways."

He indicated with his head, but Miles had already set down his rifle and was on his way. In less than five minutes he was back, with three shirts. Davey looked at the bundle of washing.

"Three?"

"Yes, I got yourn and Zeke's thur. His'll make it a worthwhile job."

Across the face of the giant Saunders came a look of utter amazement but Miles was onto the next part of his plan.

"Byfe. What's the Portuguese for: "Will you wash these; how much""

Byford thought, and then answered.

"O ira lave estes. Quanto."

Miles looked at him and repeated the words.

"O ira lave ….. what?"

"Estes – it means these?"

"Estes, right, and quanto means how much?"

"Yes. Now get over there before you forget."

With a leery look growing in his eyes, they all watched Miles go over to the girl, carrying the shirts. It seemed as though he had managed the words quite well, for she took each shirt, grinning all the while, and gave each a cursory examination, whilst Miles looked on from close range, taking full advantage. The girl spoke something, but Miles, unsurprisingly, did not understand and so he spread his hands open to show that he did not comprehend. The girl, evidently bright and knowing, shouted the word at him and they could see that it must be a number, for she spread her hands wide three times, with all fingers out. She wanted thirty, of something. Miles, for his part, understood well enough the business of bargaining but that it was not the best response for this occasion, so he pulled out a King George sixpence, the smallest silver coin of the British Realm. The girl looked at the coin, it somewhat on the small side and screwed up her face, to then shake her head. Miles pocketed the coin, to then produce a button, which he handed over. The girl bit into it with her good teeth, concluded that it was silver, then nodded, to speak words which left Miles clueless, which showed, markedly.

"Amanha. Aqui. Pouco antes de Meio-dia"

Her response was to shout it at Miles, laughing all the while. Miles, looked across at Byford.

"Byfe, Did you hear? What she say?"

"Tomorrow. Here. Just before Noon."

Byford paused, to again raise his own voice.

"And what comes next is obrigado, senhorita."

Miles face broke into a broad grin as he spoke the words, his white teeth still showing an idiot grin in his brown face as he returned to his comrades, but only after two or three glances back and the waving of his hand to the girl. Davey was less than impressed.

"Right, so contact's been made, but wher' do you think this is goin' to go? You could find yerself engaged or fightin' a whole family, afore this is done."

Miles shrugged his shoulders before picking up his rifle and then lifting the sling onto his shoulder. He gave a very self-satisfied grin to himself as he gave the girl one more long look before they moved away, but she was elbow deep in the waters of the fountain.

"Could be, John, could be, but I'm for chancin' my arm, an' don't forget, tomorrow is Friday night and dancin' in the square outside the main gate."

Still highly pleased with himself, Miles led them all back up the hill to the fort.

Friday, just after the Noon gun, found Miles returning to his barrack with three very clean shirts, two of which he threw at Davey and Saunders. He was in a very good mood.

"'S'all right. Thee'ce don't owe I no coin!"

It was Saunders who replied.

"Nothing was further from my mind."

But Miles was lying on his palliasse, twiddling his thumbs and grinning.

With the sun almost set, torches appeared around the square and the musicians began to assemble, including several from inside the fort, most wielding fiddles or squeezeboxes. The population gathered, trestle tables were set up for food and two wine barrels were set on cradles, one of white and one of red. Even in the growing dark the colourful national dress of the Portuguese stood out, the men with dark jackets, dark breeches, bright waistbands and white socks, the women with dark jackets, dark skirts and white socks, but the dark of each was offset by livid headscarves, bright shirts and blouses, but especially bright multi-coloured waist shawls. Most of the soldiers in the fort that were interested preferred to watch from the battlements, for many had tried to join in the formation dancing and, unsurprisingly, found it beyond them, them causing such a mix up that they turned the whole affair into chaos. Therefore, most preferred to watch from the best vantage point available the events of the evening, these events being the succession of general dancing, exhibition dancing formations, and professional couples describing a dance which involved a lot of stamping and arms poised in the air, and was solely accompanied by the woman of the pair making rapid clicking with a pair of shallow wooden cups held in her hands.

To begin, an ancient stood on a barrel to call the moves, the locals formed up, and off they went in a whirl of skirts, scarves, sashes and flouncy sleeves. Miles was in the square, amongst the crowd, along with a deeply anxious John Davey, Ezekiel Saunders and Len Bailey; Joe Pike was up on the battlements entwined with Mary. It was not long, but a few dances, before Miles saw the object of his desires and he went straight over and made a very reasonable fist of bowing and inviting the girl to dance by holding out his arm. She, highly amused at this exhibitionist soldier, accepted and they joined the lines, where

150

Miles, much to the relief of John Davey, essayed a good performance that caused little confusion and did not require much prompting nor pushing from the fellow dancers. At the finish Miles was stood before the girl looking very pleased with himself, when next came something that they had not seen before. The band dissolved from the stage simultaneously to the arrival of some kind of hobbyhorse, carried shoulder high by four burly men. Simultaneously, a kind of gallows arrived and at the end of the horizontal was a multi-coloured ball, but what it was made of, could not be discerned in the poor light around the square, however, it did shine bright in the yellow torchlight.

The purpose of all soon became clear, when the young bloods of the neighbourhood each climbed aboard the horse to be lifted up, so that they could, at full stretch, reach the ball with what looked like a heavy spoon. However, the bearers were not at all accommodating towards their rider. They seemed well practised and thoroughly co-ordinated in making the task of hitting the ball extremely difficult, by rocking the horse forward and backwards and from side to side, also accelerating up at speed, then stopping. All this gave their rider a much greater concern to avoid falling off, than hitting the ball. Several riders tried, but ended up looking ridiculous, clinging on for dear life, one ending backside uppermost over the front of the horse, another actually sliding under the horse's belly, before mercifully being lowered to the ground. Miles' girl was in a state of great excitement, cheering shrilly if a rider actually managed to touch the ball and laughing when they lost their seat.

Miles decided that this was a most excellent way to create a good impression and so, via much pointing and chest beating, he managed to convey to her that he was going up on the horse. The first that Davey and Saunders knew of what was about to happen was when Miles climbed aboard. Both groaned together.

"Oh no!"

The bearers, realising that someone different, even special, was aboard, carried him around the square and the soldiers watching on the battlements, seeing a Redcoat up and "giving it a go" began loud cheering in encouragement. Then the attempt by Miles began. Both Davey and Saunders had groaned aloud because each knew that Miles would try something different, they both knew that he was wiry, strong and doubly cunning. The four bearers trotted Miles up to the gallows, with no twisting and turning, their idea was a sudden stop to send Miles over the front, but that was exactly what he wanted. Before they realised the plan he had in mind, it was done. He was up and standing on the horse, very perilous, but he maintained his footing just long enough to leap at the ball and hit it a two handed swipe with the spoon. The ball broke from its mooring and partially broke apart to scatter coloured pieces of paper beneath the gallows and over the cheering crowd. Miles had shot over the front of the horse and landed on his feet, to somersault over, whilst still holding the spoon. Davey became even more fretful and pulled Saunders forward.

"Oh God, he's broke the rules. They'll beat the living Christ out of him"

Davey was wrong. Amidst much exultation, Miles recovered the ball and held it high, showering out more of the contents, then he looked around for his girl. The whole point, as he saw it, was for a lovelorn swain to get the ball for his ladylove and Miles had achieved just that, but he couldn't see her. Then it all went wrong for Tom Miles. He was immediately seized by the four burlies and many others, for his arms to then be immediately thrust into a marvellously embroidered and ornate coat, which was quickly fastened at the front, despite his loud protests. Still protesting, with much bad language, the frustrated Miles was carried to a very high chair, with two long runners on the feet, to be hoisted onto the shoulders of the burlies again, but now almost twelve feet in the air. He was then paraded around the square, looking more and more frustrated and annoyed, to disappear down one of the main "alleys" that led off from the square, followed by a jubilant crowd, all preceded by a Cleric carrying a large and highly ornate cross, the like of which his three companions had never seen. The last Miles saw of his girl, as he was whirled around the square in triumph for the last time, was her accepting the attentions of what could only be a Rifles Officer, with his green pelisse jacket slung over one shoulder, black leather straps and a very distinctive purple waistband.

Davey, Saunders and Bailey were each using a tree for support, each almost helpless with laughter. Brushing away the tears, Davey spoke to Bailey.

"Len, you'd better get in, and tell Ellis what's happened. Miles've been kidnapped to be the main attraction of some religious parade."

Almost unable to walk, Bailey left to enter the castle, while Davey spoke to Saunders, both finally getting themselves under control

"We'd best follow, see that he behaves himself and don't commit no great doctrinal sin, nor get too drunk, nor start a fight, which could be the most likely."

***

The French army was arriving, to march immediately through the city and down to the harbour. The role of the British army was twofold, firstly, to clear the streets that led onto the main road down to the harbour, this to keep away any Portuguese intent on vengeful murder and, secondly, to search each soldier for possessions not covered by the Convention. Lacey assembled his Officers and read out the instructions received from Headquarters, reading slowly enough for his audience to write it down, then return to their Companies to read it to their men, who were to undertake the searching. Minutes later Heaviside was stood before his men, assembled all in one crowded barrackroom. He carefully read out Lacey's words, sonorously and sepulchrally.

"Soldiers are allowed to retain their own weapons and personal possessions, including a reasonable amount of coin and valuables. We are to look especially for valuables that can only have come from a Church, and valuables that could not possibly have been brought on a campaign; family ornaments, for example. Whatever is confiscated is to be handed over to the Church."

He lowered the piece of paper.

"There shall cleave nought of the cursed thing to thine hand. Deuteronomy 13, verse 17."

Soon after hearing his weighty words, Number Three Company were taking their turn, all overseen by their two Ensigns, Rushby and Neape, but, in actuality, supervised by the two Colour Sergeants, Jed Deakin and Harry Bennet. Heaviside and Major Simmonds added their grave presence to dealing with the Officers, undertaking the searching of such themselves, but significantly supported by Sergeant Obediah Hill. None would allow any Frenchman to march through without his full belongings being spread out on a table. Such was the hatred towards the French from the Lisbon mob, that, after being searched, the French had to be escorted in small groups down to the harbour by British soldiers, in a ratio of two to one, or the Lisbonese would have broken into the column to tear their erstwhile oppressors limb from limb.

After the search began, it wasn't long before a substantial pile of confiscated booty began to grow from both the Officers and other ranks. Any soldier who complained found himself nose to nose with a very angry Jed Deakin, him backed up by a bayoneted musket held by Toby Halfway, then the rest of Number Three Company. Relieved of his booty, the soldier received nothing by way of sympathy other than a jerk of Deakin's thumb in the direction of the harbour. For the Officers who complained, there was little difference, simply minus the thumb but plus a Bible quote from Heaviside.

"Whomsoever the Lord our God shall drive out from before us, them will we possess. Judges 11, verse 24."

Whether or not the Officer understood made little difference, around mid complaint they received a shove from the musket of Sergeant Obediah Hill, sending them also in the direction of the next group for dispatch to the harbour.

The 105th were doing a thorough job, but all involved were becoming more aggressive and suspicious as they became incensed at the paltry items that some soldiers claimed as there own; mean and meagre objects that had plainly been looted from the homes of the poor, taking the best that had been there, such as a plain and very unremarkable brass candlestick which was discovered in the backpack of a Voltiguer, him identified by the gaudy tassel still swinging besides his shako. Deakin seized the trivial item and slammed it down onto the table, then thrust his face but inches from the now uneasy Frenchman.

"You piece of shite! Wher's the gain in liftin' such as that, other than 'twer the best and only thing in the house, an' you felt the need, so's you just 'ad to walk out the door with somethin'?"

Such was his anger that he tipped the whole contents of the Frenchman's rucksack onto the pile of confiscated spoils, then thrust the empty object back at the Frenchman. His action was immediately supported by Halfway, who pushed the offender away with the butt of his musket. The Voltiguer then let loose a torrent of French, but this was terminated by Colour Sergeant Harry Bennett, who added his weight to the argument in the form of a two handed shove. Faced with three, very angry Englishmen, the Voltiguer held his peace and slunk away to stand in the next group for escort. The men engaged in searching were growing more angry and more likely to confiscate. So, a huge French Grenadier, similarly found with a meagre brass Altar cross, was relieved of almost all he had, including some coins. In response, he moved forward to menace the smaller Deakin, but he was immediately halted by four bayonets leveled at him, three besides that of Halfway.

Albeit slowly, the French battalion was searched and the last soldier passed morosely on, lighter in his pack and hating all that was English. The pile from both men and Officers had received their last additions and those in charge stood looking at the mound of objects, mostly Churchware, but many plainly personal to the particular family they had been stolen from. They all stood in silent anger, silent until the inevitable from Heaviside, but this time, all within hearing, being faced with the clear evidence of the misery inflicted, responded with "Amen".

"If ye shall still do wickedly, ye shall be consumed. First Samuel, verse 25."

He turned to his Ensigns and Sergeants.

"Gather this up, careful now, no damage and take it into the Cathedral. I suspect you will find plenty there already, waiting for what you are about to add."

The valuables were bundled into blankets and it took almost a whole section to carry it into the Cathedral. There was, indeed, already a substantial amount spread over much of the long aisle and the transepts, but on the way out, Deakin could not remain silent at the sight of the incredible opulence all around, for, in addition to that of the Chancel and the main Altar, were the chapels and shrines all along the side walls, each containing bejeweled statues, each surrounded by golden challises and ornaments. He spoke to no one in particular, but he simply voiced his own thoughts.

"I ain't never seen the like. Not no-wher', such riches in here, an' so many poor out there."

He referred to the crowd of beggars always present at the main doors, but he was overheard by Ensign Rushby.

"Best keep such thoughts to yourself, Sergeant. Here, as in many other countries, their Religion is everything. Did we not see similar in Sicily?"

Deakin nodded.

"Indeed Sir. Yes Sir, we did."

He spoke no more, nor looked any more as they passed out of the main doors, but, once outside, Deakin threw a Portuguese coin into more than one beggar's cup.

Meanwhile, back at their Headquarters, Carr and Drake, having been summoned, were entering the room that was serving as Lacey's Office. Lacey looked up at their entrance and began immediately.

"Carr. Drake. We have a problem. General Loison has been discovered by a patrol of Number Two, or, more accurately, he found them. They were called into a house by a French Officer, and Loison was there and he surrendered to them. He's terrified, more so than most French are; it was his Division that massacred the town of Evora before we landed, and news of that spread across the whole country. He may get away unrecognized, on the other hand, so he reasoned, he may not, in which case he'd be torn apart, hence the reason why he holed himself up in a house until the main army arrived and he feels he can sneak out, lost in the throng as it were."

He sat back.

"But we still have the problem. The Convention gives the French our full protection. If he is recognised, we are honour bound to protect him, which may involve us in wounding, even killing, any Portuguese assailants. We need to get him down to the harbour and onto a ship, without an attack on his person. The problem is, he's a distinctive looking cove, well filled and well upholstered, but what's more so, he has only one arm. The Portuguese know him as the "One Armed Assassin." If they see a one armed Officer amongst the French, much adorned as they tend to be, they'll assume it's him, immediately."

He sat forward.

"Any ideas?"

Both Carr and Drake looked blankly back at him, then Drake's eyes grew wider and he grinned.

"We could disguise him as a British soldier, Sir."

Lacey sat up as Drake continued.

"Yes Sir. As an escort for the next lot of French."

He paused as Lacey continued to listen.

"Which arm, Sir?"

"His left."

"So, we put him in the middle of a rank, towards the rear of the column, with his left side, close to the centre, where a missing arm won't be noticed. Sir."

156

Lacey nodded.

"As good a plan as I can think of. See to it, the pair of you."

Within an hour they were at the house, accompanied by Sergeant Ellis, and then up into an upstairs back room where Loison was hiding. It was small, damp, dirty and dingy, with cracked and peeling plaster disfiguring an outside corner. It contained but one chair and small table and a child's bed. Carr needed but one look.

"How are the mighty fallen!"

His tone and expression meant that they took an instant dislike, one side to the other. Loison was above average height and corpulent, just as Lacey had described, but exuding deep disdain and contempt for the British before him, all of minor rank relative to himself. Drake explained the plan; it needed one simple sentence.

"Nous vous obtiendrons vous est déguisé hors en un soldat britannique."

The General's face screwed up into a look of intense distaste and offence, perhaps to preserve his honour, but Carr was having none of it. Because of this General's attitude and his heinous deeds, he was angry to a level rare even for him.

"Tell this snotty bastard that if he's a better idea, let's hear it. This is the best we can come up with to save his rotten hide. We know what he did at Evora and we saw some of the like in Sicily, a place called Catanzaro. This is how we're going to get him out, so he can shut his French mouth and swallow it. If I had my way, he'd be dangling from a rope!"

It was Drake's face that now became distorted but, in his case, from deep anxiety and no little embarrassment.

"I don't think my French is quite up to that, Sir, and, I've a sneaking feeling, that they understood most of it anyway."

Drake was correct. All the French in the room were looking daggers at Carr and he was staring back at them, one at a time. Drake intervened, speaking to the French.

"Nous obtiendrons un uniforme. Nous retournerons dans une heure."

Carr understood enough to turn and leave, followed by Drake and Ellis. Outside, as they descended the stairs, Drake asked two obvious questions.

"Can we get one that fits? He's rather rotund."

Then another.

"What about his Staff Officers?"

Carr answered the second question only.

"They can take their Godamned chances!"

But Drake was persistent.

"We should escort them down as well. It could be a good thing. Any Portuguese watching will be looking at them, not at the soldier escort, where he'll be."

Carr nodded, reluctantly.

"You're right. That's what we'll do, but we'll have the bastards in step, not wandering down like a bunch of "hard bargain" recruits!"

Ellis followed them through the lower passage and into the street, grinning from ear to ear, as he had been since they left the room. Such was music to the ears of so stern an NCO, but he was "ranker" enough to relish the discomfiture of so senior an Officer.

A large British tunic was found, in Seth Tiley's cell, where he was locked up for safe keeping, therefore he had no need of it, and the other items for a British uniform were obtained from stores. Drake's Section assembled and so, in less than an hour, they were back at the house, with Drake's men remaining outside. The three entered the room to find all as before and Ellis laid out the items on the bed and on the table and stood back, but none of the French made any move, until one of Loison's Aides stepped forward and spoke, in perfect English.

"The General wishes for some privacy, whilst he changes from his uniform."

Carr looked at him, then at the General, with all the contempt he could muster.

"What makes you think that we want to look at this tub of lard anyway?"

With that he left the room and then the two followed to sit on the stairs. Drake asked the next important question, the subject still in the possession of Ellis.

"Should we give him a musket? We brought that spare, but, I mean, can he carry it?"

Carr nodded.

"He'll have to, carried at "Shoulder arms", cupped in his right hand"

He turned to look at Drake.

"How'd he lose an arm? D'you know?"

"Haven't a clue."

Carr turned to stare down the stairs.

"Wish whatever did it had gone a bit more to the right!"

His audience chuckled, including Ellis, who was still enjoying every moment, but soon the door above them opened and out came the Aide.

"The General is ready."

All three stood, climbed the few stairs and entered. The General was now dressed in the uniform, with all the other equipment, including a shako. Being so similar to a French uniform, there was little out of place, but Carr was not going to waste the moment.

"Sergeant Ellis, get this defaulter into a proper state to join a Section of the 105th Foot, Prince of Wales Wessex!"

Ellis grinned, propped his musket against the wall, and walked forward. He then stood before the General, hands on his hips, to slowly look him up and down, whilst continuously shaking his head. His mouth was screwed into an exasperated twist. Carr and Drake had to make strenuous efforts to maintain the correct deportment, stomachs aching from the suppressed laughter, but then Ellis walked forward to pull, prod and adjust the figure that had once been a General of Imperial France. Loison went red with anger to almost match the colour of his tunic, but he had no choice but to remain submissive until Ellis stood back and gave his face a "just about" expression before turning and nodding to Carr and Drake. Carr took charge.

"Give him the musket."

Ellis held the weapon out, such that Loison could carry it with his right hand.

"Right. Sergeant. Get the men outside, formed fours, right shoulder arms, two sections, and the rear with a hole ready for the General here. In the middle of the back, that is, and tell whoever's on his left to stay pressed up to him."

He paused.

"And I want a file ready to form down the outside of these Officers. They have to be given some protection as well."

Then he looked scornfully at the group.

"Though not from my choice!"

Drake and Ellis left the room, leaving Carr to indicate that they should all leave, making himself the last. They clattered down the stairs and Loison's Aides took their places in the centre, carrying their own bundle of possessions, whilst Loison was pushed into place by Ellis, who then spoke to the soldier on Loison's left.

"Mercer! This 'ere French gentleman has no left arm, but we don't want that made obvious, so you keeps up close to 'im, to hide it. See?"

Mercer nodded and Ellis stepped outside the parade. Carr was at the front, Drake at the back and so Ellis took his place on the outside of the right files where he could see Loison. Taking his cue from Carr's previous words he ordered the parade.

"Parade. Attention!"

All feet came together, bar those of the French. Drake spoke from the rear.

"Sort them out, Sergeant."

Ellis went first to Loison and kicked his feet together, the same for his French Staff. Back in his place, he continued.

"Parade, by the left, quick march."

Within a few steps, the parade was marching in unison; the French had taken the hint that they were to behave in a soldierlike manner and, in fairness, this did not sit uneasily with them. They came out onto the main road that led down to the harbour and progressed on. Such a body of men in scarlet red soon drew attention and, soon after, the Lisbonese that were stood watching, quickly noticed the hated French uniforms, then they noticed that they were Officers, therefore some picked up anything that could be thrown. However, the British uniforms ranged on the outside, deterred any from actually throwing anything, so instead they hurled what they could; insults and curses, emphasized by shaking fists.

Then, suddenly, came a threat that they had not thought of. A young girl, pretty and very excited, ran forward to the Redcoats, carrying a flower. She came in from behind Ellis, and so there was nothing he could do to obstruct her, as she came, by pure chance to the rank that contained Loison. He was second in, and for the first soldier she tucked the flower under his crossbelt, but for the second, this being Loison, she had nothing, and so she kissed him. Luckily, the first soldier who received the flower was Byford and, with his free left hand, he eased the girl out and spoke, so that she turned her attention immediately to him.

"Obrigado, Você é muito gentil. Senhorita."

160

So she kissed him as well, then fell behind, laughing and well satisfied. Byford gave himself a glance at Loison. His face was of stone.

They reached the harbour and the French filed aboard a ship. No one noticed Loison, for, with Ellis in accompaniment; he looked no more than a member of the soldier escort. Ellis followed him to a cabin and remained outside, as Loison changed back into his General's uniform, which had been carried by an Aide. Ellis waited until the Aide passed out the British uniform to him, but the Frenchman spoke as he did so.

"The General hopes that you and him will meet again, Sergent."

The remark dripped with contempt, this not being lost on Ellis, who saw no point in his reply being delivered by proxy and so he passed the bundle to a waiting soldier and pushed violently past the Aide to enter the cabin. At this commotion the General turned round and Ellis fixed him with a stare that fully conveyed all the loathing he felt. He paused to emphasise the black look, then spoke through a jaw he had to unclench.

"Yes, I 'spect we will, an' we'll be waitin' for you. Me and my mates, atop some hill, all mindful of dead wimmin an' childs. An' if you wants to find us, you only has to look for our Colour, 'tis green. Green, like this!"

He pointed to his cuff, then deliberately barged into the Aide to better make his passage out.

***

Loison's journey to the British warships had been amongst the last for any of the French, at least a journey which could be described as organised, because the formal searching had ended, but this did not mean that there were no more French heading for the harbour. Several had remained hidden, not wanting to be searched, for they still hoarded valuables, of such quality as could see them set for life back in France. Therefore, they waited for the dusk and even deeper darkness, hoping that, in the dark, they could sneak down back alleys and board the British ships not yet sailed. When some were caught by British patrols, Lacey saw the need to send out his own men to patrol the warren of narrow streets within their area and, perhaps, find these last with their cosseted treasure. Besides these were also plain stragglers, just arriving at Lisbon, and so a patrol led by Jed Deakin and Toby Halfway had now, in the dead of night, found a few of the latter and were searching their belongings, before forming them up to march them down to the harbour.

The business was conducted in a filthy alley, lit, if such a term applied, but by the paltry candle light that somehow contrived to penetrate the curtains of nearby windows, and their own single candle lantern. Deakin and Halfway were stood together, watching Stiles and Peters search one Frenchman, a straggler it would seem, for he had no valuables.

"I don't like this, Jed. There be more abroad this night than we, all lookin' for French, but some for to do 'em in."

Deakin nodded, ineffectually, in the enveloping gloom.

"You'm right, but what can we do, but get what we finds down to the ships, keepin' 'em in one piece through the journey."

He pointed at their latest French captive.

"I'd say this poor bugger've 'ad a fright somewher'. Look at the state of him."

The Frenchman was plainly shaking and, with the search ended, he was thrust by Stiles over to the other three that they had gathered. Gratitude for his safety and wellbeing, soon poured out of him.

"Merci, merci, mes amis, mes camarades. Vous êtes mes sauveteurs."

Deakin patted him comfortingly on the shoulder, but not understanding a single word.

"Yes mate, yes. I'm sure you'm right. But you just fall in with us now, an' we'll get you down to the harbour."

He began to leave the alley which movement was taken up by the other three British who pushed their French charges forward. They wended their way for some minutes, following no route that they knew, only choosing by the need to go downhill to reach the waterside, when they crossed the entrance to an alley which joined that which they were using. From within its black reaches came a cry.

"Aidez moi. Aidez"

Then came the sickening sound of a bubbling gurgle, followed by the sound of running feet, but lessening, going away from them. Jed Deakin immediately faced the way they had come, bayonet "en guarde", whilst Halfway adopted the same pose forward.

"Alf. Sam. Get up there. See what's happened."

Their French charges flattened themselves against the wall as Stiles and Peters took their lantern into the dark depths of the alley, which was little more than a tunnel. They soon returned.

"Two Frogs, Jed. Both with throats cut. I d'reckon they was bein' took away for some extra treatment, but that shout to us got 'em their throats sliced."

Deakin looked up the alley himself to see one set of boots protruding from a corner.

"Nothin' we can do. Dead bodies is for some other cove. All we can do is get these down and afloat."

Taking his cue, Toby Halfway led them on.

Once relieved of their even more relieved charges, Deakin took his men back to Fort St. George to report to Captain Heaviside.

"Two Frenchers, Sir, had their throats cut afore we could pull 'em out. They'n down in the alleys somewher' but I'd never find the place again, Sir, it bein' such a warren down there."

Heaviside nodded.

"You did your best, Colour Sergeant. Duty done. Be of good cheer; I have overcome the world. John 16, verse 33."

Then, for Heaviside, he did something Deakin would never have predicted. He reached down besides the leg of the table at which he was sat and produced a bottle, plainly not of wine, more probably a spirit, most probably brandy, and once that was placed on the table, there appeared two tin mugs.

"Here, take a drink with me. This has been a bad business, but done well. "Fear God, and keep His commandments: for this is the whole duty of man". Ecclesiastes 12. Verse 13."

With that he poured a little into the first cup, then much more into the second which he handed to Sergeant Deakin.

"Good health to you, Sergeant!"

"And to you, Sir."

Deakin drank, then passed the cup onto Halfway, who raised it in Heaviside's direction before drinking himself, then passing it on to Stiles and Peters, who also raised it to their Captain. Heaviside then ended the interview.

"Now get some sleep. "The sleep of a labouring man is sweet. Ecclesiastes Five, verse 12."

The following day the two bodies were found evilly mutilated, but rumours were coming in of similar ill treatment meted out to the

French all over Portugal, as they made their way to the various coastal ports for their embarkation. It was spoken with much satisfaction amongst the local population that the French garrison of the border fortress of Almeida, whilst on their march across Portugal, had suffered so many men being picked off, that they had taken to the interior of a Convent until a whole British battalion arrived to escort them further to the sea and safety. Similarly spoken of, was Oporto, where the Portuguese own Lusitanian Legion had to be used to prevent the local populace from boarding the transports and slaughtering every Frenchman aboard. Therefore, as compensation, the infuriated citizens of Oporto seized and plundered the French baggage still remaining on the quayside. Perhaps predictably, few amongst the British army when hearing of this, particularly amongst the 105th, could muster within themselves too much concern and even less sympathy.

Whilst the arrival of the British had been the cause of much celebration, the final departure of the French gave cause even more so. For three days the Portuguese celebrated with dancing, feasts, illuminations and more fireworks; presumably replacements had been made anticipating the need for this second occasion. Any British Officer or soldier found in the streets was immediately whisked away to be the guest of honour at some party in a local square, or wide street. Colonel Lacey, mindful of the events that befell Private Thomas Miles, forbade anyone to leave Fort St. George, but most were content to watch from the high battlements of their castle. The said Miles, Davey and Pike included themselves in this crowd, Joe Pike with Mary permanently clasped to his arm, her condition now beginning to show. Miles' "capture" had been the source of much amusement throughout the whole battalion and he was still being ribbed for it, but, eventually, even he had seen the funny side.

"I wer' set down afore the Altar an' showered with water an' a lot of mumbojumbo, afore they carried I back outside an' I got dumped on the steps"

Also, he was not in any way put out, to be given a share of someone's rum ration in return for once again relating the story.

***

Tedium soon set in, with rancour hard on its heels. It was now mid-August and the British army, in and around Lisbon, had settled into routine, a routine too close to strict, as dictated by Dalrymple, whilst he

164

busied himself with organising a new Government for the whole country of old Portugal. Shackled under his orders to remain in and around Lisbon, almost all were growing bored and short-tempered, impatient as much with themselves as with the local population. Arguments, even fights, between themselves and local civilians were frequent, for the local merchants and shopkeepers saw this huge influx of potential customers as a business opportunity not to be repeated, rather to be exploited for full profit. On top, as was common the world over, many civilians were both suspicious and alarmed by the presence of so many 'men of war" and, in addition, the extra demand generated by these extra thousands of customers was inflating prices.

For the 105th, and their neighbours of both Rifle Regiments, the square outside the main gate to Fort St. George soon became a regular marketplace and the soldiers of all three Regiments also saw prices creeping up. Sergeant Major Gibney, mindful of the growing likelihood of trouble, organised the non-Commissioned Officers whom he trusted, to make regular patrols and this, unsurprisingly, included Jed Deakin and Toby Halfway. Being on the spot, when the occasions arose, both themselves and Gibney felt it best to deal out their own measure of justice or provide their own common sense solution, rather than let any dispute drag on officially and inevitably fester into ill feeling on both sides. On one such patrol it was not long before they heard raised voices, shouting in both English and Portuguese. They hurried around to the next rank of stalls to see one of their Grenadiers in vigorous dispute with the Portuguese stall-holder and the Grenadier was growing more angry and more liable to strike out. Deakin hurried up.

"What be the problem?"

Seeing the Sergeant's stripes, the Grenadier stepped back from the stall-holder.

"These oranges, this bloody shyster took my money, then gave I these, just gave I half the number from three days before an' one's rotten. Now he won't put nothin' right!"

He looked daggers at the stall-holder, whilst Deakin looked at both.

"That's his price. Doesn't thee want it?"

"No."

"Right, put the fruit back on the stall."

The Grenadier did so and Deakin now looked fully at the stall-keeper and spoke from amongst the few words of Portuguese he did know, spoken with all the authority of his rank.

"Devolva o dinheiro."

The stall-holder looked angry and reluctant. Deakin spoke again.

"Devolva o dinheiro!"

The stall-holder was reluctant to give back the money and lose a sale, so he removed the rotten orange and added two more. Deakin looked fiercely at him, incensed by the rotten orange first given, and he held up one finger, whilst pointing at the pile. He thrust the finger forward for added emphasis. Another orange was added, so the Grenadier scooped up the pile and departed, as did Deakin and Halfway, after a withering look at the aggrieved stall-holder.

"I don't like the way this be goin', Toby. We'd be better off out in camp, or in some barracks. This cheatin' and chiselin' will bear more trouble, 'specially when drinkin's involved. I don't like it; could end up with lads on the end of a rope."

"You'm not wrong. Haven't we seen as such afore?"

Both nodded as they walked on and, as time passed, their concerns, added to those of others, reached the ears of Lacey and he, knowing the full picture from all sources, decided that he had no recourse but to confine the men to barracks when in Lisbon and to also take his Regiment out to the Campo; the bleak, hilly plain close to the Northern outskirts of the city. This he did and there he kept his men fit and in sound wind, practicing a new firing drill, whereby the two wings of the Regiment would fire independently, controlled by O'Hare and Simmonds. He had concluded that his men stood too long at the "make ready", most were closer to four reloads a minute than three. He wanted a system that enabled them to fire more often and this was not all he wanted introduced. When back in barracks, with the men confined, Lacey asked for suggestions for a "Regimental song". Finally he decided on "Brighton Camp", although under the alternative title "The Girl I left Behind Me", it was becoming more well known, and so each Company was "drilled" in the singing of what was considered a good marching song. To Sergeant Major Gibney, the order to sing was the same as an order to polish their brass buttons – there to be obeyed.

"Tha' sings! Either a song comes out tha' mouth, or tha' teeth! Tha' choice."

Many of a deeper Religious persuasion considered that a stirring hymn would be more appropriate, but Lacey would have no truck with any such suggestion and, in fairness, the likes of Prudoe and Sedgwick harboured no profound objection, bar one deeply Spiritual Soul, that being one Captain Heaviside, who could suggest several hymns of an uplifting spiritual nature, but, to his surprise, his suggestion fell upon deaf ears.

Lacey's final choice sparked thoughts in the emotional reaches of the minds of Drake and Carr, but, inevitably, that of Drake first.

"We must write. Who knows where we could be, even next week. Have you had a reply from Jane?"

"Yes. Last week."

Drake looked inquiringly and waited, but nothing came.

"Well?"

Carr took a deep breath.

"Well. She says that there is nothing that she wishes more, but she doesn't know how we are going to overcome the desires of her Father. It seems that he would want her to marry anyone other than me; when reading between the lines, sort of thing. She doesn't actually say that, but it's there, I'm sure."

Drake looked overjoyed, but said nothing; he simply spread his hands as if to say 'Well there you are, then!'

However, Carr did not share his enthusiasm.

"Well, yes, but …….."

He paused, as if unable to voice his concern, but the words did come, welling up from his own anxiety.

"There's nothing about waiting, nor being prepared to persuade her Father otherwise, or even defy him."

Carr let out something between a gasp and a sigh before continuing.

"I don't feel encouraged."

His face took on a downcast look by two or three levels, but Drake, sat forward; he had now become exasperated.

"You've not got the grasp of this at all!"

He sat even closer.

"Look! What's important? That she says she wants to marry you is what's important. As for the rest of it, well, that's up to you and her to work something out. Between you. Together."

Carr looked up at him.

"Really, you do think so?"

Drake stood up, nodding the while, and went over to their escritoire and brought back pen and paper.

"Right. A reply, and lay it on thick, "undying love", "constant affection", "you will overcome together", all that, and lots of it, said twice, then a third time, in a different way!"

Having given Carr the necessary, he shuffled his open palms towards him, urging him to "get on with it".

Carr smoothed out a piece of paper.

"Have you written?"

Drake nodded vigorously.

"Yes, but as yet unposted. Get that done and ours will go together. I will delay mine, for which I expect your gratitude. Imagine how Jane would feel if Cecily received a letter and she did not. It would be a catastrophe beyond magnitude concerning the national romantic affairs of the heart!"

Carr was ignoring him; he had begun to write, the words being followed by Drake, who began to grin enormously.

"Good, good, exactly so. I do believe some improvement can be ascribed to you regarding the art of romantic letter writing."

Carr ignored him, he was now struggling to compose the next sentence.

Also in the building, but in a different room, Major Simmonds, Lieutenant Ameshurst, sitting with Ensigns Rushby and Neape were obeying another order from Lacey, in a similar theatrical vein to the Regimental Song. They were to consider some kind of entertainment or production to keep the men, particularly the Officers, occupied when in barracks. The four were sat gloomily around a table, heads supported by one hand, the other poking and pulling at the supply of literature that they had gleaned from amongst the more erudite members of their Officer Corps. No one was saying anything. Inspiration, even half-suggestions, were very slow in coming, in fact more conspicuous by their absence.

Meanwhile, in yet another different room, but the same building, reading was the predominant activity, not writing. Carravoy and D'Villiers were reading together a broadsheet of The Times, the paper spread between them. From time to time each pointed to a particular phrase that they felt the other should pay particular attention to. D'Villiers finished first, and sat back, thinking, whilst Carravoy proceeded to the end, then he threw his hands in the air.

"We are vindicated! At home, the Convention is regarded as an absolute scandal. Dishonourable and shameful! Just as I said to Fane, and there it is, writ large!"

He then began to read the article again, his brows knitting closer together with every line. D'Villiers sat patiently waiting for his pronouncement and when it came it was again explosive and unequivocal, born as much from the memory of the privations that Wellesley had ordered for their march to Rolica, as from the legitimacy of the furore back in England.

"Damn Wellesley! Damn "Sepoy General!" I hope he's hauled before a Board of Enquiry. Sure to be, and thoroughly censured for it, signing that damnable Convention."

D'Villiers just about managed to pluck up the resolve to argue, but it came more as a bleat than a retort.

"Well, ..."

He pointed to the article.

"....... this looks more critical of Burrard and Dalrymple, and Wellesley was ordered to sign by Dalrymple, and he did beat the French!"

Carravoy was now staring straight at him, evidently not pleased with what he was hearing, but D'Villiers, to his credit, had more to say.

"Twice! Which is more than both the Austrians and the Prussians have managed to do, over years!"

Carravoy's reply left no room for counter argument. He was at full temperature.

"Luck! Sheer luck. He got lucky. I've never heard of such ludicrous tactics as the French employed back up North. We're shot of him and he deserves all the censure and.... and ...... condemnation he receives. In good measure, I sincerely hope."

D'Villiers lowered his head and turned the paper to another story.

Coincidentally, in his Office, Lacey was also reading, or more accurately passing a letter over to Major O'Hare for him to read. Lacey sat silent and waited for O'Hare to read the brief missive that had come from Brigadier Fane. O'Hare laid the letter down on the desk.

"What choice do we have?"

"None. Send the word around. Prepare for full Ceremonial Parade."

Within the hour all letter writing, self congratulating, and producing entertainments had ceased, in order to respond to the order from their Colonel. The day following was to be a full parade. Wellesley was going home on leave and the reason had gradually become known; that he felt himself wholly superfluous with two Generals superior to himself already there, ordering and organising, three including Sir John Moore, newly landed with a fresh Division. Brigadiers Fane and Acland had both agreed that some kind of farewell was required and a Guard of Honour seemed to be most appropriate, made up from the Regiments that had served Wellesley particularly well, these being the 5th Northumberland Fusiliers, the 2nd Battalion 52nd Oxford and Buckinghamshire Foot, the 95th Rifles and the 105th Foot The Prince of Wales Own Wessex Regiment. The order was greeted by groans amongst all belonging to the four Battalions, because Ceremonial Parade meant the wearing of the hated stock, the stiff black leather collar that held their heads erect and immovable. Duty in England meant its frequent use, but abroad they were forgotten at the bottom of their backpacks, until now, when they were fished out and glowered at, before being polished.

The mid morning on the day of departure, 20th September, saw the three Regiments lined up, two deep, Colours flying. The day was chilly and blustery, unseasonable for a Portuguese day in early Autumn, but the sun made its presence felt through several gaps in the racing clouds, its bright light highlighting, then diminishing, the unmistakable colours of both uniform and Colour Party. Being the least senior, the 105th were the final Battalion in the parade and O'Hare walked the length of the battalion, impatient, but pleased with the appearance of his men, all well turned out despite two battles and much marching before, between and after. He stopped alongside Simmonds, him at an equal level of irritation and willing to give vent to it.

"Typical Army. We've cleared the French out of this country, yet there's armies still in Spain we should be taking on, all no better than the ones that we saw off as though they were no more than an irritating fly. Instead, our Commander meddles in local politics. And what's the response - hold a parade!"

170

O'Hare grinned. He had a strong liking for Simmonds, knowing him a capable and effective Officer, in addition the valuable Battalion expert on fortifications and sieges.

"What would we be doing, now, instead of this?"

Simmonds nodded. He knew that O'Hare's question was the obvious reply, but he made one of his own.

"More of twiddling and kicking. Thumbs and heels respectively."

O'Hare grinned again, but movement had caught his eye and so he hurried back to the centre to take his place behind Lacey's right shoulder. Wellesley, mounted but with just a single servant riding behind, was approaching. Lacey judged his moment.

"Parade!"

He had gained their attention.

"Parade. Present arms!"

Muskets were swung from shoulders and The Colours were lifted to sit erect in their holders. Wellesley slowed his horse to a walk, then halted before them, the fourth Regiment he had seen so far.

"Lacey!"

"Sir."

Spoken around an erect sword.

"And your One Hundred and Fifth!"

He adjusted himself in the saddle and smiled down.

"Not so Provisional now!"

"No Sir."

Wellesley nodded.

"I hope I will have the honour of serving with you and your men again, at some future time."

"Thank you, Sir. We hope so too."

With that Wellesley urged his horse onward to the quayside. There he dismounted, walked the gangplank to board his ship and then his horse joined him, but hoisted aboard using an under-sling. The Regiments on parade remained at the "present" until nothing could be seen, only active sailors releasing the vessel from its moorings.

***

The parade was the high point of the week, the following days being barely discernable from each other, a procession of clock chimes and changing of the guard, each more monotonous and dreary.

However, perhaps time did not sit so heavily upon the 105th, because their Colonel had both experience and proof of the old adage "The Devil makes work for idle hands." In obediance to this, he had his men more often than not up in the Campo, running to a mock battle, their opponents being no more than a set of whitewashed posts with whitewashed rope between, but, even though winded from their run, he required them to blow the construction apart with ten volleys, which they often did, always for longer than three minutes, but always well within four.

Carr gained permission for his command to sometimes remain out in the hills for an extra day, to learn and re-learn the skills of surviving in the open, with the late September wind and rain now beating upon then. The potential of a cold soaking added urgency to the need for solid, windproof, if not wholly waterproof, shelters, each section being taught the technique by ex-poacher Chosen Man Davey. Exercises were practiced, with One Section opposed to Two, not with live firing, but the men being moved as on a chessboard. Shakeshaft proved to be a capable player and, although often out-manoeuvred by Drake, each time was far from being a foregone conclusion, nor a quick checkmate. Miles, as part of Drake's section, would glower from behind his wall or tree at the uncouth signs sent his way by many in Shakeshaft's 2 Section, all of who took extra delight in "winding up" the easily aggrieved Miles. However, both Lacey and his two Majors were not in any way aggrieved to hear, as they marched back to Fort St. George, "Brighton Camp" being sung by the whole Batalion. It had caught on, with even the Officers singing, as they marched out before their men and all sang lustily, such that the inhabitants of the villages they marched through, often came out to applaud.

Nature abhors a vacuum and so, with the fateful decision now made, the tedium was filled enthusiastically by preparations for their choice of entertainment, now decided. So that, once back within the Fort, preparations moved on apace for the production of Sheridan's The Rivals. The young females, Lydia Languish and Julia Melville were easily identified as Rushby and Neape respectively, but Lucy the maid and Mrs. Malaprop were far more problematical until Major Simmonds volunteered for the former and then, after some persuasion, so did Major O'Hare, for the latter. Many were of the opinion that a Mrs Malaprop with an Irish accent would add some extra spice to the performance. The alternative, being a vacuum of mind numbing inactivity, gave added impetus to rehearsals that took place each evening, often long after "Lights Out." D'Villiers had found enough within himself to accept a part, despite Carravoy's disdain for the whole affair, this springing from, although D'Villiers did not know it, Carravoy offering himself for the part of Captain Jack Absolute, yet only being offered the part of Thomas – a servant. This of course he turned down and he was even more discomfited when he learned that the hero was to be played by Nathaniel Drake. Similarly, the men of all three Regiments were not left behind in creations of the arts. The musicians of all three, fiddlers, squeezebox men, drummers and fifers had coalesced into a very proficient orchestra, rehearsed by Lieutenant Ameshurst, who proved to be a more than proficient musician himself upon the erstwhile Portuguese Royal Guard piano. Many a jolly evening was spent by the men and the camp followers, with singing, dancing and the odd solo. Even Carr relented from their night-time manoeuvres in the Campo, getting his men back to the Fort to play their parts, whichever they happened to be, in whichever artistic company.

Lacey and O'Hare were now far more comfortable regarding the mood of their men. Their manoeuvring and fire drill were of the highest order and the evenings' "good old sing-song" as many described it, often had O'Hare himself joining in, in fact leading, with rousing Irish rebel songs. In addition to the upbeat in tone, the dead hand from on high was about to disappear, for they had heard that both Burrard and Dalrymple were now recalled to attend a Board of Enquiry into the Convention of Cintra and thus it transpired, both Generals boarding the fast frigate HMS Amphion on the bleak, final day of September. There were no farewell parades, merely the noting of the void now at the top and who was to fill it. The answer came the next day, conveyed in the form of Brigadier Fane.

"Moore, unless they send someone superior. Sir John Moore. Ye've heard the name?"

Both Lacey and O'Hare nodded.

"But not met?"

"No," came with the same co-ordination as the nodding, but this time both heads were shaking.

"I have."

Fanes brows knitted together.

"I think he can be described as "modern". He wrote the training manuals for my Rifles, but how it'll work out with him in overall command, weell, that's for the future."

A similar conversation on the possible overall situation of the army was developing in a Lisbon cantina, not far from Fort St. George. Lacey, in better humour himself regarding the humour of his men, had relented, partially, from confining all to barracks, to the extent that non-Commissioned Officers were allowed out into the town. By chance, on previous occasions, those NCO'S of the 105th had encountered those of the 20th East Devons, who had fought beside them at Maida, so the bar was now a regular haunt for both and the news had spread of Moore's appointment. The discussion group was formed of Jed Deakin, Toby Halfway, and Obediah Hill from the 105th, whilst from the 20th, all were Sergeants, these being Able Lines, Michael Bridger, and Tommy Vickers. All were very different in stature, from the corpulent Hill to the small and wiry Vickers, but all had the hard look in their eyes of men well versed in their trade and the thorough knowing of that which turned the odds of life or death in their favour. It was Bridger of the 20th who could speak with some knowledge of Sir John Moore.

"I was speakin' to some of the Rifle lads as was trained by him at Shoreditch. He's no flogger for a tarnished button, was how they put it. Somethin' about havin' to look to yourself, "self discipline" was the words."

The "old school" Obediah Hill put down his beaker of wine.

"Well I don't see how that works! There's no Regiment in the Army as doesn't give their Sergeants a sponson to poke men back into line if they looks like steppin' back. That's the kind of discipline that we 'as now, backed up by the rope and the lash!"

Deakin looked across at him.

"'Cept we hasn't got no sponsons!"

Hill looked back.

"Well, no, we carries a bundook, an' a bayonet, on account as at Maida we was still spatchcock Provisionals."

Bridger set his beaker on the table, he was still on the topic of his Rifleman.

"Now that's what I said, but he speaks of these Rifle lads, and any Light Infantry too, as havin' to fight independent. They 'as to sort theirselves out, 'cos often there's no Officer 'andy as to give 'em orders. So, as he put it, you 'as to shift entirely for yourself, which means that you 'as to make sure that all your kit and such is fully up to scratch. They fights in threes, so, if you'm off with just two more, in a file, no Officer, then it's fully up to you if you comes out of it."

Deakin had listened carefully.

"He's right, Obediah. 'Tis 'as he says. You saw them Rifles dancin' about, out in front when they French columns was marchin' up at Vimeiro. An' I saw, even if you didn't, how they took care of them guns, and, on top, kept them French sharpshooters back. Bein' stood with The Colours, in the middle, like you, for that I'm doubly grateful. Seems to me that any man as can come up with decent notions like that, ought to be as good a commandin' General as any."

Now Vickers leaned forward, the candlelight glistening on the scar along his jawbone.

"But we've lost Wellesley as made good use of such knowhow as that. You was with him from the off. We only come in for Vimeiro. What was he like?"

Hill and Deakin spoke the word together.

"Hard!"

They looked at each other and grinned, but it was Deakin that continued.

"Hard! Some was strung up for lootin' and he had no mind to worry about our creature comforts when he knew the French was near. Mind, I'll say this for'n. He's no man to go straight at 'em, an' damn the butchers bill. He tried to winkle 'em out at Rolica, with lads way out on the flanks. On top, that's the first General as has put me safe behind a hill 'till the last moment and that's what he done at Vimeiro. I'm no looter, so what he does to them vermin I care not, but if he does what he can to preserve my skin, then I'll take'n again, soonest, as the one to be in charge."

Hill spoke again.

"I'll not argue, but that's all by the by. 'Tis Moore now, an' we can only hope that he's a horse of the same colour, 'cos no-one can convince me different, that soon we'll be off. Out of here to find a Frog army!"

Now Lines spoke for the first time.

"What about these Frenchers? How do they measure up to those we saw off at Maida?"

Hill guffawed in contempt.

175

"No different! They can't stand our musketry for no more'n a few minutes. Just like then."

He sat forward for added emphasis.

"You saw it yourself, back at Vimeiro."

It was Vickers who replied.

"No, we never, at least not close. We was up on the hill behind you, an' never fired a shot. Our Rifles an' Lights went down to take on that Frog column on the road, but that wer' about it."

The three East Devons nodded, but Vickers continued.

"We saw not too much of the detail, hardly any at all."

Then Hill continued, perhaps with the drink talking as much as himself, the sentence borne as much from emotion fuelled by alcohol as rationality.

"Now listen, for I d'tell 'ee."

He set down his glass heavily. He spoke into silence.

"There b'ain't no Frog army anywher' that I has any fear of! Even with the odds on their side! They've got no stomach for what we deals out, line to line. None at all!"

He sat back, content with the image he had created.

"An' I'll say one more thing, there's few in this army, amongst them as stands in the line, that is, as thinks any different!"

In the silence, Deakin gathered their beakers for a final drink.

***

Chapter Five

March and Counter-march

Fane and Lacey sat side by side, both feeling uncomfortably hot in the hot room. Although the second week of October, the weather remained typically Portuguese, seasonally warm, such that many had scandalously undone the top two buttons of their tunics, the company there gathered being the highest Officers in the now named Army of Spain. So far, throughout the occasion, there had been little to occupy them other than their own conversations and the long distance viewing of several maps and charts that occupied a significant fraction of the walls around, many disrespectfully covering the portraits of the great and the good that had ruled, or misruled, Lisbon for the past hundred years or more. Many maps had long reds lines marked over them, a large one of Spain had several blue circles and the rest, fifteen in all, were in the form of lists. From a distance their titles could be read, to show that, five Divisional Commanders had three each; "Beresford", "Paget", "Fraser", "Craddock" and "Hope", this making up the fifteen. However, below the headline titles, all was indiscernable, but both the appearance and the size of each was plainly different, this determined more by shape rather than by analysis, but military etiquette required that none stand and leave their place for closer examination. The arrival of their Commanding Officer was imminent and so each remained firmly in their place, though curiosity burned within each, especially Brigadiers like Fane, wanting to see which Division they were in, whilst Colonels like Lacey irked to see which Brigade had claimed them. Their impatience was interrupted by an Aide de Camp.

"Attention!"

All stood as General Moore entered. He had been confirmed as their overall Commander the week before and he had lost no time in consulting with all his General Officers, and more besides, so the result of their planning was now close to disclosure. Moore moved, as though somehow beatified, through the bright morning sunlight funnelling through the back windows and Lacey could, at last, gain some impression of his new Commander in Chief. He saw a man in his late thirties, somewhat mournful of countenance, fair hair above a fresh face, but eyes that carried a depth of intelligence that radiated confidence into the room. Moore waited for all to resume their seats before beginning. He stood behind a desk, resting forward on arms that extended from powerful shoulders. When he spoke, it was with the faintest of Scottish accents.

"Gentlemen, you may have already heard, but I am saying this in confirmation, that, on 1st August the Spanish re-occupied Madrid and promptly set up a new Government. However, the French are now massing at Miranda on the river Ebro, little more than twohundred miles to the North of Madrid. My orders are to support this new Spanish Government and, to that end, we are going to advance into Spain."

He paused to allow the murmur around the room to subside and recreate the silence of before, then he moved on.

"We will advance as four Divisions, each along a different route, to re-combine at Salamanca; the Divisions and their routes being as follows."

He then left his table to deliver a detailed description before each map and chart. Detail was the operative word. He made no use of notes, but the depth of his talk was impossible for each member of his audience to memorise, save for the occasions when their own Regiment, or name, was mentioned. However, when all left the room, they left with the impression of being about to execute a thoroughly thought out and carefully considered plan of campaign. Apart, that is, from Lacey, because the 105th were on no list, but, at least, they were not on that headed by "Craddock", whose Division was to remain in reserve in Lisbon.

Meanwhile, down at the harbour side, two pairs of British eyes were looking into about two dozen, even more mournful eyes, these being those of mules looking out of several pens, all well filled with such animals, eyes all staring back at whichever human was regarding them critically. The word had gone out amongst the population that the British were buying animals and so hundreds of such beasts of burden were being unloaded after a journey down the Tagus. Another type of "word", not official, but via the Brigadiers privy to Moore's deliberations, was that the army was going to move and that each Regiment was to organise its own transport, which then divided out into each Company. Thus Miles and Davey, accompanied by a Portuguese Cacadore, a Rifleman of that Army, were despatched to buy six for the Light Company. Davey was there for his agricultural experience, Miles for the bargaining, the Portuguese for the translation.

178

Davey entered a pen, crowded with animals and picked out six, which were hauled out by both himself and Miles for them to then be tethered outside the pen to a rail. Miles looked at the Cacadore, whilst pointing at the herdsman.

"Call him over. Ask him how much. Each!"

The Cacadore shouted something and the herdsman came over, to be asked the important question. The answer came immediately.

"Vinte e dois. Cada mula"

The Cacadore looked at Davey.

"He says twenty four. Each mule."

Davey looked at Miles.

"Twenty four. That's 144 the lot. Sounds steep. Do we have that?"

Miles nodded, then looked at the Cacadore.

"Offer twenty."

The Cacadore did, but the result was the herdsman throwing up his hands and walking away. The two looked at each other and Davey sighed.

"It's a seller's market and he knows it. We've got to get six, we've got the money, and it's not coin from our pocket. Let's pay and get back, they're good beasts, none old and all male. We've got here early and got the pick."

Miles nodded, his eyes narrowed, he was suspicious. He looked at the Cacadore.

"Call him back."

The Cacadore did so and the herder returned, but it was Miles who spoke.

"Ask him again."

The answer came back as before and the Cacadore translated as before.

"Twenty four, each mule."

The result astonished Davey, for Miles seized the jacket of the Cacadore at the shoulders and pushed him back against the pen rail.

"You chiselin' sod! Four is quatro, which I b'ain't 'eard not once!"

He turned to Davey, whilst still pinning the shocked Cacadore to the woodwork.

"Get the herder to write it down."

Davey produced paper and pencil and motioned to the herder to write on the blank page, using some of the few Portuguese words he knew, plus what he had just heard.

"Cada mula. Quanto?"

The herder took the pencil and paper and wrote, then showed it to Davey. It said "22". Still holding the now terrified Cacadore, Miles looked as well, then back at his prisoner.

"You workin' some scheme? You gettin' a rake off, somehow?"

The result was a vigorous shaking of the Portuguese head.

"No, is not true. I say wrong. I am sorry. Si, much sorry."

Miles scowled, pushed the man harder back against the rail, then released him. Meanwhile Davey had calculated the full price and, under the watchful eye of the herder, was carefully counting out the tokens, stamped out discs of metal that had become a currency since the arrival of the British. He came to 132. The herder grinned, showing uneven brown teeth, which made the grin more sinister than benign, as Davey handed the tokens over. Meanwhile Miles was handing two of the mule halters to the Cacadore, at the same time treating him to as malignant a look as he would receive for the rest of his days. Taking two each, the three began their climb back up to Fort St. George, each pulling their pair of mules.

Lugging their charges back up through the narrow alleys proved tiring and difficult, the mules often baulking at odd, illogical, places, but they eventually perked up and almost ran forward, to enter the square with the fountain, the one which served as the local wash-house. The mules ran to drink and soon their muzzles were in the cool water. The three allowed them to drink their fill, whilst Davey nodded knowingly.

"I bet they Portugee didn't give these a drink all the way down."

Miles nodded.

"I'm sayin' you'm right."

The mules were still drinking when into the square, carrying a basket of washing, came the object of Miles' affections from some weeks previously. Noticing Miles and Davey she swung the basket at her hip coquettishly, which was matched by the look on her face, a sidelong gaze, face slightly down, eyes slightly elevated and an insolent smile on her face. Davey immediately grew very apprehensive, but Miles was, surprisingly, in a very understanding frame if mind. He motioned the Cacadore to them.

"Tell her we're soon going off to fight the French in Spain."

The Cacadore did so, and it sounded accurate.

"Ask her name."

The answer came back.

"Consuela."

Miles nodded.

"Consuela?"

The previous look had gone from her face and, calm and steady, she regarded Miles. He again looked at the Cacadore.

"Ask her to name my mules!"

The Cacadore laughed, now more at ease, and so did Consuela, but she tilted her head, thought, then replied.

"Pablo and Paulo."

In his turn Miles, smiled and nodded, then pointed to his chest.

"Tom Miles."

He leaned forward and took her hand and gently raised it to a height comfortable to hold.

"Adeus, Consuela."

She looked at him steadily.

"Adeus, Tom Miles."

Then, with her other hand she touched his face. Miles released her hand and gathered the two tethers of Paulo and Pablo then pulled both from the fountain. With a last wave of his hand he pulled them into the alley out of the square back to Fort St. George, followed by Davey and the Cacadore, each pulling theirs. Once in an alley, Davey spoke out.

"That was very gallant! I didn't know you had such as that in you!"

Miles turned to show a look of mock annoyance at such a judgment.

"Oh, I can turn it on, when I feels the need to!"

Meanwhile, above them in the Fort, a conversation of even deeper sentimentality was taking place. Moore had offered all the followers a passage back to England in the transport fleet about to leave and both Jed Deakin and Joe Pike were trying to persuade their women to take the offer. Jed was sat on a chair opposite to one that supported Bridie, him holding both her hands, but making no impact; nevertheless he persisted.

"Look, I got a bad feelin' about this one, Bridie. 'Tis pushin' on for November and we'n off into the mountains and plains, which, from what I've heard, b'ain't no pleasant a place to be come winter. An' there's French all over, many more'n we've men here, so we could be on for a retreat or such, an' that's never too good a set to be in."

Bridie continued to shake her head.

"No Jed, were stayin', all of us. I don't want to be stuck back in some rotten old disease ridden barracks, for months, maybe even years, waitin' for your return. The Lord Knows that even if things gets better, they'll not ship us back out to be reunited."

She leaned forward and kissed him.

"'Sides, who'll cook your food and boil your tea? On top, were tied in with Mrs Prudoe. Who'll look after her? She's a lovely lady, kind and generous to us all, and I'd no more walk away from her than I'd walk away from you!"

Deakin gave vent to a deep sigh and nodded, squeezing her hands before releasing them. He spoke resignedly, for he had next to no choice.

"All right."

He looked down, then up, straight into her eyes, his tone serious, even commanding.

"But I'm tellin' thee now, you do as I d'tell 'ee! Look to your kit and such. Good boots, good packs and warm clothes. And extra sacks for any food we finds along the way."

He looked straight into her eyes and nodded to signify a silent finality, 'This is how it's got to be!'

She leaned forward and kissed him again.

Joe Pike didn't even get started, before Mary clung to his neck weeping. Nelly Nicholls at least listened to the first sentence spoken by her Henry before pronouncing her own judgement.

"And, sure, who's goin' to be around to stop you makin' an eejit pickle of the whole sorry business?"

***

The 105th were amongst the last to leave. The first were Paget's Division on the 18th October, quickly followed by Beresford's, both being required to follow circuitous routes up to Salamanca, just over the Spanish border. Fane, before he departed as part of Beresford's

Division, hosted a last dinner with Lacey, Blake, and Webster, the two former now being his ex-Brigade Colonels, Webster was to remain with him, because Fane was keeping the 95th. However, he could add nothing to what Lacey already knew; that the 105th would be amongst the last to leave, probably with Moore himself, after the departure of the central Brigade under Fraser. They knew that Hope was already well out at the border fortress of Elvas, but his route required a huge loop to Madrid and then over to Salamanca; his command including almost all the army's artillery. Lacey's querying of the serpentine routes of the army, brought one simple, Scottish, response.

"Dalrymple didnae do one damn thing to examine the roads to Spain. Nae transport, nae commissariat. A more festie beastie ne'er pulled on a pair of breeks! Moore has to move, knowing little of nae use, if he's to support the Spanish. There's nae time for any discovery. Politics requires nae delay."

The dinner broke up with each wishing the others well and good fortune.

The following morning Lacey found O'Hare.

"How many wagons have we?"

"Enough for our baggage and whatnot."

Lacey looked back at him seriously.

"Get a spare. At my expense. Load it with any winter clothing you can get your hands on. Anything, blankets, boots, greatcoats. Horse blankets! Anything useful that we took from the French at Vimeiro. It'll be mouldering in some warehouse somewhere."

O'Hare stared back at him, himself now as deadly serious.

"You think such will be needed?"

"Odds on. I've been through Canadian winters and an extra blanket was too often all that stood between me and freezing to death! Cold's cold, wherever, and Spanish uplands in winter will be no different."

O'Hare turned to leave, but Lacey had not finished, and it brought a broad grin to the Irishman's face.

"And it seems the world of drama will be denied your interpretation of Mrs. Mallaprop!"

O'Hare laughed openly.

"We'll count that a blessing!"

On the morning after Paget and Beresford took the Roman road North to Santarem, Fraser was seen to leave, taking the same road before he diverted away. From the battlements of Fort St. George, the 105th had a grandstand view of the long column of red uniforms, broken by the green of the 60th, and then followed by the drab brown of their followers and their wagon train. Almost all the Officers of the 105th were up there, all in idle mode, except perhaps Ensign Rushby, the Regimental artist of some repute. He was busy sketching an old woman sat at a stall on the battlements, the table spread with various pastries and confections and she was doing a very good trade. Carr, Drake and Shakeshaft were leaning on their forearms, each in their own crenulations, eating one of her pastries and watching the six battalions inch past, seemingly a crawl from the distance of their high perch. Carr looked over at Drake.

"Who's that leave still here?"

Drake set his pie on the warm stonework and consulted a piece of paper. He had obtained the information for his own personal journal, begun the week before, and he was now up to date. He had begun at their landing in Mondego.

"Besides Craddock's lot that remain as garrison, just us and the 6th. And Moore's Headquarters, of course."

Carr nodded, but it was Drake who spoke next.

"Have you written?"

Carr pushed himself away from the battlements.

"Just about to."

Frustratingly, their own departure was delayed for several days more. Their Commander in Chief, it was rumoured, but it was mere conjecture, was involved in acrimonious and detailed correspondence with his Spanish allies regarding, not just co-operation in the field, but the amount of supply that Moore could expect from the Spanish as he advanced his army into Spain. However, finally, they were drawn up on the Santarem road, the 105th behind the 6th Royal Warwicks, the followers and baggage of both Regiments tailing behind and away into Lisbon itself. The population of Lisbon, including all those that dwelt in the warren that clung to the slopes beneath the castle, turned out to see them depart. The 105th had lived amongst them for almost six weeks and relationships had stabilised at no small measure of affection. The drums and fifes of the 105th formed up before the Colour Party and they marched off to the inevitable tune of "Brighton Camp", the popular tune soon being taken up by the 6th some 100 yards before them. The Officers, including Moore, felt buoyant and optimistic, why shouldn't they? Or so they thought, but the likes of Jed Deakin sang not a note, instead biting hard on his chinstrap.

Soon they were beyond the autumn bleak Campo, now so familiar to the 105th. The weather immediately turned cold and blustery, slanting showers coming in from across the Tagus and many rolled down the tarpaulin folded at the back of their shako and unstrapped their greatcoats from the tops of their backpacks. Come the end of the day, as the men and their followers prepared their first camp in the open for some weeks, making full use of the shelter afforded by trees, walls and hedges, Lacey and the Colonel of the 6th were invited to Headquarters for dinner. Soon the food proved to be wholly ancillary to the more important topic that Moore wanted to discuss that evening, this being the gathering of extra details of the battle of Vimeiro. Thus, most of the affair was taken with moving knives, forks, spoons and glasses around the table, interspersed with the odd mouthful of food. The 6th had not been called upon to fire a shot at the battle and so Parker, their Colonel, took virtually no part in the discussion and was able to stuff himself from the good dishes, whilst Lacey was plied incessantly with questions to add to what seemed to be Moore's already considerable knowledge. Moore remained with them but one day more, before riding on, all his entourage well mounted and their baggage all in sound and stable wagons.

Over the following days the weather did not relent, but the men were in good spirits, they were marching forward, all confident that what they did at Rolica and Vimeiro they could do to any French army they met, even if significantly outnumbered. To Jed Deakin this was all "yes, no, or maybe." What concerned him most was what could be and that 'could be' was as bad as it gets. They were marching through country only just coming back to life with the departure of the French armies who had not long marched through for their evacuation from Lisbon and, on one of the better days, they stopped close to an oakwood. Grazing on the acorns within, released from some pen either near or far, were several pigs. Realising what they were eating, Deakin ordered all of those he could and advised all that he could not, to get into the wood and gather what was a feast for such as pigs, but was emergency rations for an army. The acorns were ripe and plump and soon several bulging haversacks were lodged back with the followers. Needless to say, the messes of Miles and Byford, around a messfire far away from their Officers, dined on illicit roast pork that night and for some days after on the left over joints.

Lisbon to Salamanca was a march of 250 miles and it was pushing through the days of November before they reached the high plains around the Spanish border fortress of Cuidad Rodrigo, leaving but 50 miles more. The steep walls and deep ditches of the border fortress gave shelter for but one night only, then they were again on the long, direct, Roman road Eastwards. Typically, there was little cheer from the weather; a biting wind skidded unhindered across the open grassland and the dark clouds it often carried invariably meant a soaking, gradually returning the burnt grass to a dull autumn green.

Gloom arrived in full measure from another source. For those looking onward to the vanishing point of the straight road, they witnessed the gradual reveal of the details of a wagon convoy approaching from the opposite direction, with an Officer of high rank riding before, who appeared from distance to be a Major. Accompanied by Linfield Parker, Colonel of the 6th, at the head of the column, Lacey spurred his horse forward, to pay the required respects. Whilst Parker and Lacey were cheery at the meeting, the response in return was anything but, after the Major had introduced himself as Frederick Vickery.

"Bad news, have you heard?"

The faces of both fell as they shook their heads.

"The Spanish have been smashed at Gammonal and Espinosa! On top, Boney himself is at Burgos, halfway to Madrid."

Both Colonels immediately registered shock, but next came dismay.

"There's worse. Moore's army is scattered all over and delayed on the march. Hope and the guns are stuck down South somewhere."

His face went blank, marginally better than the yet more gloom he had to impart.

"If Boney gets over the Samosierra before Hope gets back round to Salamanca, the game's up! No guns."

Lacey and Parker looked at each other, then back at Vickery.

"What of Moore?"

"At Salamanca, watching his army arriving in penny packets. Things have not gone well."

Both Colonels nodded in agreement, then Parker asked what was of import to them.

"What are conditions on the road from here?"

Vickery looked positive for the first time since their meeting.

"Good! A good chalk road. You'll make good time."

He paused.

"What of mine, on from here?"

Vickery answered.

"Good! All through to Lisbon."

Parker smiled.

"That's reassuring. These wagons are needed back in Lisbon and then to make a return trip."

He reset himself on his horse.

"That's army life, no? Nine tenths boredom, one tenth terror!"

Leaving the two grinning, he saluted and moved his horse on, to lead the wagons of what seemed sick, certainly not wounded, on past the column, down which the news was speeding, that the Spanish were scattered to the winds, that Napoleon himself was at their throats, and that a retreat was imminent. That night, around their campfires, there were Councils of War conducted at various levels. The result of that occurring amongst the Senior Officers of both Regiments was that they had no choice but to advance on and join their Commander in Chief as ordered. Amongst the Regimental Officers, that occurring between Carr and Drake was typical, with Shakeshaft as an audience, the discussion begun by Drake.

"We've no allies! We're marched into Spain expecting to be part of an allied force and we find ourselves alone. Outnumbered by the Johnnies led by Boney himself, no less."

Carr poked the fire.

"That we are on our own there's no doubting. But ……."

The last word was forceful, then repeated.

"But, we can at least make a nuisance of ourselves and what we cannot do is just march out. This country is full of good positions to mount a defence, if needed. If his army comes together, Moore will march on."

He looked directly at Drake.

"Boney's no magician! He sends his men forward just as Junot did, following Boney's own prescribed tactics, and we sent them back after ten minutes! Moore will give it a go if he can, and I think he's right."

Resignedly, Drake threw some more wood on the fire.

Back with the followers, Deakin and Pike were asking questions to gain the reassurance they needed, but Bridie soon ended the interrogation by handing over a bowl of stew, containing meaty pork bones.

"Now hush, eat your stew! We're as well up together as any can expect. What comes we'll meet as best we can. We've the extra kit and food, so stop worryin' over what hasn't come about. When it does, we'll cope as we always do!"

They marched on, the road leading on into December, but the apprehension that now hung over the column did not dispel. That is, until they came within sight of the towers of Salamanca itself, for on the road before them was a long red column and, as the briefest of examinations through a telescope would show, all along the column was the unmistakable sight of horse artillery. Having drawn the right conclusion, Lacey turned to Parker.

"There's our guns. That must be Hope. Now we've only Boney to worry about!"

***

If the cities of Spain were classed as national treasure, then Salamanca would be a major jewel set in a Royal crown. Architecture from centuries past rose assertively on all sides, buildings of worship, learning and culture, not least the two huge cathedrals and the ancient buildings of the university. Neat streets led from the wall gates to a breathtaking square, this with its own impressive buildings, but also footed by a colonnaded walkway, similar to the cloisters of a major monastery but more inspiring still. What was missing were people, several had quit their homes, only too ready to take the time for a controlled exit before the arrival of a French army. They too had heard of the disasters that had befallen their armies at Gammonal and Espinosa and yet another at Tudela and, yet more, Napoleon had forced the pass of the Samosierra, which news had sent Moore into deep depression, partially relieved only by the arrival of the final units of his army. The upper classes of Salamanca, realising that only a small British army stood between them and the avenging French, had fled West and North, leaving many vacant buildings, too small for Companies of infantry, but perfect for Brigade Officers and those smaller still, perfect for groups of followers. Thus they settled in, the spirits of all lifted by their return to good billets after the cold privations of the march from Lisbon.

The 105th were as close to being in barracks as was possible, without that precise term being actually applied. They were in an abandoned monastery. The stout outer doors had been locked and barred, more from hope than expectation and these were soon forced to allow the entry of the whole battalion. Soon the various messes were claiming their places, these being the cells of the absent monks, but Miles could not desist from commenting on the "miserable commons that goes with being a monk." He was told to stop worrying by Davey.

"It'll never happen to you!"

Whilst the accommodation was perfect for the men, for one in particular of Lacey's Officers it was far from such and certainly also not for one of his Privates, and for a reason very much different from its physical comforts. Joshua Heaviside could lower his head at the collection of statues in the entrance hall, but he could not countenance the statue, picture and crucifix in his quarters, this being one of the larger cells. However, he called in a member of his Company and had the three removed, to be replaced by his own plain cross. That established, he fell to his knees, having composed his daily prayer within sight of the city walls. However, there was no such comfort for Percival Sedgwicke. His objections to the "Popish" imagery were immediately dismissed by Prudoe and so, in their quarters, the devoutly Low Church ex-Cleric could only bow his head and satisfy his conscience by looking away.

Now fed, the men were looking to their own entertainment. The gamblers were in full spate in the other messes, but that of the six to which Davey and Byford belonged contained no students of the dice, but they had their own vices. It was Miles that poked his head out of their cell to look down the long, bleak corridor that held the doors to their cells. He withdrew his head and spoke to the five within.

"I've heard that these places brews their own spirits. For sale, like, to earn a bit of extra coin."

It was Byford who spoke up, speaking resignedly; knowing that what was to come could only end badly.

"You are not wrong. The Benedictine Monks, as only one example, do exactly that."

Miles nodded, thus encouraged, and grinned mischieviously.

"Right. I'm off for a poke around. We ain't been here so long as to allow our Officers to see all what's 'ere. I'll take a shufty."

Davey looked at Saunders.

"Zeke. Go with him, for Christ's Sakes. Keep him out of trouble."

The accuracy of Davey's words propelled Ezekiel Saunders to his feet and both left the cell. Within five minutes they were back.

"I've found a door. A bloody great big thing and b'ain't no cell. No-one's living in thur. Could be the cellar!"

Davey looked at Saunders, his brows knitted suspiciously.

"Zeke?"

Saunders nodded.

"It's true! A large door, all iron bound. I couldn't shift it, but we got a big echo."

Davey looked at the others. They all knew what he knew, that a supply of spirit was never unwelcome, to lift their own spirits, especially on the kind of cold march they had just endured.

'Right. We'll go take a look."

They all filed out in Miles' wake, Pike and Bailey carrying the candles, to be led around several corners until they were confronted by what was, indeed, a very imposing door. However, it was immovable, plainly barred by a very secure lock behind a keyhole that was equally imposing, large enough to hide a mouse. Miles shoved the door again, plainly frustrated.

"Let's smash it in!"

The response was several frowning looks at such idiocy, but then Len Bailey spoke.

"Wait here."

He was gone but a minute, before he returned with two pieces of thin steel, one curved, the other with a hook. As he knelt before the keyhole, it was Davey who spoke.

"I didn't know that you had criminal tendencies! For good Christian boys like me and Joe to associate with the likes of you."

Bailey was poking around inside the lock, but he had enough concentration spare to make an answer.

"Parson 'an Tiley b'ain't the only King's hard bargains in this Regiment."

He continued his poking, until.....

"Tom. When I says so, turn the handle."

Miles seized the handle. Seconds more.

"Now!"

The handle was turned and the heavy door swung open, to reveal inky blackness and a powerful smell of dust and damp. Pike was pushed in first with his candle and then came Saunders with the second. The dim light fell first on rows of bottles and jars, then shelves of ancient books, all bound in leather that once shone, the pride of the bookbinders art, but now all were dull and dusty. One section contained rolled scrolls, yellow and equally grey from the dust of decades, bound with faded cord, its colour of some long gone description. Miles' eyes lit up at the sight of the bottles and he went immediately to the shelves to make his selection. Having the thinnest criminal streak of those embarked on this villainy, Byford was the last in and he stood still, looking all around in deep thought, then he went to the bookshelves to examine the titles, then to the shelves with the bottles and jars.

Meanwhile Miles had expertly decapitated one bottle and was sniffing the contents.

"Brandy! Of the best, I'll wager. These Monks knows what's good for a body, after all."

Byford had now moved to read the labels on the jars, all being of the finest illuminated script. It took him some time before he could make out the words, but meanwhile Miles had taken a deep swig from the bottle. His face showed that he wasn't sure.

"Well, 'tis brandy, sure enough, but not quite of the best."

Meanwile Byford had come to his own conclusion.

"I'd drink nothing from what's in these jars and bottles. Nothing!"

By now Miles had taken another sample, in order to make his final judgement. His face showed that he was of the same opinion, but he had heard Byford's pronouncement.

"What do you mean? Nothin'!"

Byford looked at him challengingly, even in the yellow light.

"Just that."

He had found a deep bowl and held it out.

"Pour your brandy into that. I've wiped off the dust, there'll be no harm."

Miles looked at him as though he were mad, but slowly and carefully he tilted the bottle, for the precious brandy to pour into the bowl. The bottle was tipped further, almost to the vertical when all those watching gasped. Out came, to fall into the half filled bowl, what could only be described as a finger, blackened and shrivelled, but nevertheless, a finger. Byford took the empty bottle and held it to the candlelight to read the label.

"St. Ignacio! And that's his finger!"

He waved his hand at the shelves.

"This is a Reliquary, the place where bits of Holy Saints are kept. What's in those other jars I dread to think, but that one, which you selected, brave Thomas, contained the finger of the Holy Saint Ignacio!"

Even in the candlelight, Miles was going green; the others were dissolving into various depths of laughter. It did not end, even when Miles ran out of the door and they heard his boots echoing down the corridor. Still laughing, Davey looked at Bailey.

"Can you lock this up again. I don't doubt the French will smash it in, but we should leave it as found. 'Tis important to some, that being the Monks as was here."

Bailey nodded and produced his "rods", while Davey replaced the finger and poured back the brandy, before stuffing a piece of cloth into the shattered opening. It was Byford who carefully replaced it on the shelf, label outwards.

***

Merely days after the arrival of the 105th saw developments of great import, but those in the ranks could merely look on. In the weak sunlight of early December, many of the Colour Company were sat outside repairing and checking their kit, close to the gate that led in from the bridge over the River Tormes, where, on the far side, the routes from Madrid and Cuidad Rodrigo combined. Thus they were in the prime position to see the entrance of the final element of Moore's army, his cavalry and the last of his guns. Finally, came a fantastic cavalcade of Spanish uniforms, and these were merely those that were on show, all atop fine horses. The really gorgeous uniforms, encasing the personages of the most important, were inside the gleaming carriages. The Spanish arrival drew a comment from Deakin, remarked to his friend Toby Halfway.

"What do they want? To join up? Might as well, they've damn all army left. May as well join ours!"

He was not in a good mood. Not from anything that Bridie was or was not doing, he was content that she was doing her best, but to what end? Whatever she did, under his advice could count for nothing when viewed against the possibilities of what could come. He had no experience of being on a campaign where he had the additional responsibilities of a family of followers and being a natural "worrier", it was this that he found so depressing. Halfway detected Deakin's subdued temper and continued to check the stitching on his pack, nevertheless he felt inclined to make some attempt to lift his good friend's mood.

"How're things between your Bridie an' our Chaplain?"

Deakin looked away from the parade of gold, light blue and scarlet.

"With our Chaplain, not much. He'd rather they weren't about, but his wife is a very gracious and helpful lady, that I'll hear nothin' against. Why, she's just let out a dress for Joe's Mary, she's beginnin' to show, and that's what I call real Christian kindness. She's teachin' the youngers their letters and Scripture and that can be no bad thing, for their pure betterment, if nothin' else. That's how I sees it. As for the Chaplain himself, well, that's a different story."

He left it there and began to examine the integrity of his own haversack.

The arrival of the Spanish did little to lift the mood of Captain Lord Carravoy, sat at the window above this main road, sipping white wine, whilst D'Villiers wrote home.

"Moore has got us into a dreadful pickle. Even coming this far, our heads are now in a French noose, it slowly closing. What chance do we have now, against the French, they've already enough men to annihilate three Spanish armies? If they concentrate on us, led by Napoleon, what odds would you give for us?"

D'Villiers looked up from his writing. His aristocratic confidence had been boosted by his not unsuccessful experiences so far; some, if not most, being quite praiseworthy, or so he judged it. This new self-belief gave him the assurance to disagree with his superior, if ever so slightly.

"We aren't the Spanish. We're a victorious army, so we can hardly turn around and go home without firing a shot! I believe all is not yet lost, that if we hold together we can still do the French some damage and still get out of this."

Carravoy looked angrily at his companion. He had not received the full agreement that he expected.

"That may be so! But only if we go no further. The Spanish are a lost cause. Better to get out now and try elsewhere."

Carravoy turned back to the window. Recent orders had done nothing to mollify his ill mood.

"And heavy baggage to remain here, that's what I've heard and on good authority. None on the march! That means not even alternate nights on a decent bed!"

He re-examined the wine bottle. But an inch! He turned to face the door.

"Binns!"

Binns was halfway through the door anyway and spoke his own words first, they being of far greater import than any request from either of his two charges.

"Message from the Colonel, Sirs. Meeting at his quarters in 15 minutes."

D'Villiers reached for his coat, whilst Carravoy drained the last dregs from the bottle.

***

Lacey stood before his assembled Officers, half standing, half sitting on a table. There was no room left for him to sit behind it; it had been pushed back against the wall, and most of his Officers were standing. Lacey was holding a stiff paper in his hand.

"We have been made part of General Bentinck's Brigade. He studied the paper.

" 4th and 42nd."

Lacey's brows knitted together.

"The 42nd, they're Highlanders, but the 4th? Do you know?"

He was looking at O'Hare, who nodded.

"King's Own! And the 42nd are the Royal Highlanders. Very fashionable! We are in illustrious company, but Bentinck is unknown to me."

"And to myself.|

Lacey looked up to his audience and to the one most likely to be his best source of information.

"Carravoy?"

Carravoy looked from Lacey to O'Hare then back again.

"Yes, I know him. He is the second son of the Duke of Portland. His father and mine are good friends, often coming to our estate to hunt and shoot."

Lacey was interested.

"So, you know him socially."

"Well, no. And yes, but I doubt he'd remember me, being that much older."

"Sort of character?"

"Sorry Sir. I had too little to do with him, to form any opinion.

Lacey nodded, then continued with the second purpose of the meeting.

"The Spanish are insisting that we support them. Madrid is holding out, so Moore considers that he has no choice but to advance on, to Sahagun. We move on the 11th, one week's time. Get the men ready, everything checked and thoroughly examined. We are not in for an easy time."

He paused and examined their faces, all of which remained blank.

"All heavy baggage to remain here."

A slight pause.

"Dismiss."

As they all filed out, Lacey raised his hand to signal O'Hare to stay. With the pair alone, Lacey found two glasses, a bottle of good French, then poured two large measures. He pushed one across to O'Hare and the two exchanged a knowing look, then, without a word, Lacey raised his glass in O'Hare's direction and drank. O'Hare studied his Colonel and drank from his own glass, allowing the deep warmth, now spreading in his stomach, to distract his deeper concerns over what he had just heard. He replaced his glass and left, without a word more.

The following day, Lacey again called for all his Officers, but this time into the Church. The rumour was that their situation was ever more serious, so that even Heaviside had his mind distracted from the Catholic images that abounded on every wall and in every alcove. With his Officers now gathered before him, Lacey stood up, and silence fell immediately. He took a deep breath.

"Madrid has surrendered."

He quickly spoke further, leaving no room for any conversation.

"But Moore is pressing on to Sahagun, with no change of plans, and I think he's right. By co-operating with the Spanish army of La

195

Romana to our North, we can operate around the fringes of the French invasion, and perhaps, if the Spanish performance improves, do it such damage that they pull back. Marshall Soult is to our North and Moore intends to move against him."

Now he did pause, but no-one spoke. All waited for their Colonel to say more.

"We still move out on the 11th as planned. I want each man to have an extra blanket and we'll carry extra boots in our baggage train. Enough to replace what will surely fall apart on the winter roads.

He paused again, to straighten himself to his full height, for added emphasis.

"I don't need to tell you how tough this is going to be, and, if the very possible happens, it will get even worse. I refer to a retreat. A winter here can be as bad, or worse, than anything back home. I leave it to you all to get your men ready for a campaign in severe weather, and what matters most is boots, blankets and greatcoats. Any spare that we can gather, from any source, we'll take with us. And for the followers too! What they provide for the men, and not just hot food, is too valuable to risk. They motivate the men to hold together. Though I wish they were not, the followers are with us, and their suffering will prey on the men's minds."

It was Carr who raised his hand.

"Sir. All the extra will be extra weight, that needs be carried on the march."

Lacey nodded.

"Yes, but we still have our mules, and the extra wagon, and I have an idea of my own that may help."

He turned to O'Hare.

"Major, can you gather the followers in here this evening?"

O'Hare looked perplexed.

"Yes, Sir. When?"

"7 o' clock would be perfect. Also, when we marched in, I noticed several stands of larch. Get some harvested, will you, twelve to fifteen foot lengths. Have them brought here."

What he did not say, which was preying on his own mind, was that Moore had spent the last week in a state of utter indecision, which was most likely to result in an immediate retreat. He could only hope that his Commander in Chief had made a firm decision and was determined to carry it out with all the power at his disposal.

The 11th was a gloomy day, wind, cold and rain, and if no rain, always the wind, but usually the two combined. The 105th stood their place at the forepart of the long column, all secured as best they could against the zero comfort of the day, all with an extra thick blanket roll atop their knapsack, it being a roll of two blankets, and all wearing greatcoats. Even so, the wind carried the cold down their necks and up their sleeves. Bentinck, their new Brigadier, rode impatiently up and down the length of his three Regiments, of which the 105th was the centre. Before them was the 4th The King's Own, behind was the 42nd The Royal Highland and no one made mock of their kilts, the sight of their bare knees sent all into shivers.

From not too great a distance, Bentinck could be taken for Wellesley, a long nose, an inexpressive mouth, but up close he resembled a Public School Headmaster, disdainful and disapproving, which emotion he bestowed from high upon all that he came near, as he turned his horse, here and there, before and beside his column. Lacey had not warmed to him as he had to Fane, but was content to reserve judgement. Bentinck's pirouetting of his horse was only stilled as they began their march, out from the sheltering elegance of Salamanca and into the bleak weather of the endless uplands of Western Spain. At the Eastern Gate stood a hooded figure, black from head to toe, sprinkling Holy Water from a silver bucket, mumbling incessant prayers, whilst a young boy shivered beside him in choir uniform, his blue hand waving a huge incense burner, whose issuing smoke blew not over the marching men, but up and over the walls to be lost against the blue and iron grey of the clouds over them all.

***

It took not one day before rumours began to circulate; there had been a cavalry action up ahead; Boney himself was commanding the French facing them; there had been another Spanish defeat. The march forward took them through a town called Rueda, where some captured French cavalry sat miserable in the square, guarded by some Riflemen, but no more signs of conflict were seen and the rumours subsided as the army marched onwards. Each evening, Deakin and Pike took themselves back to the followers to check on their welfare. At first Joe Pike carried back a portion of his own rations to add to that for Mary, but he was soon told to stop, partly by Bridie, but mostly by Nelly.

"You're a marchin', fightin' sojer, ya gombeen eejit! You eat, Mary needs you whole, hearty and strong, not some weak-kneed, worn out pune!"

However, what reassured Joe Pike the most was the fact that Mary was riding in the Chaplain's wagon, as insisted upon by Beatrice Prudoe. Even Jed Deakin himself began to feel marginally more reassured. He was worried, as Carr had been, about the weight of the extra clothing and food that the followers were asked to carry, but felt much better when Bridie showed him the contraption that she had made back in Salamanca under the tutelage of Lacey, some days past now, in the church.

"The Colonel saw them used when he was in the Americas, used by the natives. He called it a 'travois'. You load it up and all you have to do is pull, like a sledge. 'Tis much easier than a heavy pack and the poles slides easy over the mud, like the runners on a sledge."

Both Deakin and Pike returned to their messes and slept easy.

The following days merged one into the other, this reinforced by the unchanging weather; rain and wind, cheered occasionally by a watery sun that warmed their backs and threw their shadows onto the damp chalk of the long, straight roads. A week after leaving Salamanca they marched into Mayorga and gratefully into warm billets once more. The population of this innocent Spanish town was even more sparse that that of Salamanca and for the same reason; the rumours that arrived from the East telling what happened when the French marched in. In his billet, Heaviside found a map and brought it to the building that rejoiced in the unwarranted title of the Officer's mess. With the map spread on the table, before the meal, all stood poring over it, tracing the route of the past week and locating their position at Mayorga. Major Archie Simmonds straightened up and looked askance at the map.

"We seem to be circling Madrid, but the radius is getting wider."

Carravoy was in no good mood, this intensified by his personal privations on the march. He felt no need to defer to rank, especially to the Junior Major.

"And to what end, I say? This is taking us further into the mountains. That may protect us from the French, but it just leaves us here freezing and not firing a shot, but dying anyway, of cold! And, from what I hear, La Romana's Spaniards aren't worth the boots they're stood up in."

O'Hare had entered the room and, standing at the back, he had heard all.

"From what I hear, Charles, they have no boots at all, yet they're ready to put up a fight!"

The gathering around the table parted to allow him through. He looked at Carravoy who showed no embarrassment, more like anger at the counter argument, nevertheless his Lordship waited for O'Hare to say more.

"Look at the map, Charles. What they call this, what we're doing, is manoeuvring. True we are edging away, but that prevents us being cut off from the coast."

He placed his finger on the two important points.

"From the port of Santander, or even Corunna, if needs be. Meanwhile, our cavalry, although unpredictable in a fight, are well mounted and are out gathering intelligence of Johnny's whereabouts. It's right that we support the Spanish as long as there's anything to be gained, and I'll wager we'll not be surprised by anything appearing behind us. Further to that, at Alaejos, the town we passed through after Rueda, we received a captured French dispatch, telling us exactly where the Gentlemen are!"

He moved his finger over the map.

"I believe Moore's got it about right. If we combine with La Romana, then we can go after Soult, who's monkeying about somewhere about here, between us and Burgos."

At that he made himself look cheerful, very deliberately.

"And Baird's coming in tomorrow, with 9,000 men. That brings us up to 33,000, enough, I think, to give a more than adequate account of ourselves. And our Brigade will be in his Division, the First Division. At last we are number one!"

He looked around, changing the expression on each face to match that of his own.

"Right! What's for dinner?"

It was D'Villiers who answered, companionably, which discomfited Carravoy even more.

"Chicken, Sir."

"Chicken!"

"Yes Sir, and some rather discreditable white, and a pile of red beans, name unknown."

All laughed as they took their places around the uneven "table", comprising a collection of furniture, which included an uneven sideboard.

Meanwhile, until relieved at dawn, the British pickets shivered in the dark. A picket of the 105th Lights, that included Ellis, Byford, Saunders and Bailey huddled around a tiny, illegal fire, built inside a barn, with each, as a pair, taking it in turns to return to the front door to stare out into the inky darkness. The rain had turned to snow and the ground had frozen, which had made marching to their present position easier than over mud, but the cold now chilled their bones. Moore had given orders that no picket was to light a fire, but the barn was undamaged and the back windows, facing back to the army, were soon blacked out with sacking found in a half loft. It was Ellis and Saunders who at last saw in the dawn which coincided with the arrival back to them of an advanced picket of the 95th Rifles. With the first arrival of significant daylight the fire could be built up and soon it was, to quickly give a cheerful blaze, enough to brew tea. The Rifles broke their journey to gather its warmth, opening their greatcoats to expose their frozen legs and chilled bodies.

As the Riflemen stood stamping and sharing their tea, Ellis began the questions.

"Where you back from?"

"Some Abbey, a mile on, called "Melgar Abatho".

Saunders perked up.

"Any brandy? Holy wine?"

The Rifleman laughed.

"Think we didn't look? If there were, we didn't find none."

He took a drink of the scalding tea.

"The place were cleaned out. Statues, Altar cloth, everythin', but the place weren't wrecked, like. We reckoned that the Monks worked out what was comin', loaded it all up an' shot off! Up to the next estate of theirs filled with Holy Joes. They'm always willin' to see each other right, when times is a bit fretful."

Ellis had no interest in the possibility of plundered spirit. He wanted more of what they seen.

"What of the Johnnies?"

"Saw plenty of fires, and there's horses. Plenty of 'em, you could smell 'em on the wind. Johnny's got cavalry up before us."

"And infantry?"

"I'd say not. Too few enough campfires."

At that point a Rifles Officer strode into the barn, very much looking the part, even though bundled up in a dark green greatcoat, his distinctive sword curving back behind him. All came immediately to the attention. His first words concerned the fire, which he pointed to.

"How long's that been lit?"

It was Ellis who answered, carefully choosing words from within his long campaign experience.

"Not long been built up, Sir. After dawn, Sir."

The Officer, plainly not of too many years in uniform, missed the subtlety of the reply and motioned to the three Riflemen.

"Right. That's Soult up ahead, so, you three get back to the battalion. Things are happening."

The Rifles Officer was correct, but what he said did not apply to the 105th, nor any other infantry. Not an hour before the departure of the Riflemen from the barn, with the slow dawn just complete, there came past two Regiments of Hussars, moving at a fast trot, led by their Colonels and their Commander himself, Lord Paget. Within the hour came the sounds of battle, a heavy conflict, but not of long duration. Two Brigades of Baird's First Division, came hurrying forward along the road, the first being Warde's two battalions of Foot Guards, arriving merely minutes ahead of the messenger to Ellis' picket from Carr, that they should return. Then came the 105th as part of Bentinck's Brigade and the picket added themselves to the column. Soon they were all past the Abbey of Melgar Abaxo and within sight of the town of Sahagun, plainly the site of the recent conflict. It had been a cavalry action, short but fierce and the result was that the French had been thrust back beyond the town. The battlefield had already been cleared; the dead laid out in rows, already stripped, the soles of their bare feet displayed towards the passing column. Many veterans amongst Warde's Guards were as concerned as Deakin regarding what the future may hold and cavalry boots were the best-made boots in any army. Nevertheless, Miles saw no harm in taking the chance to leave the ranks and secure three thick horse blankets from the pile, Ellis turning a blind eye; he had one himself. Deakin and Halfway did the same, as the Colour Company came up and passed by the diminishing pile. Deakin passed comment to Halfway.

"Seems our cavalry 'ave come up to the mark! Seen off three, four times, their number."

Halfway nodded.

"Rather them than me! Fightin' on horses is too far to fall!"

Both smiled and slung their rolled blanket over their shoulders.

The 105th settled into Sahagun with the same speed and expertise as they had at Salamanca, moving into warm billets, with the 105th all indoors; everyman, for all had now taken their turn on picket. Supplies were arriving and their stew was soon hot, albeit the recipe being of a most varied culinary combination, including horsemeat from the Sahagun battlefield. For warmth against the freezing night, both the men and their followers were crammed into the various buildings allocated, to the extent that there was little floor space for anyone to walk across. However, they made the best of it, and there was little argument, even, surprisingly, when Tom Miles and Nelly Nicholls found themselves in the same room. A truce had been struck, caused mainly by the need for all to consume the hot food, then to sleep. Between these two events, both soldiers and followers sat on the rush floor attending to their possessions, either equipment or weapons, or both. Joe Pike sat with his arms around Mary, Jed Deakin sat with Bridie and her children, whilst around them was the usual collection of soldiery, all members of the messes of Pike and Deakin. The latter was, as usual, fretting and worrying about Bridie and the children, to the extent that he inspected each, their clothing, boots and packs. Sinead, the youngest, then Kevin and Patrick all passed muster and required little improvement that called on the experience of "Uncle Jed", but Eirin, now in her late teens had prejudiced her clothing more towards its appearance than its durability. Her coat of a mid-green, of which she was so pleased when she first acquired it, was now showing its age and original lack of quality; it was now wearing thin and was much patched up, with mending of the mendings. Deakin shook his head.

"This won't serve! Not no more."

He turned to his chief scrounger.

"Tom. Get over to the supply wagons. See what's there. See if you can get a coat of some sort for Eirin yer."

Miles looked up from the ministering of his own greatcoat, to looks daggers at Deakin, then at Eirin, but he laid aside his sewing and picked his way out of the room, but fifteen minutes later he returned empty handed.

"There's nothin' there, Jed. Least not nothin' that they'll give up, they bloody griptight bastards as is watchin' over it all. "Not yet" they says, for now. Any for the ranks is forbid."

The two shared a blank look, but then Miles went over to his own bundle of possessions to release from the bindings one of the horse blankets, French and light blue with dark blue and gold edgings and the distinctive "8'eme N" embroidered in two corners. He took it to Eirin.

"Here, perhaps you can make summat out of this?"

He held it out to Eirin, but it was Bridie who took it, quickly examining it for possibilities. Eirin looked up insolently at the slightly embarrassed Tom Miles, holding her head to one side, a cheeky smile across her face. She came straight to the point.

"You gone sweet on me, Tom Miles?"

Embarassment and bewilderment quickly overtook Tom Miles, but not for long, and so, sending back looks of confusion and ill temper, he quickly returned to his own place and his own affairs.

The following day, 22nd December, some Spanish returned to the town to trade and sell. Miles' discomfiture continued when he found himself losing a bidding war between himself and a Hussar for a small flask of brandy. The Hussar had extra booty looted from the recent battle and Miles was outbid, by a silver button. Thoroughly out of sorts he returned to his billet to be greeted by Ellis shouting for all soldiers to parade outside, full kit, ready for inspection. Outside, his was pored over by both NCO's and Officers, then all were ordered to fall out and obtain double campaign rations, including rum in his water flask, which did something to repair his mood. All through the day, the men sat idle, back in their billets, expecting the order to "form up" to come at any time, but it did not. Instead the call to collect mess rations came and so their martial state gave way to one closer to domesticity. This gave Bridie and Nelly Nicholls more time to work on Eirin's new "coat" and the blanket was eventually turned into something useful, Nelly scowling at Tom Miles all the while, but him ignoring all, sat there in a state of readiness, greatcoat off but draped around him, pack and rifle close by, busy eating his stew and chewy Spanish bread. The next day, the 23rd, the order came to rest all they could. They would be marching through the following night, to surprise Soult's men by coming out of the dawn.

Well rested from the previous night, the men slept little, but lounged in their buildings, gambling, singing, talking, and arguing. Not so for their Officers, whom Lacey had called to his Headquarters, and he wasted no time.

"Moore is pushing on to link up with La Romana. The whole army is to advance, and the First Division will be in the lead, including us, with Warde's Guards at the very point. We and they are to take the bridge at Carrion, nine miles ahead. We march at dusk."

He looked at Carr.

"Your Company ahead, Henry, we may need you out, quickly."

He turned to Carravoy.

"Yours next, Charles, ready for an assault."

He turned himself to the map on the table, his curt briefing now ended.

"Make your preparations. Dismiss."

They filed out of the narrow door, not sure to be either pleased at the prospect of action, or apprehensive at advancing yet further. Whilst the thoughts of the majority vacillated between the two and in the mind of Charles Carravoy the latter was uppermost. Back in the small room he shared with D'Villiers and Ameshurst, he threw his gloves angrily on the table.

"This is absurd! He's relying on La Romana's rabble to actually turn up and count for something! And the rest of the Spanish Army down South, if there is any "rest" worthy of the term, to hold at least part of Napoleon's vast forces down near Madrid!"

D'Villiers and Ameshurst looked knowingly at each other, not because of witnessing another of Carravoy's frequent rages, more because they knew that what he was saying was very close to the truth. They sat quiet, attending to their own affairs of cleaning and checking.

The afternoon dragged on, with even Ellis running out of reasons to inspect and question. Then, finally, as the grey December light died through the windows of their billets, they filed outside to form fours on the road. The followers came also, to stand, cold and anxious to say their last words to their men again about to march off to war. Mary stood close to the column, holding Joe's hand, Deakin and Halfway teased Eirin about her coat with no sleeves, and Nelly Nicholls told her Henry to make sure that his stupid head remained on his shoulders.

Come full dark they marched off, soon beyond the town and out onto the dirt road, crunching over the frozen soil and ice. It was utterly dark, so that men found the need to place a frozen hand on the pack of the man in front. There was no sound, bar their crunching footfalls, hour after cellar dark hour, on the frosty earth, the rime providing the only guide to their onward march. Suddenly a horseman galloped past, his rank or regiment impossible to see. Then came the order.

"Halt."

The men did so, their frosty breath now lingering within their ranks. Next came astonishment at the order that came from their own Officers.

"About face!"

Utterly bemused, they turned to face the way that they had come.

"Forward. March on."

No-one could say why, but Gibney was able to say what.

"We're goin' back. Back t'Sahagun."

At the rear of the column, that which once had been the head, Bentinck and his entourage were beating on the door of a small hovel at the roadside. The group included Lacey who entered with the others, in past an astonished and apprehensive Spanish peasant, who was immediately ordered to light a candle. This done, Bentinck opened the despatch handed to him by the Aide-de-Camp, him now long gone, having delivered by word of mouth the important order to halt and return. Bentinck read the note out loud.

"Return to Sahagun and prepare for retreat. La Romana can muster only 8,000 men and one battery. Napoleon has left Madrid and is marching North towards us, with all the force at his disposal."

Bentinck re-folded the note and put it into his pocket. He then left the building without a word, followed by all others, who hoped to find their horses in the dark.

They finally filed back into their billets during full daylight, watched by their astonished followers. Tired, most immediately slumped down, divested themselves of their full marching kit and began to prepare food. Those who did go outside were promptly quizzed on their return, 'what had they learned?", especially Jed Deakin when he came back from returning The Colours to Lacey's Headquarters.

"What's the rumour Jed? What's goin' on?"

The whole billet fell silent. Deakin had halted in the middle of the room, the sign that he had something to say.

"I got it off Lacey's Sergeant Clerk, Bert Bryce. We've got to get ourselves out of a hole, an' that means retreat. Boney's on our case an' comin' up with all he's got. On top the Spanish b'ain't worth a row of beans. We've got Soult still on our doorstep and The Man hisself hurryin' up to join 'im. We'n full outnumbered and have to pull back."

The next question came from several quarters.

"Where"

"Bert didn't have no certain information, but his guess, an' I'd say he was right, was to Corunna, a port up North. It looks like a full evacuation."

Those last words were taken as being his last words and the billet dissolved into chatter, much in angry tones, but Deakin had more to say. He shouted at the top of his voice.

"Now listen. Listen! All of youse!"

At such an order from such as Jed Deakin all fell quiet. Deakin took a deep breath.

"I'm not goin' to make this sound easy, not in any way. We've got a retreat comin', in winter, and over mountains. It don't get any 'arder. We could be movin' any minute, so, once you'm fed, look to your gear. Particular to your boots! What you've got on your feet decides if you can march an' whether you can go lookin' for food. If you can find a spare pair, carry 'em, chuck somethin' out if you 'as to. Take only what'll get you out of this."

He looked at Tom Miles, not a popular man, but well known as an experienced veteran.

"Tom. Be there anythin' you can say, as might help?"

The answer came immediately, calm and grave, spoken as Miles stood up.

"Rain don't just make you wet and cold, it adds to the weight on you. If you can get a piece of tarpaulin to cover yer shoulders that sends the rain off, all to the good, but if not wax up a piece of cloth to put in the same place. Failin' that wax up your coat, to shed the rain off."

He took a deep breath, the next coming from his scrounger's experience.

"The first to be left behind is supply wagons when the horses packs up. Their covers is tarpaulin. Try to get a piece."

With that abrupt ending, Miles sat back down to attend to his own affairs and even Nelly Nicholls sent a kindly look in his direction, which he returned with an understanding nod. All throughout the billet was alarm and anxiety for the future, but the many tasks for preparation at least occupied their thoughts and partly shrouded their fears and soon there came the smell of melted tallow as many took Tom Miles' advice, the candles used being of the finest, looted from nearby churches. Soon came the order they had been expecting: Parade at Noon. They did so and quickly marched off, noting gratefully that they were not part of the rearguard, now marching up and past in the opposite direction, a long column, the whole of General E. Paget's Division and the two Light Brigades, Robert Crauford's and Alten's King's German Legion, comfortingly supported by Lord Paget's now celebrated cavalry, after their astonishing victory at Sahagun.

***

They had been marching non-stop for a night and a day, over roads identified across the landscape simply as a ribbon of barely passable half frozen mud. Now a weak Noon sun peered down upon them, but it afforded no warmth through the gathering clouds. They had been sent on a more Northerly route compared to the Divisions of Fraser and Hope, which meant that theirs, Baird's, was faced with crossing the river Esla at Valencia de Don Juan, by means of only a ford and two ferry boats. It was Christmas Day, but none had noticed, save a few Officers. In their Company at the front of the 105th, Carr passed around his hip flask, with the words "Merry Christmas" but he could not make them sound anything other than deeply ironic. The flask went on to Ellis and Fearnley, both grateful for the strong spirit. In the Grenadiers, Carravoy, too, felt the need to make some form of acknowledgement of the occasion and in the same way, the flask was circulated to D'Villiers and Ameshurst, but what was passed further on down to the two Grenadier Company Sergeants came from their individual Captain's own supply. In Number Three Heaviside also passed the flask but also accompanied by the inevitable.

"Unto you is born this day in the city of David, a Saviour, which is Christ the Lord. Luke 2, verse 11"

Having received the flask back from Ensigns Rushby and Neape, he took another swig himself before passing it onto Deakin, who delivered his standard reply.

"Yes Sir. Thank you Sir. I'm sure the lads all sees it that way, Sir."

Deakin passed it onto Gibney, who, having drank, sank his face deeper into his collar.

There had been a sudden thaw; the road was being churned into mud and so anyone beyond the first battalion had to contend with the sucking loam, now churned up, which threatened to pull off their precious footwear, but many of the followers, taking the advice of the more experienced, bound their feet with straps or even rags, to hold their shoes in place. The soldiers produced their parade spats, which should be gleaming white for such an occasion, but they now held on their boots and shed away the water. Soon they were indistinguishable from boots and trousers, each man having mud up to his knees.

The thaw was worrying Lacey and O'Hare.

"This thaw will raise the river at Valencia, Padraigh, let's hope by not too much."

O'Hare's answer was not on that subject at all.

"Who's behind us, holding off Soult?"

Lacey turned in the saddle to see him and give a clear answer.

"Anstruther's Brigade, that's the 20th, 1st 52nd, and the 1st 95th"

"The 20th. Aren't those the lads that came up beside us at Maida?"

Lacey nodded, this emphasized by his bi-corn, worn "fore and aft".

"The same."

He paused.

"Also Lord Paget, or a least some of his, and in him I have every Faith. His two Regiments saw off twice their number at Sahagun and so, I strongly suspect, any French cavalry will be wholly wary of the sight of a Hussar's pellise."

O'Hare chuckled .

"I hope so. I'd like to think that I could win battles just by the cut my tailor puts into the uniform I wear."

They both laughed and then busied themselves with unstrapping their cloaks at the rear of their saddles. It was beginning to rain again.

Valencia de Don Juan proved to be a bottleneck in more ways than one. Firstly, the streets which led down to the ferry and ford were narrow, but, at least for those waiting in these streets above the river, there was some shelter afforded by the walls of the houses and the over-hanging thatch. Secondly was the river itself, with its ford and ferry. Strangely, the entire town was on their bank so all buildings were available to provide some degree of comfort for the followers of the 105th and the protection was doubly welcome as they pressed against the walls of cottages, watching the water dripping from the overhanging thatch above them. One line was composed of the broods of Bridie and Nelly Nicholls and all were now munching army biscuit, not now so crisp from the rainy march of the past 24 hours. The Chaplain's wagon was before them on the road, waiting its turn, either for the ford or the ferry, and in it was Mary, cold but at least not wet. Sedgwicke was at the reins, sat just out of the rain, partly under the wagon canopy. A horse blanket was over his head, which at least covered his shoulders and part way to his knees, this keeping off the occasional, and wayward, gathering of raindrops that the odd gust of unkind wind sent slanting under the short awning above him. His Chaplain was also inside, content to be in the company of his wife, but not so sure about the company of Mary, but he relented slightly on discovering that Mary carried a Rosary and that she knew her prayers very well.

Sedgwicke looked across at the huddled figures against the cottage wall. He did his best to sound cheerful.

"How are you, ladies? How are you coping? Reasonably well, I hope."

Nelly Nicholls was the nearest and it was she who answered.

"Parson, me darlin'! Now, sure, things could be worse. We're all sufferin' from the cold but we've enough commons as to keep us all goin'. How might things be with yourself?"

"Well enough, Mrs. Nicholls. I must be thankful that I am not required to walk, as you are, so I'll make no complaint."

"You've got the truth of it there, Old Parson, Lord Save Us all."

She paused.

"Now, is there anythin' that you might be knowing?"

She was asking about rumour and Sedgwicke did have an answer.

"There's talk that we'll give battle at a place called Astorga."

"And how far might that be?"

"Two days, would be my estimate."

He saw her nod her head knowingly, then he asked further.

"How are your children?"

Nelly looked down at the six shapes grouped between herself and Eirin Mulcahey.

"Sure aren't they holdin' up well, but 'tis the Lord's Truth that the three youngest is feelin' the pace of it all."

Despite the rain, Sedgwicke leaned out of the wagon a little more.

"Then can I offer that they sit up here with me for a while?"

He pulled a blanket up from under his bench.

"Up here they can rest and perhaps be a little warmer."

"The Saint's Blessings be upon you, our Old Parson. They'll come up, if you're certain now."

Sedgwicke nodded and the three, Sinead Mulcahey, then Violet and Trudy Nicholls, clambered up over the wheel to sit on the vacant space and huddle together under the extra blanket. The noise alerted Chaplain Prudoe, under his own pair of blankets, but him very disinclined to abandon their warmth, for however short a time, to peer outside the canopy.

"Sedgwicke! What is taking place out there?"

"Just a measure of re-organisation, Sir. Sorry to disturb you."

Sedgwicke was gaining the art of dissembling as well as any veteran, only with a better vocabulary.

Meanwhile on the edge of town at the river, all was feverish activity. The guns were surging over the ford, the water above the wheel hubs, but not enough to apply enough pressure to move them with the current and into danger. The two ferries were in constant use, but the usual ferrymen were by now exhausted and sheltering in their hut, or counting the money earned. The necessary manpower was now provided by soldiers, all anxious and urgent to get themselves across and to the relative safety of the other side. Therefore, there was no shortage of willing hands to pull the hand rope as it ran through the guides at the side of each ferry, nor of willing hands to pull back that which was empty, using an extra pair of ropes attached to the back of each vessel.

The 105th were the next and stood patiently waiting their turn. Conveniently each ferry could take one Company, and the Lights would be next after the Grenadiers. John Davey, always suspicious and wary, broke ranks to examine the lashings of the guide-ropes on their side, not only were they used to propel them across, they also held the ferries safe in the stream. After careful inspection he returned, nodding his head and grinning at the exasperated looks he received from his comrades. Within half an hour all were across and volunteers now did the same for their followers. However, the baggage wagons were required to use the ford and they entered the water to find it surging up against their sides. Most gave a lurch at the pressure, but their wheels held when they were seized by the grooves worn by the countless traffic that had traversed the smooth stone over the centuries. However, within the Chaplain's wagon all was suddenly not well. The water, higher than the floor of the wagon, bubbled in between the floor and the sides and soon all on the floor was wet through. Mary and Mrs. Prudoe, sat on their bench, lifted up their feet out of danger, but the Reverend Chaplain found both himself and his precious blankets now very wet, cold and uncomfortable. He heard children's giggling from the front, but was too discomfited with his own situation to pay it any further attention.

Having crossed the Esla the Brigade made camp. Once his Regiment had crossed, Lacey turned himself to the business of the day, this being a written command from Bentinck, this having come from above him, from Baird, Bentinck's Divisional Commander. Lacey stuffed it in his pocket and turned to O'Hare and Simmonds.

"We are ordered to drop behind, to form a link between our Division and Anstruther. The 4th and 42nd will move on, at dawn. How far back he wants us, I've no way of knowing. Closer to which, himself or Anstruther?"

He paused.

"How far back is Anstruther?"

He looked at the two and received no answer, not surprisingly, for there was none that either could give. He exhaled loudly through his nose.

"Right, we'll hold here for tonight and half tomorrow. At dawn get the followers on; get them closed up to the 42nd. At least there, they'll have some protection from marauding cavalry, but I'm worried about them getting their rations. Send a loaded wagon on with them, and our Chaplain. He can take command. Now, sleep!"

With that he took himself off to his tent, and his two Majors off to theirs. With the night came unwelcome activity. Many of the 105th, seeing the empty village across the water, saw their chance for plunder or valuables, or firewood at least, but this possibility had not escaped the attention of RSM Gibney. Merely a few hours into the night, he assembled a line of the NCO's he trusted, these including Deakin, Ellis, Hill, and Fearnley, all positioned to meet any returning miscreant from wading the river with contraband. Those that had gone back merely to gather firewood from the stock within the cottages were allowed through, whilst those with any item that could be described as valuable, even if the intention was merely to smash it for firewood, received a punch on the jaw from Gibney and orders to return it, then report back empty handed or receive another stinging blow. However, Seth Tiley used his formidable strength to ford the river higher up and avoid the line. He had ransacked but three cottages and found but a few coins, but he regarded that as being a satisfactory result for little risk and effort.

Soon after the following dawn the 105th found themselves alone, both their fellow Regiments and followers having now gone. Lacey spread his men in their Companies along the banks and waited for the hands in his half hunter to show Noon, or for Anstruther to appear on the opposite bank. What did appear at around 10 o' clock was a small squadron of Hussars, who plunged straight into the ford and crossed. They halted and their Officer went straight to Lacey.

"Cornet Miles Cornish, Sir. 18th Hussars."

Lacey nodded

"Lacey. 105th. What do you know?

The Cornet leaned forward in his saddle.

"We've just seen off a couple of squadrons of French. Yesterday, Sir. Other than that I know little. I am to get myself back to Benevente, where the main column is."

"Anstruther?"

"Some way back, Sir, but I could not say how far. Yesterday we heard sounds of fighting, but that's all."

He paused, clearly having more to say, but not information.

"I have some wounded, Sir, please can you help?"

"Of course, our Surgeon is in the trees there. And get yourselves some hot food!"

The Cornet saluted and returned to his men, who were having their own conversation with the Light Company, Davey in particular, who had immediately noticed their wounded.

"When did they come up on you?"

A dark faced Sergeant gave answer, Light Cavalry pigtails prominent either side of his face, but his reply contained no aggression.

"Yesterday. Over 100, but we gave 'em a caning. His cavalry don't seem to have the stomach for it, but his infantry is different, so it seems. They've got Boney behind 'em and that always makes a difference."

Davey nodded.

212

"The rearguard? Seen anything?"

"There's fighting, we hear it every day. So far, us and them have kept the Johnnies back, but they'll come."

He took a drink from Davey's offered canteen and was evidently pleased to taste the rum. Thus more warmed, now both inside himself and also towards Davey, he said more.

"Anstruther will be with you before night, and watch out. He'll be needing all he can get, and you, hanging about here; you'll get pulled in!"

He handed back the flask and followed his Officer, riding to the rear of the 105th's position. Within an hour they were gone, leaving Lacey to again study the hands of his watch, both seeming to be moving in reverse. At 11.45 he gave orders to assemble for march, but at 11.55 an Aide de Camp came riding up, plunging through the ford with more urgency than that shown by the 18th.

Lacey looked at O'Hare.

"Here it comes!"

The Aide de Camp reigned in his panting horse and saluted.

"Sir. I'm hoping I've found the 105th."

Lacey nodded.

"You have."

The young Officer's face broke into a broad smile, before relaying his message, both written and by word of mouth.

"General Anstruther, Sir, wishes you to march on, to the next defendable position and hold there, Sir. He leaves that decision to you and to hold until he comes up. Sir."

However, his next words would severely limit Lacey's choice.

"He expects you to find one and hold it before nightfall today, Sir.

Lacey opened the note and found the words to exactly echo what had just been said.

"How far back is the General?"

"About two hours march, Sir. His nearest to you are the 1st 95th. His others are further out, perhaps another hour, each."

Lacey nodded again.

"Well done, Captain. Get yourself some refreshment before you return."

The young face changed to surprise.

"Oh, I'm not to return, Sir. I'm to get on to General Baird."

"Baird, you say? What have you for him?"

"That Soult is operating close to the North of here. We are mostly against cavalry, but Soult has moved his whole infantry towards La Romana, who's holding the bridge at Mansilla. Sir."

As the Captain moved off, Lacey looked at O'Hare.

"Mansilla! As I recall from our map, if Soult gets over, he's just short of being across our route back."

O'Hare nodded.

"I'll get the men going."

He departed, at an urgent trot.

The defensible position, when they came upon it, was a bridge over a stream. It was not a raging torrent, but it had carved itself deep down to create steep banks composed of either hard clay or soft rock. The 105th filed across the bridge straight into defensive positions, but Lacey's first order was for them to prepare hot food. The Grenadiers and Light Companies were closed up at the bridge, either side both up and downstream, along the bank, with the Lights supported behind by the Third Company, these with The Colours. This gave the likes of Davey and Miles a chance to talk with Jed Deakin and Toby Halfway formed up behind them.

"Alright, Jed? How's things?"

Deakin lifted his head that was leaning on his hands grasping the staff of his Colour.

"John, Tom, Zeke."

He ceased his naming and smiled a greeting to the rest, these being Pike, Byford and Bailey.

"Could be worse, and it won't be, not yet, as long as rations holds out. What do you know?"

"Not much more, bar there's French cavalry feelin' for our backsides."

Deakin nodded.

"Knowin' that, I feels better stood yer, with that stream in front an' some commons about to arrive, soon, I hopes."

Miles now spoke up.

"You'm right about the food, but for one, I've had enough of sloggin' our way back through this mud. I'd not complain if we was to stand and fight this crew of Johnnies as is pushin' us out of one place after another. They'm nothin' special. We took 'em easy enough at Vimeero!"

214

Deakin nodded sagely. Such outbursts were not unknown from such as Tom Miles, but he knew he had a point, shared by the vast majority of the army. He levered himself off his flagstaff.

"You'm not alone in thinking' that, Tom, not by a long chalk."

The hot food arrived, mostly beans and something green, but they made no examination, merely wolfed it down. Now, they had the simple and tedious task of merely holding their positions, but fires were lit and they benefited greatly from their warmth. They had been there little more than an hour, before the cry came from a picket positioned beyond the bridge, shouted on the run, as they came back.

"Stand to! Stand to!"

All rushed to form their firing lines, but what appeared was a column of Riflemen, all in green, not the recognizably friendly red, all men formed in fours, trotting along the road and soon to reach the bridge. They crossed the bridge and halted on the road, to the right of the Grenadier column. Both Carravoy and D'Villiers studied the panting men, all dirty and dishevelled, many with uniforms parting at the seams, many tolerating minor wounds bandaged up, but all, nevertheless, were tough, intimidating soldiers, with their Baker rifles clean, maintained and ready. There was nothing of the societal army about these; there was no purchase of Commissions into their ranks as there was into those of more fashionable Regiments. They embodied the hard, dangerous world of life on the edge; of trusting themselves and each other in the extremes of combat but yards from the enemy.

Leading them at the run was a young Captain and he left the road as his men crossed the bridge but they remained on the road, holding to their column. He went immediately to Lacey.

"Captain Brotherwood, Sir, 1st 95th."

"Lacey. 105th. Who's behind you?"

"We believe French cavalry, Sir. Saw them from a distance. If they are coming for this bridge, I'd give them fifteen minutes before you see them."

Lacey nodded.

"And the 20th and the 52nd?"

"The last I heard the 20th were to the North, the 52nd to the South, Sir. It's reasonable to say that they'll not come through here. My 95th are divided between the two."

"General Anstruther?"

"No idea, Sir, I've not seen him for over two days."

Lacey sighed and looked at O'Hare."

"That bridge should come down. What do we have?"

"Nothing! Not even picks and shovels."

"Right. We hold until night, then sneak off, leaving fires. That'll give us a few hours grace at least. I'd say there were two more hours of daylight."

O'Hare nodded his agreement.

"Hold positions?"

Lacey nodded, then looked at Brotherwood.

"Give your men a rest and some food. We'll hold here. Do you have rations?"

Brotherwood drew breath to answer, but Lacey continued.

"No, keep what you have. Draw some from our wagons."

He pointed.

"Over there."

Brotherwood saluted then ran off to his Sergeants. Lacey went straight to Carr.

"Cavalry may well be on their way."

Carr looked puzzled, so Lacey answered.

"French!"

Lacey allowed the word to sink in.

"Send one of your sections upstream, the other downstream, for a good half mile. Check that there are no other crossing points."

Carr saluted and ran off to Drake and Shakeshaft. Lacey took himself to Heaviside.

"I'm sending the Lights off scouting. Bring yours forward into their place."

Number Three Company had been stood guarding the bank for but ten minutes before their own pickets came running back again to the bridge, their urgency conveying its own message. Strangely, with all stood to, all was silent, to be broken by a single word shouted at a distance, in French.

"Arretez!"

A French Officer, of what seemed to be light cavalry, had appeared at a bend in the road, about 300 yards off. He rode forward but a few yards more in order to improve his view, then stopped. Many telescopes were on him from amongst the 105th, including Lacey's, who lowered his and remarked to O'Hare.

216

"He'll not try here. He'll be off to the flanks, to see what he can find."

Carr was with Shakeshaft and by now they were almost half a mile downstream. If anything the streambed was becoming deeper and beyond where they were, where the stream continued on, the ground was falling away steeply and the stream became a succession of rapids. He was just contemplating returning, when a soldier ran up.

"Sir. French cavalry, Sir, on the other side."

The soldier pointed to where Carr should run and he did so, just as his men began to open fire. As Carr arrived, he saw one riderless horse, and the rear end of several more with the backs of their cavalrymen riders. They had seen all they wanted to and, with the far bank held, they wanted no more of where they were. Carr stood with his men, waiting some minutes, before ordering their return. As they marched back, they heard the sound of more cavalry beyond the trees. They stood to, but the sound passed. Back at the bridge they found all space crowded by their own cavalry with the unmistakable figure of Lord Paget himself talking to Lacey, O'Hare and Drake. Carr walked up and waited his turn. It was Lacey who addressed him.

"Captain?"

Carr saluted.

"There is no crossing below us, Sir. We saw some French Hussars, and downed one at least, before they made off."

Carr saluted formally and left, accompanied by Drake.

"What did you find upstream ?"

"A crossing, yes, but deep and easily defended. We saw no French."

Carr nodded, then both saw and heard movement. The 95th were deploying along the bank and their own Regiment was now forming up on the road, prior to resuming their march. Carr looked at Drake.

"Right. Plainly, we're moving on."

Much was about to happen, not least Lord Paget leading his cavalry back over the bridge to the French side. Before he left he leaned over in his saddle to Lacey, who immediately came to the attention.

"Colonel. I'm informed that you are the 105th. I'm told that you had a bit of a set to at Maida.".

"That's true, Sir"

Paget nodded.

"Pleased to have you along."

Lacey's reply was accompanied by a brisk salute, punctiliously returned.

"Thank you, Sir."

Their eagerness to leave and continue their retreat was giving the 105th both focus and energy, and soon their column moved off, leaving the Riflemen in possession of the bridge. The Light Company were the last to leave and all would have been a tranquil transition were it not for a disgruntled Rifleman, with holes in both his trousers and boots, feeling the need to make a comment to the passing 105th Lights.

"You can clear off now then, you lightweights. We'll take care of this."

Unfortunately, this was within the hearing of Tom Miles, whose own inner character immediately came to the surface.

"Leave it to you? Just like it worked out at Brilos? You thought different back then!"

The Riflemen frowned deeply, a big question forming in his mind.

"That was you?"

Miles fell out from his Company, just enough to display the lurid facings of his uniform and point to his cuff, for added emphasis.

"It were. The 105th Wessex, a wearing of the green, but perhaps you didn't notice, on account of you bein' too busy lookin' out for what was comin' up from behind!"

***

The sign said "La Baneza", it clinging to life as an object of significance, its crosspiece bearing the word at the oddest of angles, it's post closer to the ground than to the vertical. That it had not been taken for firewood was a miracle, perhaps attributed to its soaking state from the incessant rain, probably because dry furniture looted from abandoned houses made better kindling. The rain had continued throughout the previous night and into that morning. At this point roads joined and the 105th, as part of Baird's Division, were now on the same road as that of Hope and Fraser, these termed the "main army". They were now the last in Moore's retreating column. The road had been churned into little short of a quagmire and it was obvious from the briefest examination of the fields to the side that many Officers had taken their men onto these whenever they could and there had also beaten the dead pasture into liquid mud. Inevitably wagons got stuck and men were required to fall out and run back to extricate their precious supplies using little more than shoulder power and a rope attached to any fixed point. However, what did impress Lacey and O'Hare was the ease with which their own pack mules negotiated the clinging mud, their hooves coming out of the deep mud with ease, as though they were on a paved road.

What did worry both Senior Officers were the signs they came upon, on each side of the road, of an army losing its discipline. Every building within half a mile either side of their route was wrecked and looted, especially for wood to make bivouac fires, and so nothing remained; doors, floorboards, furniture, even window frames had been stripped out. Distraught local inhabitants stood cursing the soldiers in the column, whilst trying to secure their homes from the wind and the cold using nothing but straw bales and sacking. Lacey had sent on a Lieutenant with a message to Baird that they were now close to the main column, but a reply came back that Astorga was crammed full and could hold no more. They were to remain outside for one more night, then march in, by which time Hope and Fraser would have marched out. Another night in the mud and rain, just above freezing, did not appeal, but, this close to Astorga, a main town, there was at least a decent selection of buildings, some more wrecked than most, some more dilapidated than others, but all could provide some form of roof that gave shelter from the continuous rain.

They marched for anther mile, then Lacey called a halt. It was almost Noon. The 105th, now ordered to halt, spread themselves around such buildings as could be found and cooked a meal from their dwindling supplies. Using all the tricks learned from his time as an "old campaigner", Morrison managed a hot stew for his three Officers of the Light Company. Bearing the pot and three dishes he pushed at the stricken door of their billet, wondering how many more times he would be able to do that before it parted company with its doorframe. Having shouldered it aside, he entered and was about to make some cheerful remark to his three charges, but the words died in his throat; all three slept the sleep of exhaustion upon the earthen floor. He left quietly and returned the meal to the stew pot, still suspended over the fire.

However, no sleep, at least not yet, for Jed Deakin. He was stood outside at a well built, but now wrecked, cottage, it's contents spread outside and spilling over the threshold, for there was no door remaining. He was examining the chimneystack from amongst the scattered furniture outside. Toby Halfway was with him.

"Get Tom Miles over 'ere."

At a summons from Jed Deakin, Tom Miles duly arrived, saying nothing, but with a questioning and irritated look on his face. Deakin pointed to a small door, almost at the top of the chimneystack, set into the stonework.

"We needs to get you up there!"

Tom Miles still remained surprisingly silent, but his questioning expression became one of exasperation. However, Deakin was looking around.

"Come 'ere"

Miles and Halfway followed Deakin into what was left of the barn. All wood was gone, but of harness there was plenty and many loops of rope.

"Toby, get some hefty lads, Zeke Saunders, the "Twins", and such."

With Halfway gone, Deakin was measuring Tom Miles for size, but by now Miles had had enough.

"Now just what spatchcock idea be you dreamin' up?"

Deakin ignored him. He had found a horse's collar and laid it on the ground.

"Step into that."

Miles did as he was bid and placed both feet inside the collar. Deakin then worked the collar up Miles body until it was just under his armpits, but would go no further. He nodded, well satisfied. He then began knotting together the lengths of rope and securely attached one end through the top rings on the collar. He then inclined his head towards the door and Miles followed, trailing the remaining rope. The "hefty lads" had arrived, all immediately grinning at the ridiculous appearance of Miles. A weight was attached to the trailing end and Saunders threw it over the apex of the roof and then it was brought round to the chimneystack. There was plenty of spare and so Deakin thought again. A length was detached and tied to Miles' belt, as a "control". With no more words, Miles was unceremoniously hoisted skywards. "Knew you'd end up at the end of a rope, Tom", came from somewhere, but Miles was more anxious about the fact that he was describing circles against the wall. However, eventually he was at the right level, but too far over, so Deakin himself pulled him over with the control rope.

"Right, open that door and see what's inside."

Unsure of his security within the collar, Miles was clinging with one hand to the rope above, but with his free hand he reached inside and his face changed to one of pleasant surprise.

"Hams! Three, perhaps four. This must be a smoke box."

Deakin became impatient.

"I knows that! That's why I sent thee up there. Now unhook 'em and send 'em down."

There were, indeed, four and each ham was carefully dropped into the arms of Toby Halfway. Deakin walked over to examine each, ignoring the way that Miles was dropped by the "hefties", who playfully released the rope a little too early to give Miles a less than comfortable landing, which gave rise to a bout of evil cursing which continued even when Saunders hauled him to his feet. Deakin had decided.

"One for each mess, the last two for my Bridie and all, when we gets into Astorga, tomorrow.

***

The dawn broke with reluctance, as if it were suffering from pangs of conscience that its day was simply another that was about to inflict yet more misery on the gathering of humanity spread within its dismal daylight. The 105th were already formed up in the first gloom that could be termed any point of daybreak and they marched on and, just within the span of the day, they were in Astorga, with their followers waiting at the side of the road, holding torches or candle lanterns. The "falling out" was a warming affair, a respite from the cold and wet of yet another exhausting march. Even Nelly Nicholls wrapped her arms around her Henry, then beamed like a bride at the sight of the ham. That night, united within family firesides, a good meal was enjoyed, a good stew of vegetables and ham, thickened with biscuit, for the followers had been better supplied than the 105th themselves. Jed and Bridie sat together, him plying her with anxious questions.

"Clothing and boots. How're they?"

Bridie clasped her hands between her knees and smiled at him fondly.

"You're not to worry now, we've extra. We've not used the stuff as was took up after Vimeiro, we can hold that back. And that travois thing that the Colonel showed us is a Godsend. We can take much more and the poles slides easy over the mud, and they stops us from falling over."

Deakin nodded, satisfied, then started on his next topic of worry.

"How're the children?"

Now her hand went onto his arm.

"Fine. We're keeping them warm and fed. Fine!"

The last word was spoken with increased volume, and drawn out for added emphasis. Deakin nodded again, for the final time, at last, but it was Bridie who had more to say.

"And, at the top of all, Mary's feeling better. She's riding on a mule called Pablo."

She pointed to the shape, not far off, the ears most prominent. Deakin leaned away from her, the better for her to see his look of astonishment.

"How'd you get that?"

"The drover ran off."

"And didn't take the mule?"

"No. It was loaded with bags of cartridges. No use to him. So we loaded them onto the other mules and stuck Mary on top."

Now Deakin did smile and looked at his "follower wife" with admiration, before kissing her fondly.

However, the night was not peaceful. Fraser and Hope's Divisions had not wholly quitted the town and word was spreading that they were not to turn and fight, which was a bitter disappointment. Moore had ordered the retreat to continue, contrary to the rumour that had spread since Benevente, that at Astorga he would, indeed, offer battle. O'Hare had not heard formally, but he learned such from a Captain of the 6th Foot who was out in the tumult of the night, trying to restrain his own men. The mayhem had extended back through the town to the ground of the 105th and so O'Hare ordered Gibney to mount again their own guard to prevent their own men from marauding. However, the guard found themselves fending off drunken soldiers from a variety of Regiments, trying to plunder the buildings they were occupying. The Captain was shouting himself hoarse, but to no avail. His men paid him no heed, reacting only to blows from Gibney and his men as they fended off with heavy blows any lurching pillager that

222

came too close, this being the only effective form of discipline able to be enforced all round and about. O'Hare had gone to the Captain and touched his arm, but the result was an ill-tempered tirade, fuelled by the desperation of his circumstances.

"Unhand me, damn you, before I shoot you where you stand!"

The anger turned to shock as he recognised O'Hare's symbols of rank.

"Sir. My apologies, Sir. I thought you were one of my men. I can't say how sorry I am, Sir."

O'Hare had greater worries.

"It's of no consequence. Now, tell me what's happening."

"Much of the army has mutinied, Sir. This place is full of heavy baggage and munitions, rum included. They've got at it and this is the result. They're sick of the retreat and now they know that we are not to make a fight of it here, as they think was promised. If I spoke honestly, Sir, I'd say that they have a point. They feel there is nothing behind them to be in any way afraid of."

"Did La Romana hold Soult at Mansilla?"

The sudden change of topic took the Captain by surprise.

"La Romana's here, Sir. In Astorga with us."

"All his men?"

"As I've heard, Sir, but no kind of army at all, more a band of beggars, and they've brought typhus with them."

"So there's your answer, Captain. Soult got past them at Mansilla, meaning that we are close to being outflanked, by him. He could be across our rear, as we speak. Also there's the small matter of Napoleon himself planting his feet in our footsteps. Moore has little choice but to retreat further before Soult cuts him off. So, when you finally regain control of your men, tell them that, will you? In fact, I will make that an order."

O'Hare's firm tone had its impact. The Captain saluted and returned back towards the town. O'Hare walked over to Gibney, holding the centre of their defensive line. By now most of the drunks were prone, more from blows inflicted, than the final effect of the alcohol consumed.

"Sar' Major. Form up a section of men, men who can take care of themselves. Get into the town. This riot will spoil more stores than it distributes to where it's needed. See what you can rescue, food and cartridges get priority, but after that, there appears to be some rum, in copious amounts. I can trust you?"

"Sir!"

The volume and tone told O'Hare that it would be more sensible to doubt the dawn of the following day.

"Recruit some Officers. Pass on my instructions to you as an order to them. I'll take over here."

Gibney saluted and hurried off to the nearest building where he knew men of the ilk he required were billeted. It contained Ameshurst's Grenadiers and, after that, he gathered Carr, Drake and Heaviside, then soon to be joined by other Officers, all armed at least with a pistol or their sword. With 50 men they set off into the town, into a scene of mayhem. Drunken soldiery in several gangs, bellowing and fighting, roamed the streets, many drinking rum from their cooking kettles and canteens. Carr, Gibney and Heaviside led the way, ignoring any drunk that fought with a fellow in similar condition, but pointing accusingly at any that was assaulting a Spanish civilian, the pointed finger sending an avenging NCO to drag off the offender and beat him senseless. Heaviside inevitably spoke what they all felt.

"The drunkard and the glutton shall come to poverty. Proverbs 23, verse 21."

Carr was one of the few that heard.

"That is certain sure, Joshua, looking at this, and what's to come after. Nothing more certain."

They soon reached the town gate that pierced the ancient walls and entered to progress on and find an even thicker and more dangerous crowd, gathered around the place that evidently commanded their most attention, these being storehouses some way in from the town walls. Ranged along the road were many abandoned carts and wagons, their draught animals dead, and their drovers run off. Heaviside commanded their men to draw up four wagons and they pulled them to the gaping doorways. Roughly dividing their men into three they entered, but only two groups found any supplies such as they wanted, this being blankets, shoes, flour and biscuit, although much had been hauled out and was now trampled into the mud of the roadway. They guarded their gains with fists and pistol butts as they hauled their gains in their carts, including even some barrels of salt meat and rum, back to the gate and then through, after hauling out senseless drunks who barred the way.

Outside the gate, houses were still numerous on either side and screams and shouts of both triumph and distress came from both the doors and windows. Soldiers were emerging from most doorways, carrying all manner of useless articles, but one figure was unmistakable, especially to Cyrus Gibney. He strode forward, delivered a killing left hook to the point of the chin, then a slamming blow to the side of the head and finally, for good measure, a downward blow to the back of the head as the figure slumped forward.

"Tiley! Tha' whore's left overs."

Although very drunk and stunned, Tiley was able to raise himself on his hands and knees, but Gibney placed a boot just below his armpit and shoved him over. By now Carr and Ameshurst had arrived.

"Tiley, Sir. Seth Tiley. Was a Grenadier of ours, now a deserter."

Carr looked down at the prone figure, noting the drunken piggy eyes, narrowed with hatred, but with enough cunning to make no attempt at escape.

"Is he one of yours, Lieutenant?"

Ameshurst answered immediately.

"No Sir, one of D'Villiers."

Carr nodded and turned to Gibney.

"Tie his hands. Use his own belt if you have to. Then use another to lash him to a wagon."

Soon all was done and the carts were finally hauled back to the 105th, but on the journey none expressed any sympathy for their prisoner, Seth Tiley. Any drunks found wearing the green facings of the 105th were forced to drink salt water and, as they vomited the rum that had rendered them senseless, Gibney pronounced judgement.

"Tha's'll thank us come mornin', that tha's able to take tha' place int'line. Crauford's comin' on behind, and he'll hang thee for a deserter, or leave thee for Johnny Frog, who'll just smash in tha' scull!"

***

The full light of dawn saw the 105th and their followers marching through a town desolate and despoiled, civilians on either side attempting to remake a home using what furniture was scattered in the roads and alleyways, but too little of it remained and all too damaged, for every building beside the main road had been broken into and looted. Many soldiers, the less drunk, had dragged themselves to their feet and were staggering on, lurching on before the marching 105th. They were the last Regiment of the main body through the town before the rearguard came up, but many of the intoxicated were still too sunk in stupor and remained in such state beside the road. The far towngate was soon passed through, with most men and followers in good heart, for all haversacks contained something and most had an extra blanket, and many again, new shoes. However, all spirits sank a little when they saw the heights in the not too far distance, high and covered in snow, the ridge of Monte Teleno, behind which lay the high plains of the Vierzo. Also, unknown to many, the Regiment had suffered its first deserter, although for the second time in the case of Seth Tiley. Thrown into a cellar, a stiff leather belt that bound his arms could not be drawn tight enough to long resist the strength of such as he. He freed himself, used his immense strength to force up the slanting doors that gave access from the yard and disappeared into the night, but not before bludgeoning a sleeping soldier and relieving him of his blanket, haversack and bayonet.

***

Chapter Six

Retreat

With every step up the Monte Teleno the rain transformed itself into thick snow. Its twelve mile climb sapped not just the energy, but also the spirit, for every horizon, glanced at hopefully from a face most often turned away from the blizzard, turned out to be just another false crest, to reveal yet another climb, wending mockingly away towards the white horizon. Equally depressing for the 105th, forming the last of the main column, were the continuous signs of an army falling apart. All along both roadsides, they found abandoned arms and equipment, by men who had discarded all, to lighten their load, retaining only that which could carry food or keep them warm. After but a few miles out of Astorga bodies lay in the snow, some now frozen to death, these being drunks, too debilitated by the night's excess, who had collapsed from cold and hunger, whilst others had rapidly consumed the drink they carried, only to sink into a renewed state of helpless coma and succumb to the cold.

Almost as soon as they left Astorga they had come to the pass of Manzanal and soon after that, the pass of Foncebadon, and almost every soldier, who felt that he carried within himself enough of the General, looked at both to judge them as the impregnable positions that they were. At both, it was obvious that a single battalion could hold off an army for days, so why wasn't Moore standing to fight? Thus the overall standing of their Commanding Officer fell to a new low. Officers such as Lord Carravoy, voiced their opinion to any that would listen, especially to his captive audience of D'Villiers and Ameshurst, whilst such as Carr and Drake retained their thoughts to themselves. Such as Deakin, Halfway and Gibney likewise held their peace, thinking merely 'good luck to the bloody rearguard!'

With the first "fall out", Deakin and Pike had hurried back to the followers, but they found them in good heart. The care that had been devoted to their clothing and boots was being repaid. The reply to Deakin's anxious enquiry was to be told to drink his tea! All were bundled up, fed and warm. Lacey, O'Hare and Simmonds equally played their part, tirelessly riding up and down the column, pleased that the men were holding their places within their Companies, with Officers and Sergeants also holding to where they should be. O'Hare, unsurprisingly, proved especially adept at passing on encouragement, which brought a laugh out of most who heard.

"Ah, come on now, boys, and don't be spoiling your throats with the demon drink! Sure, now isn't there tea, buns and a Carol Service waiting for you all at the top?"

With their arrival at the summit, achieved as the light was dying, came their arrival on the plain of the Vierza, bleak, open, treeless and wild, broken but by a few low stone walls built upon earthen banks for extra height, such was the dearth of good building stone. Likewise, wood for fuel was very hard to find, therefore campfires were built with the absolute minimum and stocks of kindling were guarded against theft as though it were the family treasures. The one saving grace was that the march to Astorga had taught all the brutal truth of exactly what was facing them, these being all those possessed of sufficient hope, buttressed with the necessary determination. Bar the worst malcontents determined to soon desert, almost all heeded the advice of veterans, it coming both from amongst the army and the followers, to " leave nothing uncovered and then just set your face into it, one foot after another." Crossing the Vierza took three days and the conditions across it kept the army together, not for reasons of high morale or discipline, but simply because there was nothing anywhere to rob and pillage, beyond the occasional one room, stone hovel. Thus all kept together; for there was no alternative that gave any hope for survival. Many survivors, when arriving home described the toil up the Monte Teleno and the crossing of the Vierza as amongst the worst of the retreat, no other part was more exhausting, for the cold, the ice, the wind and the absence of fuel and shelter created the harshest and most unforgiving conditions of the whole march. They attributed their survival to the fact that, at that early stage, they were still well clothed with sufficient food. Most hovels encountered were dismantled down to the bare walls, but the inhabitants were not maltreated, if such a term were not misapplied, as their home was pulled apart before their very eyes.

One such home was not so lucky. The vanguard of the army, in this case wholly the opposite of such a noble and heroic term, was made up of various gangs of deserters and criminals and one such was led by Seth Tiley. He had attached himself to just such a collection, battering its erstwhile leader senseless and then usurped the throne. Knowing that they had to stay ahead of the army, for the Provosts were now at work with the hangman's rope, they were always the first to any place of habitation. So it was now, that Seth Tiley held his bayonet under the chin of the daughter of one pleading crofter, whilst his wife and other children screamed hysterically, but not at such a volume as to drown out the harsh tones of Seth Tiley himself.

"Alimento! Alimento, pronto!"

The husband knew that surrendering their meagre stock of food would be a death sentence to some, if not all, of his family and so he continued to supplicate before the implacable thug that was Seth Tiley, constantly repeating his plea.

"Ningún alimento, por favor, nosotros moriremos de hambre."

Tiley had not the faintest idea what the man was saying; only that food was not being produced. In response he thrust the bayonet up a little higher. The girl screamed as blood trickled down the blade and Tiley shouted louder.

"Alimento! Alimento, pronto!"

The wife yelled something at the man, his face fell and he then motioned them towards the yard, where a door was uncovered of its snow and opened to reveal some sacks and some dried vegetables. As his gang emptied the store, Tiley punched the man in the face for his reluctance to divulge the food sooner. Now better provisioned, his men made off into the snow, still following the road, with three of his gang carrying extra looted items, which they soon threw away as useless.

Similar was inflicted on the first town off the Vierza, this being Bembribre. When the 105th entered the town, it looked as though it had been fought over as the lynchpin of a battlefield. All the houses were wrecked; all furniture and fittings taken for fuel and some buildings were even on fire. The streets either side were full of abandoned wagons, or their sketchy remains, them having been pulled apart for fuel and also there were many dead animals, butchered down to the skeleton. The men of the 42nd of Bentinck's Brigade, two hours ahead of the 105th, had tramped on through, giving but small glances to the pitiful inhabitants attempting to make sufficient shelter from canvas and blankets, these stretched over bare walls in the place of ruined roofs. However, for the 105th, here they would spend the night.

The appearance of a battlefield was added to by the dead drunk soldiers lying in the streets, but, incapacitated deserters and marauders as they were, they were not alone, for many yet remained on their feet and were still yet prepared to rob and scavenge from any humans within range. Lacey knew this and ordered a double guard on both their camp and that of the followers. He chose this instead of obeying his instinct which was to round up all deserters and march them on as prisoners, but he also knew that he would be placing his own men at risk, not only from injury, but of deserting themselves. However, no such good sense existed in the thoughts of Reverend Chaplain Prudoe. He took himself and his wife further away, to the Church, there to find better shelter and there to better enlist the protection of his merciful Lord. Chaplain's Assistant Sedgwicke was horrified at the risk he was taking and felt impelled to speak out. He knew what would almost certainly happen.

"Sir, they'll take our mules, Sir, then we'll have to abandon the wagon!"

Prudoe turned to Sedgwicke and spoke indulgently.

"Not so, Private. My rank will protect us, as will the good Lord, I feel sure. I cannot imagine any of our soldiers robbing a Chaplain, their Man of God."

Sedgwicke wrung his hands together, it was useless to argue, but, even in the gathering gloom, he could see the wraithlike gathering of scavengers slowly approaching. He could only say one thing, as Prudoe entered the church.

"Sir. It would seem that such soldiers are here already. Perhaps you could exert your authority at this moment, Sir?"

Prudoe paused from ascending the steps and looked back puzzled, looking down at Sedgwicke, stood at the bottom, who had nothing more to say, for the deserters were already running forward, and so Prudoe descended and placed himself before them.

"Halt! I order you. I am both an Officer and a Man of God. I order you to halt."

As an answer he was bowled over and the men ran on to cut the traces of the mules and drag them off. Then, others arrived to ransack the wagon. Within minutes all was scattered on the frozen ground, books, papers and the slates used by Bridie's children, their letters still visible, perfectly formed. Sedgwicke stood looking at the disaster, also Beatrice Prudoe, stood at the church door, her hands beside her face in horror whilst her husband sat shocked and astonished. Unwounded, but heavily bruised in his pride and self-esteem. As Sedgwicke walked to the nearest drunks to gain backpacks and knapsacks, he heard two shots as the mules were slaughtered. He returned to the wagon to gather such of his own possessions as he could find, after handing a knapsack and haversack to Mrs. Prudoe to enable her to do the same. He had but brief words of consolation.

"It was inevitable, Mrs. Prudoe. Sooner or later the mules would have died anyway, of starvation and work; in this cold."

She nodded dumbly, but he had more to say.

"We should now get ourselves amongst the followers. They are probably better able to care for us than we are for ourselves. I do think that best."

She nodded, then she did her own packing, making tortured selections regarding what to take, then she helped her stunned husband to his feet.

***

Bembribre held no fond memories for any of the 105th. Those stood on guard had to spend every minute of the first night on careful watch, each, several times, being required to fend off deserters with the threat of their bayonets or even the discharge from a musket. However, the threatening figures of Gibney, Hill and Saunders kept many at bay. The various messes, united with their followers, were sat huddled in the wreckage of buildings, but at least they had better fires, using wood gathered from the wrecked wagons, which eventually included The Chaplain's. The morning and the rest of the day brought small comfort and even less to their Colonel. An Aide-de-Camp came riding back with a strong cavalry escort and went straight up to Lacey and handed him a note. Lacey immediately took umbrage.

"Captain. This army of ours has not yet been brought to so parlous a state that Captains fail to salute a superior Officer!"

The Captain jumped to the attention and saluted.

"Yes Sir. Sorry, Sir."

Lacey glowered at him and opened the note. Having read it, he looked at the messenger.

"No reply."

The Captain saluted and remounted his horse to ride off, continuing down the road, seemingly to reach the final rearguard.

Laccy looked at O'Hare.

"I want to see all Officers, in there, in fifteen minutes."

He was pointing to the church, with his Chaplain's possessions now even more scattered and blowing in the icy wind. Within ten minutes all Officers were in place, sat under the one section of roof remaining. Holding the note delivered to him, Lacey stood up, within the silence already there.

"We have been re-assigned to the rearguard, now called the "Reserve Division'. That's us, our old friends the 20th, the 52nd, 91st, 95th and 15th Hussars. All led by Lord Paget. It would seem that he has asked for us specifically. Crauford's Lights are leaving the army to be embarked at Vigo, not Corunna. Once they were the rearguard, now we are."

Having read what it contained, he lowered the note.

"I am not going to speak grandiloquently about honour and privilege. To be part of the rearguard for this retreat will be the toughest trial that any of us have faced, including Major O'Hare and me. In addition, we will have the followers with us, I cannot send them on, there is now too big a gap between ourselves and the 42nd, thus there is too great a risk that they will be robbed and murdered and left to freeze and starve."

He took a deep breath.

"Which brings me to my major point. This army is falling to pieces. On top, we are being pursued by an army of considerably larger size. I, for one, want to get home, I do not wish for my bones to remain here, to be scattered to the Four Corners when the snows melt. If we are to get home, we must hold together, as a Regiment, able to take care of both ourselves and our own, and fight out way out. So far we have held together; our losses have been minimal, but morale is low, I don't need to tell you that. Whether or not this Regiment gets back to England lies in your hands.

He paused, then continued, with each "you", when spoken, being significantly louder, to convey the added responsibility.

"I speak of how you treat and how you deal with your men. If they see you enduring what they must, if they know that you are concerned for their welfare and that it has your highest priority, they will stay with The Colours and we will reach Corunna. So, let that be your watchword: "First, Your Men". They come first, you second. Look to their needs before your own. Join their campfires, share their worries, advise where you can. If we fall apart, then fall indeed we will. Together; we may just stand."

He stood for a second, studying the blank but serious faces.

"Dismiss."

All left the church and dispersed through the growing gloom with a different mood to that which they had harboured when they entered it. To those such as Carr, Drake, Ameshurst, Shakeshaft and Heaviside, such instruction was plain and obvious. In fact Heaviside went straight to the fireside of Deakin and Bridie, his Colour Party, with his Bible, and then read from heroic passages, having passed out tea from his own meagre stock. He needed no light to see the words, he knew them by heart anyway, and he needed to finish what he had begun the previous evening. His audience sat listening intently, as the deep, sepulchral, and thespian tones of Heaviside rolled around the bare walls of their meagre dwelling, and all fully appreciated the gesture of his joining them in such a concerned and comradely manner.

However, for such as Carravoy and D'Villiers, such was almost beyond countenance, certainly beyond experience. They had performed their duty at affairs laid on for the tenants of their Estates, but to actually socialise, in conviviality, with their men, all being both their social and military inferiors, would be something entirely new in their privileged lives, if it were to happen at all. D'Villiers fully appreciated the dilemma that dwelt within his Company Commander and tried to make light of it by describing what they were being asked to do in terms that were not too onerous.

"Well, I see no problem in walking around and asking how things are, sort of thing."

Carravoy turned on him

"And if they say, "bloody awful", what then? Do you give them your own shoes and blanket? A drink from your own spirits? What power do we have to put things right? Moore has got us into this God Awful mess, let him get us out of it. Let him provide the men with what they need. I've nothing spare, even if you have."

He turned to walk away and gave his parting shot.

"I don't know what Lacey is talking about!"

With that he stomped off, calling for their servant. D'Villiers stood for a while, enveloped in the cold silence of the chill evening, broken but sparsely by the sounds of the camp around him, mostly the crackling of campfires. He sucked in a deep breath and studied the desolation, so obvious even in the weak glow of the same meagre sources of light and comfort. What Lacey had said was as pure and obvious a truth as any he had heard; they came out as a whole Battalion or they would not come out at all, and that depended on the men. He drew another deep breath and sent it out into the darkness as a white cloud that hung long in the air, similar to the conclusion that now hung in his mind. He went to the ruined houses occupied by his section and up to the first campfire.

"Hello men. Can I get a bit of a warm from your fire?"

Two immediately made room for him and one cleaned out a beaker for some tea.

***

The following day saw two arrivals, one welcome, the other decidedly not so. Paget came in with the Reserve Division, this being welcome; whilst arriving from all points and routes came a host of stragglers, some sober, but many staggering drunk. Where they still managed to obtain supplies of drink remained a mystery. Paget paid them no attention, but immediately called a conference of his Colonels, using the church that Lacey had used the previous day. Paget came in last and, as all stood, Lacey saw the change in his new Commanding Officer. The strain was telling. He stood erect, looking every inch a cavalry commander, tall and slim, and fine featured with dark hair, slightly thinning at the front, but his eyes and cheeks were deep sunk, from worry, cold, lack of sleep and poor food. He stood before his Colonels, intending to be brief.

"All followers are to be sent on, from all Regiments. We are the rearguard, we cannot be slowed in any way. They can wait at Villafranca, two days on from here. It is a major depot containing fourteen days rations of biscuit, flour and salt beef and pork, also hundreds of barrels of rum and ammunition. Secondly, I intend to make a stand beyond this town. I am informed that there is a more that halfway decent position not a mile beyond us, and I intend to give M'sieu a reverse to buy ourselves some time. Prepare your men. Those are your orders. Dismiss."

However, Lacey stood, greatly concerned.

"Sir. The followers will be exposed, alone on the road, prey to any group of deserters who see them. It could cause desertions from our own men, deeply concerned for their safety. They should at least see them depart with some form of escort. Sir."

Paget looked at him, a level gaze from the tired eyes, but not unreceptive.

"I hear you, Colonel."

He thought for some seconds.

"I will detach two squadrons of Hussars, to escort them for one day. That should get them over Cacabelos bridge and not too far from Villafranque. How does that sound?"

Lacey nodded.

"Thank you, Sir. The best that can be done in the circumstances."

Within an hour the followers were all assembled, a long column from the five Regiments. Prudoe, as an Officer, and Sedgwick, as a non-combatant, were to go with them. For the very brief time they were allowed together, sadness and worry were the common emotions, not least from Jed Deakin, but Bridie scolded him for yet more of his continued fretting. He watched them move off, him slightly less concerned by what Bridie had said, but not so Joe Pike studying Mary. She was still on the mule, but her condition meant that she could barely keep any food down, thus she departed with little inside to sustain her.

The 105th, having rested, were the last to leave Bembribre. There were now hundreds of drunks in the town, for those who had already been there when the 105th first entered, had now been joined by as many again. Perhaps it had entered their befuddled minds that the presence of the rearguard meant that there were no more of their army between them and the pursuing French. Knowing that they were the last to leave, Lacey sent his Lights and Grenadiers on first, to push out of the town any that could walk, but few were gathered up. Within an hour the 105th took their place on the very far right of Paget's battle line on a steep ridge, 15 minutes march out of the town, with the 52nd inside them on their left.

For the whole of the morning they had nothing to contend with but the wind and the cold, but at midday they saw three columns of cavalry, the first of the pursuing French. Their entry at one end of the town saw the rapid exit of many of the inebriated from the other, soon to be pursued across the open ground by the French cavalry, who were seen not to be Light Cavalry, but instead the feared Heavy Cuirassiers, their steel breastplates obvious, even in the weak light. The next act of the drama was wholly sickening to those watching from the British positions, made more so by the watchers being so utterly helpless to intervene. The Cuirassiers were plainly veteran at pursuing broken infantry and each cavalryman took their time to carefully overtake each victim, to strike, not at their back or even their neck, but to deliver a downward stroke back into the face. Time after time, man after man was felled to roll writhing in the snow, clutching the wound. However, such a blow was rarely fatal and many rose to run to the shelter of Paget's line, blood streaming from a deep cut that had in some cases destroyed eyes and removed teeth. The Cuirassiers never took themselves within range of Paget's line and, being content with the damage they had done, immediately returned to Bembribre. The 105th, as did the other Regiments, stood looking at the tens of bodies lying prone and dead in the snow and listening to the collection of moaning wounded, now sat behind the British line, but they could do no more than shout insults and curses across the acres of red stained snow. As the evening gloom closed in, Paget, face even darker with anger, had the wounded collected and marched off into the growing twilight, while his men held their allotted ground through that evening until the deepest darkness, then they too marched off, leaving meagre campfires to deceive the French, now thoroughly in possession of Bembribre.

On the road Carr and Drake marched side by side, each with heads held deep within turned up collars, each thinking if the cold they could feel in their feet, was just cold or, more worryingly, seeping water. However, Drake's character, as usual the more cheerful of the two, found something to say.

"I feel moved to wish you a Happy New Year. A very healthy and prosperous 1809."

The sound of a strangled laugh came out from behind Carr's deep collar.

"I thank you for that, and will banish any thoughts I may harbour of sarcasm and irony. In return, I wish the same to you."

Sharing the humour brought the face of each up from within its protection, as Carr continued.

"Have you made any plans between you? You and your Cecily?"

Drake opened his jaw and held it open as if thinking of something to say.

"Well. Waiting for my Captaincy may well appear to be a little superfluous. We could well set a date on our returning home from this little sojourn. This Peninsular War, as some are now calling it, looks set to last for some time, so, the idea of a wife at home, building the nest as it were, appeals more and more with every day spent within this burnt or freezing country."

He took a deep breath.

"When we get home, I feel sure we'll be sent out here again, after not too long a time. I like the idea of calling myself a married Officer. Also, I'll have the right for her to accompany me, should we choose!"

Carr nodded and this time his reply was heavily ironic.

"Oh what larks indeed! I feel sure that Cecily will be overwhelming in her thanks to you for bringing her to such a welcoming clime as this!"

At that, both laughed and trudged on, then their attention was drawn to their left. The Lights and Grenadiers of the 91st were hurrying past, on to Villafranca, as the Light Company Captain informed them as he came up to them, to try and secure some much needed stores from what were there and then await the arrival of the Reserve Division to claim them.

The morning of the 2nd January saw the Reserve Division passing over the bridge at Cacabellos and it was not long before it was clear that Paget was going to offer combat yet again to the French. All his command were over the bridge that spanned the river, the Cua, except half the 95th Rifles and a squadron of the 15th Hussars. Paget sat the road on his horse, ordering his battalions to their places as they came to him off the bridge.

"Lacey. Get yours to cover the guns up there, and send your Lights back over. The 95th need some support."

Lacey looked at Carr, who had heard all, and was immediately wheeling his Company off the road for them to turn and retrace their steps. With the 91st completely over, Carr led his men back, to a defile that was being held by the 95th, with the 15th on their horses some way forward, which placed them closest to any arriving French. He positioned a section each side of the road, where it emerged from the defile, in reserve of the 95th. He looked behind, over the Cua, and was within sufficient distance to see the six blank muzzles of the battery, covering the bridge, and also to pick out the green facings of his own Regiment in place beside them. All along the far bank he saw the red of their supporting battalions, but before him, he could see nothing of an Officer in command of the advanced guard of the 95th, he could only assume that one would either arrive, or present himself. That created a small worry, but he held it in check and instead he continued to look forward down the road, standing with Shakeshaft beside him.

"Well, that's set, then. All we need now is the French!"

The same thoughts were in the mind of Jed Deakin as he stood beside Ensign Rushby in the Colour Party; Halfway, Stiles and Peters, beside and behind. The cold seemed to silence all sound, save the jingling of the bits of the gun horse-teams, stood waiting patiently behind and gnawing hopefully on some dirty straw. Deakin looked at the narrowness of the bridge, then at the number of British on the far side, including cavalry, and felt some unease, but he was comforted. He could say, at least, that he was on the right side and so he rubbed his hands up and down the wood of his musket to gain some warmth. Across the other side, Drake was looking at his watch, having lifted up the cover to fondly look at the inscription therein. As the watch tinkled 1 o' clock, he had urgent cause to snap the cover shut and thrust it into his pocket.

"Stand to! Stand to!"

The shout was coming from somewhere ahead, but what mattered now was obedience and, all around, his own men were unslinging their weapons to hold them "at the ready". Further down the defile, events were happening quickly; and for the worse. French cavalry had arrived, seemingly in strength, and they had wasted no time in charging the single squadron of the 15th Hussars. Rapidly, all was descending into chaos. For the 15th, far ahead, to remain was suicidal, and they were withdrawing at the gallop. The Commander of the 95th, finding his targets masked by the retreating 15th, felt no confidence in the ability of his own men to hold such a large body of horsemen and he was also withdrawing down the defile, chased by both the 15th and the French cavalry. From the other side of the road, Carr was now shouting himself, his voice desperate!

"Withdraw. Fall back!"

Shots could be heard from the 95th, but too few to be effective, for the French were masked by the galloping 15th. The only option was to reach the bridge and get to safety. Drake saw all of his men away, then took to his heels himself.

It was a run of several hundred yards and soon Carr's men were mixed with the 95th and all soon overtaken by the 15th on their horses. There was an unreal period of time, after the 15th had passed them, a time filled with fear and terror, for them, the running infantry, all knowing that the next horses they heard would be carrying French cavalry. Some men had already been knocked to the floor by the 15th, but Ellis was on his feet, running with Saunders and Byford. Davey, Pike and Miles, were just ahead, with some Riflemen before and after. The bridge was 200 yards in front and, when Ellis looked behind, the French were less that, and mounted. On top, the French cavalry who were heading straight for the bridge had already overtaken them. Despite his laboured breath, he managed an order, to take them off on a wide tangent.

"Go right. The river! Get to the river!"

They ran on, then the horse came up beside him and, incongruously, he noticed that the cavalryman had filthy boots, then came what he knew was coming, the swish of the heavy sabre from above. Ellis held his rifle horizontal and took the blow just before the trigger guard, then he thrust the barrel between the rear legs of the horse. The horse staggered but did not fall, but Ellis had bought seconds, which brought him nearer to the river, but then he stopped. It was of no avail; himself, Byford and Saunders were cut off, themselves and three Riflemen, because before them had appeared at least half a dozen horsemen. Between the horses legs he saw Davey, Miles and Pike leap into the river, but himself and those with him were now lost. The nearest cavalryman pointed his sabre at the six, all now grouped together.

"Se rendre, maintenant."

Ellis did not need a translation.

"Drop your rifles. Lay them down."

The other five followed his example and their weapons now lay in the snow. The cavalryman must have been of some rank, for he pointed to three of his comrades and then he led the others off to the bridge, leaving the disarmed infantrymen guarded on three sides and awaiting their fate.

At the bridge, Henry Carr was seething with anger, at an intensity unusual even for him. Men would be lost because of a mismanaged shambles; someone had miscalculated badly. Close to the bridge he screamed at his own men to form a firing line and soon a Rifles Officer copied his example and some Riflemen joined on, wherever they could. All immediately stood at the "make ready". Carr saw French cavalry above the heads of the last of the retreating 95th and 105th, which was good enough for him.

"Present!"

The rifles came level.

"Fire!"

About 40 weapons barked out and Carr saw several French cavalry go down, before smoke obscured his view.

"Back! Fall back!"

The bridge was now but yards behind them and they had bought time for the last retreating British. He joined his men for the final sprint, expecting to be overtaken, but no horse came. Over the bridge he sank to his knees, supported by his sword, and looked around at the chaotic melee of his own men and the 95th, but he had no time for rest. A Major was yelling at the mass of fugitives.

"Clear the road! Either side."

Some stood in confusion at the vague order.

"Move, damn you. Move! Any side."

The urgency was justified. The French cavalry commander had come through the defile with two Regiments, one had scattered in its pursuing the British advanced guard, but the second was about to charge the bridge. They were already formed up, four abreast, 300 yards away from it. Deakin looked from his vantage point. He knew what was about to happen, but he spoke softly.

"Damn all damnable Officers and their damn ache for glory."

The 105th were stood squarely upon the road, in a column two Companies wide, three men deep. The remaining Companies supported behind. Deakin heard the Battery Commander dispassionately begin his orders, from just yards in front, with three guns just each side of the road.

"Stand by."

The French cantered up to 100 yards, to then spring into a gallop to force the bridge. Deakin could barely bring himself to look. As the first ranks closed together to meet the bridge, the cannon fired. The round shot ploughed through the whole column, their sickening passage being marked by the arms, legs and bodies of both men and horses being flung into the air. Whilst the guns rapidly reloaded, the following ranks of horsemen actually managed to cross the bridge, to be halted by the company volleys by ranks of Deakin's own Regiment, supported by both battalions positioned along the banks. The brave cavalry could get no further and within half a minute the once proud Regiment of Horse was a bloody wreck, with men and wounded horses staggering away

240

from the butchers block that was the road and the bridge, some horses pitifully trying to escape with but three legs. Another discharge from the guns wreaked yet more havoc but it mercifully ended the affair, this being signalled by the sound of a French recall.

For Ellis, Saunders, Byford and the three Riflemen, nothing had changed. They still stood under guard, their weapons still at their feet. One of their escorts kept his eyes fixed upon them, but the other two were watching events at the bridge. However, Ellis was studying the riverbank and he saw what he hoped, a faint shade of red through the frost covered reeds. He spoke softly to his fellow prisoners.

"Stay alert!"

His nearest cavalryman, looked sharply at him, on hearing the words, and lowered his sabre to point at Ellis, but he raised his hands in surrender and shook his head. The cavalryman was not appeased and continued to stare at Ellis.

Davey and Pike were hauling Miles up onto the bank, Davey keeping low, as his ex-poacher's instinct told him, and pushing down Pike if he threatened to get too high. He peered through the reeds to assess the scene and then spoke softly to his two companions.

"Get loaded."

A faint reply came back from Miles.

"Already am."

However, Joe Pike was rapidly going through the motions. Davey ground his teeth with impatience, for the French withdrawal was beginning, but Pike soon joined them, gently pushing his weapon through the reeds. Davey gave his orders.

"I'll take the far one, Joe, the nearest, Tom, him as is left."

All three took careful aim, but Pike and Miles waited for Davey to fire. He did and they, also, carefully pulled the triggers. Davey's cavalryman made off with but half a head, Pike's toppled backwards, dead, on top of the prisoners, whilst Miles' fell wounded from the saddle, to be dragged some yards by his horse, his foot caught in the stirrup. Wasting no time on any of them, Ellis picked up his rifle, which was quickly copied by the others and they began running, with no sound, to the river. The three there had reloaded as the escapees leapt over the reeds to descend immediately to the deepest part of the streambed. The telltale smoke had thankfully drifted away. There were other targets but Davey thought it foolish to reveal where they were, cavalry were all around, so instead they silently viewed events as best they could through their cover. Ellis joined him.

"Can we get back to the lads?"

"That I doubt. Take a look."

They were on a bend of the river, which gave them some view of the bridge on the French side and it was a scene of heavy skirmishing. French Dragoons had come up and were disputing the issue across the river using their short carbines, more were arriving and, what was worse, heading straight for them, presumably to try to cross the river. Ellis did look, but briefly, to see that Davey was right. He looked behind, hoping to leave the river on the British side and circle back left, but, as chance would have it, behind was a steep river cliff; to climb it would take too long and the Dragoons would have them like flies on a wall. He took but one look.

"We're off, up stream."

Crouching low, the nine splashed off through the freezing water, heading away from the fighting.

Back between the guns, the 105th were placed too far back to become involved in the skirmishing between the Dragoons and the British, but it was plain that the French were making no headway against the 52nd and the 95th holding the British bank. Suddenly, there came the sound of loud cheering from the Rifles as one man ran back to their ranks, having taken himself right up to the river bank, but Deakin paid it no more attention. He stood watching the skirmishing between the two sides, eventually being required to stand as observer for over an hour, but then came more developments. General Moore himself arrived to sit his horse behind the guns to the right and view for himself what the French were about to attempt. They were massing for an infantry attack across the bridge and over the bloody remains of their own cavalry, some still struggling in agony. Deakin groaned and cursed again. He looked at his two Ensigns, who looked as horrified as he felt and it was not long before they heard the same sounds that they had heard at Vimeiro, the rhythmic drums and the fervent shouts of "Vive l'Empereur!" Again the Battery Commander calmly gave his orders.

"Load grape. Double! Stand by with case!"

Deakin watched the artillerymen run back to their caissons to obey and then return with the canvas bags of heavy grape shot and the steel cylinders of case shot. Each gun received a double charge of grape shot, these being heavy balls, each over an inch diameter, then all stood ready, this signalled by the arms of the six Gun Captains raised in the air. By this time the French column was at the throat of the bridge, but the 52nd and the 95th were waiting for the cannon to fire. When it did come, it came as a single blast, to rip the column apart at the first discharge, then they added their fire. The column maintained the attack for barely seconds before breaking and running from the hopeless task.

Bickering continued across the river when Voltiguers came up, but the fighting died with the daylight and Moore ordered a withdrawal. When they formed up, prior to marching away, the Light Company of the 105th called their Roll, Sergeant Fearnley in the place of Ellis. Carr counted the names not answered, 11 in all, and took himself elsewhere, deeply angry once more.

As they marched away the news passed through the ranks of the whole Reserve Division, the Rifleman that they had seen running back had shot a French General, the one in command, or so many hoped. The Rifleman's name was Tom Plunkett and he was now the toast of the Division. Deakin sniffed and clenched his jaw, picturing in his mind what that same French General had ordered his men to do. He thought again to himself, a very consoling thought, "No bastard Officer ever more deserved a hole through his head, and that's a fact!"

***

Meanwhile, as the Reserve Division was leaving Cacabelos in the growing dusk, Chaplain Prudoe was leading the followers into the outskirts of Villafranca and there was a deep apprehension growing within him, from two sources. Firstly, the troubling expectation of continued life within the donkey he was riding, the wheezing that was emitting from the frozen muzzle of the beast made him fearful that each breath would be the last, but secondly and mostly, what he could see and hear from Villafranca. As at Bembribre, it sounded and looked as though the town were in the centre of a battlefield; it was on fire and the sounds of shouting and shots were all too clear above the crackling of the flames. Instinctively, his fear caused his hands to tug the reins and the donkey immediately stopped. Beside him was Sedgwicke and, just behind, were his wife, Bridie, Nelly and Mary, she still on their mule. Prudoe felt obliged to exercise his position of leadership.

"Private! We cannot take the women and children into there."

However, his powers of command were immediately exhausted. Sedgwicke waited for more but none came; instead Prudoe was alternating worried looks both down at Sedgwicke then at the burning skyline of Villafranque. Finally, something occurred to him.

"Private. Whatever is happening in there, you can blend in, mingle, as it were. Take yourself forward and take a look to, ….er."

He trawled his mind for the correct military terms.

"…… assess the possibilities."

Sedgwicke looked at him aghast at what he was being asked to do, but an order was an order. He saluted, re-set his packpack and marched forward, with some deep apprehension of his own. The first humans he encountered were a group of citizens fleeing into the fields, some with wounds on their heads and arms. The next was a staggering soldier, with a mess kettle full of rum, who invited him to take a drink. When Sedgwicke politely declined, the soldier took a swing at him, but missed, mostly from poor aim and partly because Sedgwicke had time to dodge the course of the obvious "haymaker". The soldier then overbalanced, both himself and the rum then spreading across the frozen snow.

Soon he could feel the heat from the burning buildings, but the amount of conflagration was not as bad as he first feared. What did fill him with horror were the sights on the main square and road. Most obvious were what had been, columns of infantry, now almost in chaos, with men running from the ranks into buildings, being chased by Officers with drawn swords and actually striking at them. Next were the scores of drunken soldiers, staggering down the road or lying in total stupor, yet still managing to drink from their canteens and other forms of receptacle. Amongst them were some dead civilians. A minor detail was a soldier, slumped forward, held by his bonds around a post, obviously lately shot. Sedgwicke stood for a full minute, for such a length of time was required for his eyes to convince his mind that he was not looking upon some illusion, nor was he asleep and gripped by some nightmare. He was looking at a riot, a near mutiny, but somehow Sedgwicke brought himself back to reality. The fugitives from the ranks were running into and out of three particular buildings, there were fires made of barrels in the road, incongruously supervised and guarded by the Provosts, and the great majority of the town was still standing, not ablaze, but certainly wrecked and ransacked. He turned and retraced his steps, more rapidly than he had entered, noting that the first drunk had disappeared.

He immediately reported to Prudoe, who spoke first, his face plainly showing the fear he deeply felt.

"Is it safe?"

Before Sedgwicke could answer Nelly Nicholls piped up from just behind Prudoe.

"Ah now, Parson, darlin'. Are you thinkin' that we can all go in, and get behind some walls, and under a roof, even, perhaps?"

Sedgwicke was in a quandary as to who to answer first, a woman or his Officer, but he found an answer to fit both.

"I think it is safe to go in, yes, but only some way. At the centre is taking place, what I can only describe as a mutiny, but it seems confined to there."

He took a breath to look at both faces.

"I think it may be possible to gain some provisions. I think I identified some buildings that may be acting as stores for such."

Prudoe was disappointed. He was hoping for information that would enable him to order "no further", but Sedgwicke gave him no choice regarding advancing, but he did have a choice regarding who would be first.

"Sedgwicke. Lead us on, as far in as you think prudent."

The good Sedgwicke saluted, then walked on, the whole column of followers tailing behind. In his head he made calculations of how many houses would be needed to shelter them all and he used that to determine when they should halt, rather than the maximum safe penetration of the town. Now satisfied, he turned to Prudoe.

"I think here, Sir."

Prudoe dismounted, saying nothing, but the followers needed no orders, they immediately dispersed into the houses on either side. Soon there were shouts and insults as drunks were hauled to the doors by groups of women and ejected into the road. Bridie and Nelly Nicholls, having completed their required casting out, soon had all in order in the room of a decent house, a room with a fireplace. Mary's mule was brought inside and given a straw mattress to chew on. Nelly took herself over to Bridie.

"Did Parson not say, that there could be provisions further on in?"

Bridie nodded, but Nelly continued.

"Sure, but we have to try."

Bridie nodded again, then spoke.

"It could be dangerous."

She paused.

"Not could be. Surely is!"

Nelly nodded.

"Right, but I've an idea."

She turned to their children.

"Youse! Set up the rest. Me and Auntie Bridie are out to see what we can find."

Soon they were advancing into the town. Both were wearing red jackets pulled from drunken soldiers to make them less conspicuous, two haversacks around the shoulders of each and a stout chair leg held besides their skirts. Instinct told them to keep to the sides of buildings and soon they were at the scene that had so disturbed Sedgwicke. With stern gaze and wholly unshaken, Nelly Nicholls assessed what she could see. Nearest was a fire, tended and guarded by Provosts. They were bringing out barrels to feed it, whilst beyond, there still pertained a scene of riot and mayhem. Nelly waved Bridie to follow and they approached the Provosts, who, noticing first the red jackets, levelled their muskets at them, then their faces turned to puzzlement as they noticed that all else was female. Nelly spoke first.

"Good day to you, Sirs. We've just got in and we're from the Hundred and Fifth. We was wonderin' if there was any chance of a bit of provender from anywhere around this place. We've children and such to feed and was hopin' for something to help us keep our bodies and Souls together, God willing of course."

The nearest Provost was typical of the type, surly and scowling, dislike and suspicion of all others not such as himself always his first reaction.

"Moore's orders are that all stores are to be destroyed! All of it. Nothin' to be left for the French."

However, their Sergeant was of a more sympathetic nature and had some admiration for two women who had ventured into such mayhem. He spoke with a London accent.

"Hold up! Hold up! I don't see the harm. There's good grub in them barrels what's burnin' off to waste. Ordered to be burnt an' not given out, like."

He walked up to Nelly and Bridie.

246

"That's what's wound the lads up into such a state. An' they've got a point, as I sees it!"

He motioned them forward and used the bayonet on his musket to roll two smoking barrels off the fire. With the butt of his musket he smashed both open to reveal smoking joints of meat, seemingly pork, amongst the packed salt.

"There you go, girls. Pick out of those what you can."

The two started forward to use their own knives to spike the joints and thrust them into their knapsacks, but the Sergeant hadn't finished, he turned to one of his men.

"Smith. Fetch out a sack of biscuit. And flour, if there's some."

In the time that it took Nelly and Bridie to fill their knapsacks of the joints of meat, Smith returned, with but one sack.

"No flour, Sarge, 'tis all scattered, but here's biscuit."

He handed it to the Sergeant who handed it to Bridie.

"Compliments of the Provosts, girls."

He stood pointing, back out of town.

"Now get along, or you'll get me shot, just like that silly sod over there."

He used his thumb to point to the executed soldier, but neither of the women was listening, it was Bridie who was speaking.

"May the Saints bless and preserve you, Sir. Do you have a name?"

"Jacob Lederman is who I am. If you're ever in the Isle of Dogs, ask for the Lederman's. Anyone'll tell you the way, and our door'll be open to you."

It was Nelly's turn.

"And if you're ever in Roscommon, ask for the O'Mara's and say you've seen Nelly, as became a Nicholls. You're a Saint, so y'are, and a good man. I'll get my Henry to shake your hand, if ever we gets the chance!"

With a wave from all three Provosts, the two followers departed, keeping to the shadows, ever wary, cudgels at the ready, but they were not disturbed and soon reached the shelter of their room to unpack the precious contents of their haversacks.

***

The 105th marched in during the late evening of the next day, the 3rd January, but for them there was to be no overnight rest. Paget's orders, received on the march, were to cook a meal, then march on, through the night. To "cook a meal" were the operative words. All in the Reserve Division had become aware of the task allocated to the Lights and Grenadiers of the 91st, and so it was in high anticipation that they watched Ameshurst's section of Grenadiers march further into the town to secure the promised rations, for which they were now in great need. Carravoy considered the idea of accompanying his men himself, an idea toyed with but briefly, because Binns had secured a good billet, too warm and too dry. Therefore, from his own initiative, he sent Cyrus Gibney in with Ameshurst as weighty support, "after all, who would be better at dealing with devious soldiery?"

What they saw at the practically lifeless warehouses put disgust on the face of Gibney and dismay upon that of Ameshurst, for all was now lifeless, bar the now common inebriates, prone or reeling and seen in all directions. The buildings were empty of what had been promised, except for a covering of smashed and trodden biscuit across the floor and what had been flour, but was now brown and damp, mixed with the bare earth. At the entrance to one warehouse stood the forlorn figures of the Captains of the 91st Lights and Grenadiers. Both felt it their solemn if wholly unpleasant duty to "face the music" and tell those of their Division what had happened. Approaching the pair, alongside Ameshurst and Gibney, came other Officers of the four other Regiments of the Reserve Division, two Captains of the 52nd in the lead. The dejection on the face of the two Captains of the 91st told all and but one sentence was sufficient, from the older of the two Captains.

"Our men joined in the looting!"

It was Gibney who reacted quickest.

"I'll take some lads, Sir, an' make a search. Should find summat!"

Ameshurst looked at him, utterly dejected, but grateful for a positive reaction.

"Yes, Sar' Major."

He nodded morosely.

"I'll take a look around myself."

The "look around" of both revealed nothing of any significant value, except dried or charcoaled meat at the bottom of various fires, these caused to be examined when Gibney recognised the remains of pork barrels. He set all the men he had to guard the remains, this appearing more as piles of ash than the source of any provender. Thus it was with but scrapings that they returned to their Regiment, carrying mummified meat and some meagre bags of biscuit, this gleaned from the floors of the warehouses.

Meanwhile, emotions were equally in deep jeopardy in the room occupied by Bridie, Nelly and Mary. George Fearnley had given himself the unenviable task of informing Mary that Joe was missing. As he entered their billet, the unique event of him arriving alone told Mary that something was wrong. Lying on a blanket on the floor, she lifted herself onto one elbow, fear and anxiety layered over her tired face.

"Where's Joe?"

Fearnley gave no answer, but walked forward to be nearer and offer some words of comfort, but not hearing the answer she wanted sent her into paroxysms of weeping. Both women rushed to her and Fearnley spoke softly the words he wanted to say. Deakin and Henry Nicholls stood helpless in the background.

"Now, it's not as bad as it seems, Mary, they're missing. No-one saw......"

He paused, not wanting to use the word "bodies", but it was too much of a struggle.

"...... well, their bodies."

He found better words.

"We didn't see them dead!"

Bridie looked at him.

"Them?"

Fearnley nodded, now encouraged.

"Yes. He's missing with John Davey, Tom Miles, Sarn't Ellis, Zeke Saunders and John Byford. If they were dead, we'd have seen them, across the river, where it happened. They could be prisoners, then alive, or they're out somewhere, trying to get back."

Nelly nodded at him, then rubbed Mary's shoulder.

"There now, Mary, there's hope if ever I heard it. He's in the company of five good lads, if ever I was a judge, including that scut Tom Miles. Now there's a survivor if ever I saw one, and you know what they say about bad pennies and he's one of the worst!"

But Mary was inconsolable and buried her face in the pile of clothing that served as a pillow. Nelly continued to rub her back, to afford some consolation, whilst the children all looked on, almost as desolate as Mary herself. Deakin looked at Henry Nicholls.

"We'd best get out an' see what rations've come back."

Nicholls nodded and both left, with George Fearnley, leaving the women to do the best they could with Mary, speaking words of hope that they had no evidence for. It wasn't long before both men returned, both as desolate as the atmosphere in the room and both empty-handed. Bridie looked at Deakin stood wretched, him shaking his head.

"There's no rations, Bridie! 'T'as all been ruined by the army, as've come 'ere before."

Bridie's face changed, it became cheerful, which puzzled Deakin greatly.

"'Tis all right, Jed, at least for now."

She crossed the room to their haversacks and opened one.

"Before you came in, we pulled some salt pork and beef from the fires. Some Provosts let us, so we've meat, at least, and a little biscuit."

Deakin crossed to the four bundles and opened each.

"Provosts you say! The first I've heard of them cold hearted beggars showin' any measure of Christian charity!"

Bridie sprang to their defence.

"Well they did, and they fetched out that sack of biscuit."

She pointed to the 10lbs sack propped against the wall, then it rapidly dawned on Deakin that his followers were better provided for, than any soldier's mess in the whole Regiment. He didn't need to think too long.

"Right. Count these out."

The haversacks were emptied and the pieces counted, to total 43. Deakin gave his verdict.

"These has to be spread over."

He pushed four apart.

"These you keep. Get the rest around the followers and they'll give what they can to their men. He looked at Bridie.

"Have you still got them acorns?"

Bridie nodded.

"Let's 'ave a look."

She went over to the bundle unloaded from the travois.

As was in that sad place, so also in others. In the hours left to them before marching off into the night, many were about their own melancholy business. Chaplain's Assistant Sedgwicke was around the campfires of the men and followers, with a pot of the pork grease that he had found at the base of many fires in the town, rubbing it onto any frostbite or open wound. Henry Carr was dining on horsemeat and biscuit, washing it down with sour wine, which did nothing to wash away his sorrow and anger at the loss of his men. He blamed himself for doing nothing when he saw the danger. Neither Drake nor Shakeshaft could console him, after all, neither could speak from the weight of experience, but the arrival of Padraigh O'Hare did help.

"These things happen, Henry, it's the game we're in. We have to set ourselves to live with it, or give it up. That's the trade we've chosen. It's all a part."

He poured some brandy into four receptacles and picked one up for himself, as then did the other three, and all drank. There was no toast.

***

The cliff provided little shelter from the wet discomfort that fell incessantly from the limitless black of the sky above, sending snow, then sometimes rain, but mostly a mixture of the two. Davey and Miles had managed a fire that had spit, failed, then found new life as the wood dried. It was now a good blaze as they dried their boots, still wearing them, after the soaking in the Cua. The three Riflemen, found to be called Alfred Verrity, Jeruel Spivey and Harry Newcombe sat alongside the six of the 105th, to feel it's cheering warmth, all sat close up within its yellow light. Beside him, Ellis had four haversacks spread on the ground, their edges close together to protect the food spread upon them from the mud beneath. It was a meagre gathering from the haversacks of them all, of raw meat, root vegetables, biscuit, stale bread, dried fruit and acorns. All disconsolately studied the artist's subject for a "still life". None more so than Ellis, for there was barely enough for one day's march, let alone the prospect of several days, before they could hope to circle back to their army, whilst still keeping ahead of the French. He sighed and began the process of sharing it out and it took but a little time for each portion to be stored in the haversacks of the nine. Ellis was securing the straps on his, when a musket barrel appeared over his left shoulder, accompanied by several others that came into the circle to complete the ring, signalling immediately that they were surrounded. The muskets were Charleroi pattern, French, but the men who held them pointed threateningly at the chests of the nine were anything but, for, at that moment came an imposing figure into the firelight to stand between Tom Miles and Jeruel Spivey. He was tall, but what they saw first were stout French cavalryman boots, then heavy pantaloons, much buttoned on the outside. They disappeared up into a round ball of sheepskin, but the impression was not one of corpulence, rather of rugged strength. The ringlets and a luxurious moustache beneath a peaked hunter's cap completed the image. He spoke two words; neither were questions.

"Inglés. Venga!"

The movements of the musket barrels gave all the translation that was needed and the nine stood up and gathered their possessions, hastily donning their equipment. As they moved off, one of their captors added fuel to the fire, leaving it as a more than adequate decoy.

They progressed on, hour after hour, through narrow clefts and thick woodland, always holding to a narrow path, defined poorly by its surrounds, but clearly shown in the slush by the footprints of those marching ahead. The pace the Spaniards set was good Light Infantry, but always upwards, and gradually the idea of them being captives faded. For one thing they were allowed to keep their weapons and for another, whenever a wineskin was produced it was given first to them. Relations improved with Byford giving the correct response, "Gracias, Senor", eliciting the reply, "De nada." Miles was feeling more secure by the minute and passed his own cosily familiar verdict on Byford's linguistic ability, "Book learned bastard!" This feeling grew further within him when one of the Spaniards tapped him on the arm and offered what looked in the dark like a strip of leather, but proved, at the first bite, to be some kind of dried meat. Miles spoke the reply he had learnt from Byford and received the same reply from the Spaniard.

It was a bleak and fitful dawn that saw them enter what appeared to be an old quarry, which impression showed true with the growing light, containing many deep overhangs that harboured rough huts and shelters. It was well populated, all genders and all ages were up and about with the dawn, all busy with some task, cooking, chopping firewood, examining clothing and bedding, sorting through stores of food. The leader indicated where they could lie down, under an overhang on a spread out pile of bracken. The nine sat, not wishing yet to lie flat, instead wishing to examine the camp. The most obvious conclusion was that a whole village had moved up into the hills, into this disused quarry, to build shelters amongst the rocks and overhangs.

They had but little time for detailed examination, because the leader was now standing before them.

"Coma, entonces duerma."

Ellis waited until he had departed.

"What's he say, Byfe?"

Byford stretched his mouth in uncertainty.

"Well, Spanish is close to Portuguese, so I think he said, "Eat, then sleep."

Ellis nodded and soon the first part of Byford's translation proved correct. Some stew arrived in large earthen bowls. It contained a lot of beans and green leaf, but it was accompanied by flat bread, one for each bowl. The nine took the bowls eagerly, ignoring the heat on their fingers from the hot dishes and soon the food had disappeared. Three women arrived to collect the empties, one a girl in her late teens, who made a beeline for Joe to collect his bowl and spent as long as was on the right side of not too obvious, studying him, until embarrassment did pull her eyes away. This did not go unnoticed by Ellis.

"You keep any thoughts on that wench out of your head, 'fore you ends up with the knife of some lovelorn swain in through your ribs or some Father marchin' you off to some Altar."

Joe Pike looked shocked and dejected at the same time.

"It's Mary, my Mary, that I'm worried about."

Ellis nodded.

"We all knows that, boy, but she's in as good a set of hands as I can think of, Nelly Nicholls and Bridie Deakin, with Jed not far, neither. Keep that worry out of your head, we've enough of our own."

All within hearing nodded their agreement and reassurance, but Ellis was changing the subject.

"Now, some sleep, but I'll take the first watch. These coves seems affable enough, but it takes more than a bowl of stew and a pile of bracken to win my trust. Keep your rifles slung right around you so's they can't be took off you easily. There's a few here as wouldn't say no to such a gun as a Baker, that's for sure."

The rest of the group, bar Ellis, unstrapped their blankets, spread them over themselves and immediately fell into deep sleep. After three hours, a woman brought Ellis a beaker of a bitter infusion to drink, but it was hot and, to him, surprisingly refreshing. He woke Davey who then stood watch, giving Ellis a short period of rest. The others woke soon after but Ellis was allowed to sleep on; then his prediction did come true. Three of the band came to Davey and pointed to his rifle. Davey nodded.

"Byfe. What's the words for good gun?"

Byford sat up and spent a second in thought.

"Fusil bueno, I think. They used the word fusil on our march here."

Davey looked at the three and smiled.

"Fusil bueno. Baker."

One held out his hands, obviously wanting to take the gun, another placed a collection of knives, jewellery and a French pistol on the ground. Clearly they wanted to trade. Davey shook his head.

"No, my fusil."

He beat the flat of his hand back against his chest.

"My fusil."

Byford helped.

"Mi fusil."

Another piece of cheap jewellery appeared, but Davey still shook his head. The three became angry and spoke in loud, rapid Spanish. Davey looked at Byford.

"What they say."

"I have very little idea, but I did hear the word "hospitalidad", which I can only surmise means that we should be grateful for their hospitality."

Davey looked perplexed, not from what had been said, more from what he needed to say in return.

"Can you tell them, that if I have no gun, I get shot by our army?"

"I'll try."

He thought for some seconds, while the three irate Spaniards continued to glare at Davey. Eventually Byford composed something.

"No fusil, inglés dispara."

He placed his own rifle behind him, then sent a stiff index finger into his own chest.

"Bang! No fusil. Inglés dispara."

The three looked only a little less irate, but their faces changed when they heard the voice of their leader.

"Déjelos sólo. Consiga acerca de su trabajo!"

The three slunk off as the leader approached, accompanied by what appeared to a scholarly gentleman, clearly no weakling, but with a gentle face. When close enough he addressed Davey and Ellis, in broken but understandable English.

"I am Antonio Sempri and he is Georgio Mangara. He is leader, of us."

At the sound of his name Mangara nodded his head, but Ellis and Davey felt it better to show proper manners and so they rose and shook the hands of both. This seemed to please Mangara, who then spoke to Sempri.

"Georgio wishes to know why you are leaving Spain."

Ellis nodded, and spoke in simple terms.

"Too many French come. Napoleon leads them."

Sempri translated and Mangara nodded, then spoke himself to Sempri who translated.

"Is this not coward? You should stand and fight."

Ellis nodded, not annoyed.

"We will be back. If we live, we will come back, more of us, and fight some more."

Ellis waited for the translation, then continued.

"But it is no use if we are all killed or made prisoner. Then no-one will come back, England will have no army."

Sempri translated, Mangara nodded, seemingly understandingly, then he spoke further, for Sempri to translate.

"Georgio says that we are "guerrillas". We fight the "little war" against the French. You could join us until your army comes back."

Davey looked at Ellis who slightly shook his head, but it was Davey who spoke.

"We thank Antonio for the honour he shows us, but our honour means that we fight with our own comrades with our own army. We are not deserters. We fight and perhaps die with our own amigos."

Davey had included one of the few Spanish words he knew, but how accurate was the translation they had no way of knowing. However, the answer seemed to please Mangara. He made a fist, beat it against his chest, then thrust it a short way towards Davey, a stern expression on his face. Davey now thought the time to be right.

"Can you help us return to our army?"

Sempri translated and Mangara nodded as he replied for the translation.

"Yes. But tomorrow. Today we go again to the French."

The answer was definite and brooked no argument, no-matter how disappointing, but then Mangara spoke for Sempri to point to the three Riflemen.

"Antonio would like those to come, with their good guns. Of this Baker, we have heard."

Ellis looked around to Verrity, Spivey and Newcombe, who had been listening, but Sempri was speaking further.

"Their green makes them good guerrillas!"

Ellis turned again to the three.

"They want you to go on a raid with them, against the French. It would help our cause if you did."

It was Newcombe who spoke. He was a Chosen Man, like Davey, but Ellis didn't much like him. He gave out a superior attitude somehow, seemingly disdainful of mere line of battle soldiers, but nothing had been spoken as yet.

"As you say, Sarge. That's us, we're Light Infantry, trained for just such."

Ellis nodded, but Davey touched his arm.

"These "guerrillas" could get us a coat, so's we don't wear red. We could go too, win their favour even more."

Ellis looked back at him.

"Careful John! If we're caught, out of uniform, we'll be killed and not clean. These three will be fightin' as they always do, an' the French knows 'em as skirmishers."

Davey nodded.

"You're right. Didn't think of that."

Ellis looked at Sempri.

"They're yours to command. Teach 'em a few tricks, will you?"

The joke meant that everyone parted laughing and in good spirits. At Noon some more stew arrived and with it consumed, the three Riflemen tailed onto the end of the long line of guerrillas leaving the quarry. The six of the 105th watched them go with mixed emotions, they were off to fight the French, but as a part of God only knew what kind of hairbrained scheme, but Ellis answered their thoughts.

"These guerrilla lads have survived so far, so they must have some idea what they'n about. Cause for Faith, I'd say."

He paused until the last had gone, then looked back at his remaining five.

"Meanwhile get all ready for a tough march. We've lost two days on the lads, who've been pullin' back all the while. We has to overtake, while still goin' what'll probably be the long way round, led by these villains."

The afternoon was spent in sleep or checking equipment, especially boots and straps. With even that exhausted, they sat idly under their shelter, counting the times the girl walked across their front stealing glances at Joe. With the day now died and the gloom of evening almost full, the party returned, accompanied by much rejoicing. Shouts of triumph came from the men and those in the quarry replied with their own cries of obvious delight. They had brought back much booty of all sorts, boots, clothing, equipment, harness and weapons, which looked like cavalry carbines, and all were invited to take their pick. All was dumped around a central pole and some kind of circlet was hung on it, but the smiles of Ellis and the others died when the three Riflemen came back to sit amongst them, to sit dejected with their faces down and facing the ground. It was Spivey who finally broke the silence. He had shown himself to be the most gentle natured and thoughtful of the three.

"I never wants to be part of anything like that again, never!"

His head dropped down again. It was Joe Pike who asked.

"What happened?"

Spivey raised his head and took a deep breath, to let it out as a long sigh.

"They saw a cavalry patrol, some way off, but comin' our way. We 'ad to run over a hill to make the track in time, but we did and got on both sides. They dragged a dead tree over the track, just round a bend, so's the Frogs wouldn't see it from way off; they'm good these boys. Then they shot 'em to pieces from front, back, an' sides."

He took another deep breath to form a deeper sigh.

"That was one thing, what came next was another. Wounded or not, all prisoners had their eyes put out before they cut their throats."

He inclined his head towards the pile of booty.

"See that, … that garland thing? That's their ears. They cut off the ears of every man as a trophy."

Total silence met his description, broken finally by one word from Miles.

"Jesus!"

Miles had more to say.

257

"This is one stinkin' war. These, our side, is as bad at doin' what we saw at that Alcobatha place, afore we closed with the Frogs at Rolica.

The thought entered Ellis' head regarding what the French did to obtain food, but it seemed lame and inadequate, so he held it to himself.

Few spoke for the next hour, instead they sat watching the guerrilla band poring over the booty, selecting what could be used. Their victory must have been regarded as a major triumph for soon a bonfire was lit and the carcass of a goat set to roast before it. Soon after, there was dancing, drinking and carousing involving the whole community. Ellis looked at the scene, then at the eight of his command.

"Come on, we can't just sit here like the poor relations at a wedding! 'Twill look bad. Grit your teeth, get to the fire and take a drink. Look cheerful. Whatever, a troop of French Dragoons is dead, them as we've no more need to worry about. How it was done won't be the first nor the last evil thing as is done in this blighted place, afore anyone's finished!"

Still carrying their rifles, slung from shoulder to hip, they rose and joined the throng. Mangara himself roared a greeting and brought over a wineskin, which he drank from in amazing fashion, holding it way above his head at arms length for the wine to squirt down into his open mouth; somehow he was able to swallow and retain an open mouth at the same time. Judiciously, the nine managed to drink short bursts of wine from the wineskins that they had been given, before choking to the roaring laughter of Mangara. Soon pieces of carved goat were arriving and the drinking abated slightly. When the goat was but a suspended collection of bones, the dancing started and it was not long before the girl arrived to stand before Joe Pike. Horrified, he looked at John Davey.

"Go and dance, boy. Where's the harm? Mary will understand, and we has to keep these sweet. Can't afford to give no offence."

Joe was dragged off to dance for the whole session, only to return when the entertainment took a more serious turn. There was to be wrestling. Two well-built youths came to the centre to be supervised by Mangara himself. He posed them for the start, right arm over their opponents left shoulder, their own left hand on their hip. Mangara dropped his hand and the bout began. It was a clumsy and brutal affair, short lived, with one getting the other in a headlock and near throttling him, until Mangara told him to stop. However, the affair then took on an even more serious flavour. A young version of Mangara strode forward, topping his Father's height by a head and exceeding his breadth. He was stripped to the waist as Mangara held up the wrestler's arm, yelling something. Davey looked at Byford.

"What's he saying?"

"My son!"

A challenger came forward, older but perhaps more experienced. However, he was despatched in double quick time by being barged over, then pinned to the floor. Flushed with such an easy success, the son looked around for more challengers and his gaze lighted on Zeke Saunders, to be followed by his finger. Mangara clearly agreed and motioned Saunders out into the ring. Ellis looked at Mangara, then at Saunders, then back and nodded, grinning like an ingratiating showman, but indicating that Saunders needed to take off the top of his uniform. He then pulled Saunders back from the ring. One thing he did know was that Saunders was once the undefeated wrestling champion across the whole of Wessex and much beyond, but he needed a word.

"Zeke! Lose! We needs these to get us back, what we don't need is any one of this gang with any kind of grudge. Put some moves on him, show him that you are a good opponent, no kind of dunce, but lose. These aren't the folks from back home, these is different. They lives by the feud!"

Saunders nodded as he removed his shirt.

"I hear you, Sarge. Problem is, how do I make it look good?"

Ellis slapped him on the shoulder as he entered the ring to a cacophony of cheering. This was to be a bout of a quality that they had not witnessed in some time and the result, if going with their man, would pass into legend. Mangara set the two, and then dropped his hand. Instead of pulling away, Saunder's ducked under his opponent's head to seize his right arm and pull that arm into an arm lock. The arm went up his back as Saunders pushed him forward and crowd fell silent as their man fell to the floor, his face pushed into the dirt from the pressure of his arm twisted up his back. However, Saunders slid forward to place his head besides the right ear of the Spaniard, just enough for him to reach up and get a hold upon it. The hold was weak, but, when Saunders felt the pull, he allowed himself to be hauled over, releasing his own hold. The crowd roared at the show of strength and

Saunders rolled away to return and circle his opponent. The Spaniard raced in to take Saunders in a bear hug and then take him to the floor, but Saunders twisted so that he landed uppermost, then from his position of advantage, he forced back the Spaniard's head into the dirt to break the hold. As the Spaniard's hands slid away, Saunders stood up and waited. The next charge came and Saunders met it perfectly; with his right arm under the Spaniard's left armpit and he used the speed to throw him across his outstretched right leg to land sprawling on the floor.

Saunders met three more attacks with a scientific answer, then he stood to gauge his opponent; was he angry, frustrated, petulant, or what. "What", seemed to be the answer, he remained ready to fight on. Though plainly not as skilful, he was not prepared to give best at all easily. Saunders judged the moment and 'Fair enough', came to mind. This time he advanced himself, looking for the move he wanted, shifting one way, then the other. This caused the Spaniard's arms to come up to where he wanted them and Saunders dodged between and under to seize the Spaniard's left leg at the top of the groin and try to lift him. The Spaniard instinctively leant forward and Saunders hoisted him almost off the ground, bar some weight remaining on the Spaniard's right leg. Then he did what Saunders hoped, he pushed up on that leg to gain height above Saunders. The Spaniard remained upright and Saunders deliberately leant back so that the Spaniard's weight was now bearing down onto Saunders head and chest. The soldier held the young man there for an agonising two seconds before toppling backwards with the Spaniard landing on top. Seemingly winded, Saunders made no move as the Spaniard seized his wrists and pinned him to the ground, whilst Mangara bellowed the count of seconds. Hearing the number he wanted, he leapt up from Saunders prone body to caper around the ring, arms raised, bellowing himself. Saunders rose to his feet and dusted off his trousers, to find his opponent stood before him with a wineskin, which he tipped above Saunder's head for the wine to emerge. Saunders did his best to swallow and drink both at once, but eventually the wine cascaded over his face. It was of no matter. Mangara's number one son was holding his arm aloft and Mangara himself was slapping Saunders on the shoulder, yelling repeatedly.

"Bravo Inglés! Bravo!"

The crowd were cheering at full volume, as Saunders, in the English way, offered his hand for his opponent to shake and this he did, before raising it high again. Ellis chose his moment and walked forward with Saunders shirt and tunic, a signal for him to gracefully leave the ring. As he came back to his companions, Byford gave his verdict.

"Plainly, Zeke, there's a place for you in the Diplomatic Corps."

However it was Miles who answered.

"Yes, if he could spell it."

Saunders gave no answer as he drank from Miles' canteen."

**

Beyond Villafranca there was both similarity and contrast to what had gone before. Beyond the ravaged town was a country similar in every way to the desolate Vierza; the depressing contrast was that this time there was so little food to eat and little fuel for each day of their march. The cruel wind carried the cold; at that height there was nothing to divert it from its chosen prevalence, always from the East, to numb the right side of their faces and penetrate any gap in their clothing that was increasingly coming to pieces. If there was not mud, to soak and numb their feet, there was ice that used the damp to penetrate chilling cold further beyond their blistered feet, up their tired limbs, chilling to their knees and beyond, to then further exhaust their hungry bodies. Each day, each step became more of an effort, as Moore's army, now barely worthy of such a description, trekked across the highland, each day's progress less than that of the day before.

The 105th, still in the Reserve Division and still with their followers, held their column. Gibney, Hill and other trusted NCO's marching out beyond, at the sides, there to deter any deserters falling out, but they were not called upon; either their presence deterred any such action, or the men of the 105th were content to stay with their Colours, seeing this as their best chance of survival. The fate of those that left the column was all too obvious, both ranged along the roadside or hanging from a convenient tree. Lacey expected every minute to be called upon to deploy and turn against their pursuers, but the hours dragged as slowly as their feet and no order came. Instead, regularly every hour, it seemed they found a cavalryman either shooting his horse or butchering the carcass. Harbouring their own precious supplies, Bridie and Nelly made a trade for some of the meat, using acorns or biscuit as currency. During the night stop, the several joints of meat was boiled in their pot on communal fires, wood was too scarce and precious to be given to each family or mess. The result was wrapped in frozen grass and then bound further in cloth. Nelly Nicholls pronounced on the result, as it was stored on their travois.

"That'll keep in this cold for a good week or more."

This was spoken to all that could hear, but she was studying Mary. She could hold food down a little better these days, but she was growing weaker by the day. The poor food and incessant cold were taking their toll. They were keeping the mule alive for Mary to ride on and they all took turns at pulling the travois, including Beatrice Prudoe, her husband maintaining a place ahead, where he would not have to witness his lady wife performing such labour alongside such labouring fellows. Not that it now mattered so much, his thoughts rarely strayed from his own dire circumstance, for he was as much a bundle of wet cloth and blankets as any of them, with a piece of torn sheet holding a felt hat on his head, the knot tied incongruously under his chin, this now a lot less well upholstered than when they had first landed at Mondego. However, on the 4th January, the 105th were halted by Paget's orders, and the followers sent on, with Prudoe and Sedgwicke included. The ex-Cleric and convicted felon that was Percival Sedgwicke had increased in regard throughout the battalion to the equal of any looked upon as "a pillar of the Regiment". Using what peasant remedies he had learnt when he visited the low hovels of what had once been his Wiltshire Parish, he gave whatever help and advice he could, keeping cheerful and being welcomed at any fire, usually with "Hello Old Parson. Come and have a warm!"

With the breaking dawn of the 5th January, Heaviside was shaken awake to open his bleary eyes to find himself looking at a well appointed Major, well clothed and well shod. Not waiting to observe the niceties of standing and salutes, the Major spoke immediately.

"I'm told by your Colonel that you are the Captain of the Colour Company?"

By now Heaviside had managed to roll out of his blankets and was stood to attention.

"Yes Sir, that's correct."

The Major leaned forward, a gloved index finger touching Heaviside's chest in time with the words, the other hand holding the reins of a good, and obviously healthy, horse.

"I want your Colour Sergeants, your NCO's and any other trusted men to make up a half dozen. I expect you by that tree in five minutes."

He mounted and rode off in the direction of the stark and haggard feature that he had singled out and Heaviside hurried off. He went directly to Jed Deakin and shook him awake. Deakin tried first to move his head and was grateful that this time his hair was not frozen to the ground, he had made his bed amidst the relative warmth of an abandoned battery of Spanish guns, then he looked up at his Captain. Heaviside spoke immediately.

"Gather Bennet, Halfway and three others as soon as you can. Meet me on the track, by that tree."

Deakin hurried off to carry out his Captain's orders and soon himself, Heaviside and the squad of men were stood with the Major, who quickly looked over the ragged collection, then issued his order, it being very simple and directed at Heaviside.

"I'll not need you, Captain!"

As Heaviside saluted and departed, hoping for some breakfast, the Major issued another simple order.

"Follow me."

They all formed into pairs and followed, led by Deakin and Halfway, who was even more curious than Deakin.

"What's all this about, Jed?"

"Search me. All I'm hoping for is some provender at the end of it."

They marched uphill for half an hour before entering a cutting that had a steep slope on one side, above the road, but on the opposite side was the edge of a sheer drop. Waiting on the road were two stout wagons, their covers all in place and each with a team of four oxen waiting at the front, in harness but all eight plainly in a state of utter exhaustion, their muzzles barely above the thick ice. The party marched up to the first wagon and halted when the Major raised his hand. The Waggoner must have heard their approach or had been looking back through the cover, for he dismounted and came to the rear, where, stood before the Major, he saluted and waited, but not for long.

"Unhitch the teams and take them on."

The Waggoner disappeared and then the Major looked at Deakin.

"Two men, to pull the cover back."

By inclining his head, Deakin motioned Stiles and Peters onto the nearest wagon and the cover was pulled back to reveal a collection of small chests in five rows, five deep. Deakin made the calculation and came to 25. Each chest was about 12 inches by 12 and 15 inches long, held shut by a small clasp with a small lock through it, but no sooner had he finished his examination than the Major was motioning him to the cliff edge, to point down to a sharp projection on a shelf about 20 feet below.

"All these chests are to be thrown over and each must hit that outcrop, so that they smash. Is that clear?"

Deakin did his best not to look puzzled and amazed, but he managed a simple salute.

"Yes Sir. Very clear, Sir."

The Major stood back.

"You first!"

Deakin stood for a second, before going to the wagon to take one chest, that had been placed on the edge by Stiles. It was heavy, but the first thing he noticed was the army stamp burnt into the lid, this being a crude copy of the badge on his shako. He carried the chest towards the cliff edge, but before he arrived the Major gave an order.

"From above your head."

Needing no orders, the others of the work party had taken themselves to the edge to watch. Feeling the weight, Deakin hoisted the chest above his head, before dropping it accurately onto the outcrop. The result was a crash as the chest burst apart, then a cascade of silver spread out, down into the chasm, to soon be lost against the backdrop of the snow at the bottom, hundreds of feet below. Shock and surprise was plain on his face as Deakin looked at the Major, who was now prepared to give an explanation, now that one was so obviously needed.

"This is the army's treasure. It must be scattered in case it falls into the hands of the French. Moore's orders. Continue!"

Forming a queue, the soldiers approached the wagon, to each pick up a chest and throw it down, the sound of the crash on the rocks making a regular rhythm. The Major stood back to supervise.

Deakin was horrified; this was a King's ransom being thrown over a cliff, to be picked up by the Spanish in the Spring. It could be given as a bounty to those who held to their Colours, given in handfuls to all who still remained with the column, as a reward, which would ease their life back home, but instead it was disappearing as a silver spectacle down hundreds of feet of cliff. Something snapped inside him and his mind began to work.

"Sir, this wagon's a bit perilous, Sir, and like to roll back, with the team gone, Sir. We needs some rocks behind the wheels."

The Major nodded and, over four journeys, Deakin fetched four sizeable rocks from the slope and lodged each behind a wheel. As the jettisoning of the chests continued, Deakin sidled up to Halfway.

"Tobe. Drop one onto the shelf, so's it don't smash. Just breaks and stays thur. Then call 'im over."

Halfway's next chest brought the required result.

"Sir. One's lodged on the shelf, Sir. Part way down."

The Major came to the edge and peered over. There was indeed a chest on the rock ledge, barely shattered. Now annoyed, the Major looked back at the wagon and saw what he wanted, a towing rope, hooked on the outside, used to pull the wagon out of the mud when required. He fetched it himself, seized the nearest soldier and tied the rope around him, using a very seamanlike bowline knot. He handed the end of the rope to a group of soldiers, then looked at the one encircled by it.

"Over you go!"

Although deeply fearful, the soldier sat on the cliff edge and lowered himself over, whilst the others took the strain, with the Major looking down to supervise that no coin found itself into the soldier's pockets. Meanwhile Deakin was giving his own orders, to Peters.

"Sam! Get under the wagon an' dig a hole in the snow. Use yer brummagem."

Peters looked at the Major, concentrating on events over the cliff, then at Deakin, then he dropped off the wagon to draw his bayonet, then crawl under the wagon to hack a hole in the deep snow and ice. After another glance at the Major, Deakin pulled a chest off the wagon down onto the snow and, using his own bayonet, quickly levered of the weak hasp. He lifted the lid, then, after again checking on the Major, began shovelling coins into Peters' hole. He risked ten double handfuls of the heavy silver pieces, then stopped, to close the lid. As Peters shovelled snow back over the small hoard, Deakin took a stone from a rear wheel, and smashed in one corner of the chest, up to the already ruined hasp. Peters had scrambled back aboard the wagon, hauled up by Stiles. The stone was quickly replaced before the sound registered with the Major who immediately turned around, instantly suspicious, to see Deakin gathering coins off the snow and placing them back in the chest, along with a lot of snow to make up the bulk. The Major came back, very angry and suspecting theft.

"What the Hell happened here, Sergeant?"

"Sorry, Sir. Dropped one. Fingers too cold, Sir."

The Major didn't wait, but thrust his hands into both Deakin's side pockets, to find nothing but a few acorns. Remaining suspicious, the Major looked at Deakin, then under the wagon, then close around. He looked at Stiles and Peters.

"You two, empty your packets!"

Both pulled out the lining to reveal nothing. His distrustful examination was broken by Deakin lifting the chest and taking it to the cliff edge, then casting it over.

"Thought that one could go over anyhow, Sir, bein' as it's already smashed open!"

The Major looked daggers at Deakin, but this was broken by the soldier reappearing from the edge, having now been hauled back up. The Major was angry, convinced that something had happened, but there was no evidence, which added to his temper.

"Get the rest over, quick, damn quick, and get it right. Another poor throw like that and you're off to the Provosts, the damn lot of you."

However, he felt happier as the army's fortune was rapidly consigned to the cold air off the cliff and both wagons were empty. The Major gave his last order.

"The wagons go over and you go back."

The wagons disappeared over the edge together, to land with a crash that could be heard even from their height above it. The Major mounted his horse as Deakin formed up his men, then turned and gave a salute, which was not returned; the Major simply walked his horse away. Out of the Major's earshot Deakin gave his orders.

"March off. Slow."

The small party shuffled back down the track, with Deakin giving frequent looks back at the Major. When he had disappeared around a bend, Deakin grabbed Stiles and Peters and pulled them over to the slope above the road. Halfway took the party on. Deakin waited two minutes, in case the Major should obey his suspicions and return, but they saw nothing. Deakin walked out to the centre of the track, waited a while more, then motioned for Stiles and Peters to run back with him. Soon they were between the wagon tracks and soon Peters found the soft snow under which lay the hoard of coin. This was scooped out, and then the hole carefully examined to find three more silver pieces hidden in the ice. Holding the sides of their tunics steady, for the weight was causing a wild swinging, they ran down to catch up with Halfway and the others. There was an immediate share-out, each man receiving twenty-eight silver guineas, enough for a small cottage or a patch of ground back home, but the brief joy died in an instant when they regained their Companies to be told by their Captains to prepare for a forced march, as ordered by Moore. Deakin went straight to Bridie to stand helpless. There were no preparations that could be made; there was nothing left to bring into use for such a march that was not already employed and about to give out and fall apart.

***

There is misery that comes from the evidence of your own eyes, yet there is a misery beyond that, the level of misery which comes as hope dies away. That day's march, followed by that of the 6th January proved to be the worst of the whole retreat. Moore ordered forced marches for 36 hours, which brought a result almost as deadly as though he had brought about the battle that almost all his army craved. The 105th were to take their turn as the final rearguard and Lacey handled his men in an orderly retreat, one wing of the battalion holding a strong position for the other to retreat past and hold the next. The French Dragoons that they saw made no attempt against them, content to wait for their own infantry to come up, but by then the 105th were long gone, disappeared into the mist and snow of the uplands.

Being the last through, the sights that greeted them were harrowing to even the most hardened veteran. Worst to see were the frozen remains of followers, frozen in abandoned wagons, as were the wounded. In addition were the able bodied but exhausted soldiers, those who should have been able to hold their place at a normal pace, but instead sat by the roadside in misery and the agonised curse that the 105th heard most was a curse to their Commander, who would not let his men stand and fight. That their own followers had been sent on ahead was a mixed blessing. Even halting to confront the French, the pace that they needed to set to keep ahead would have proved impossible for their women and children to match. These were out of sight, but not out of mind, and the groups of dead followers they came across were carefully studied as they marched past. However, frozen faces, covered with frost rime, soon lose the familiar that was once loved and cherished, therefore they could do little but march past, heads down, hearts hardened to the pitiful cries of all the dying; soldiers, women, and their children, although some were hoisted, if still alive, from their dead Mother's arms to be carried inside soldiers tunics and rotated round, when the carrier became too weary himself from the extra burden. Drake and Shakeshaft felt the misery and despair that this caused worse than Carr, who clenched his jaw and slogged on, occupying himself with the military business of conducting the rearguard for the Regiment. This caused him to often look back down the track, marked out in the white snow not as a black ribbon formed by their passing, but by the continuous dots of red, where a soldier had given all up.

On the morning of the 7th they caught up with their own followers and their own mourning began, for their own families had not escaped the deadly embrace of cold and its cruel henchman; hunger. Several of their number now remained back on the hard road, casualties of the past two days. Deakin went first to Bridie, immensely relieved to see her still alive and still caring for the others, although he was shocked to see her sunken eyes and dark cheeks, the signs of ensuing starvation, but she immediately turned his alarm for her and the children aside and took him over to Mary.

"She's lost the baby, Jed."

Deakin looked in horror from Bridie to Mary, but his eyes finally remained on the girl. She was lying on the travois, face turned away and covered with a blanket. There was neither sound nor movement. Bridie touched his arm.

"She's not got enough in her even to cry, Jed. She's alive, but full borne down. I don't know how she'll go on."

Deakin looked at her, a tear emerging from his eye, which Bridie wiped away.

"The child?"

"A girl."

"What did you do?"

"Mrs Prudoe wrapped her up and we buried her in a churchyard. We think it was a place called Cerezal. Parson told us the name when we got him to say some words. He's good like that."

She looked down on Mary's prone figure.

"I told her it's for the best. It was her or the child, or both. The baby born would not survive anyway. Not in this. I told her, but it did no good, not then, but perhaps later it might."

Deakin nodded, but now it was Bridie's turn to cry. She leaned forward to place her forehead against his chest, she was now fully sobbing.

"There's worse, Jed."

She almost broke down completely.

"I've three children, Jed, but I can only keep two alive. If I try to feed three, they'll all die."

She spoke no more agonised words, but sobbed uncontrollably.

Deakin levered her away from him, to hold her at arms length.

"They're wearin' that drummer boy kit? An' it don't make no difference?

She shook her head.

"Which is the weakest?"

"Sinead."

Deakin nodded once more; Sinead was the youngest. He lifted Bridie's face.

"Now listen! She'll not die, not whilst the lads is near."

He turned to Halfway, sat nearby on a log, unshaven, horribly dirty, a piece of torn blanket wrapped over his his shako and his boots held together by a variety of bridle harness. A piece of wagon covering was draped around his shoulders, giving him the look of a down-at-heel undertaker.

"Tobe! You heard all?"

Halfway nodded, then stood up.

"They'll not die, Missus, not none of 'em! We've lost one child, we'll not lose no more!"

That night extra arrived in the pot for the children, albeit mostly acorns and horsemeat, but that night each slept in the arms of a soldier, these being Halfway, Stiles and Peters, wrapped in horse blankets. The next morning Deakin gathered yet more of the thick blankets, cut a slit in the centre and each child of his own and also Nelly's then wore a broad poncho for the day's march.

***

The order to continue the retreat did not arrive, at least not immediately. Moore was content that the main French army was some way behind, their forced march had added to the distance, if also to the long list of his own dead. The Reserve Division were to hold where they were and, at Noon, the main army marched off, the 105th's followers with them. The Reserve would catch up the following day. Deakin, Halfway, Stiles and Peters watched them go. The children looking like mobile playing cards, escapees from their place in a "house" made of such, but all three now looked better, as did Mary. She seemed to have recovered somewhat, especially when Sinead joined her on the mule and she took responsibility for the small child upon herself.

The 105th were ranged across a hillside, seeing practically nothing, for all was shrouded in a deep fog, although this did nothing to stop the snow, which continued to fall, not thickly, but enough to soon cover their backs and shoulders. Being unable to see anything, an eerie silence reigned, each soldier feeling the need to hear what he may not be able to see. Each Company was required to send a picket down into the valley, some 200 hundred yards below them. In the case of the Colour Company this was Stiles and Peters, both now sat on a horse blanket, leaning against a tree, saying nothing, staring at the odd shapes, seen eerie and distorted by the swirling fog from over the stream, all gathering their share of the falling snow. Their visibility was but yards beyond the stream and so, like the veterans they were; their muskets were easily to hand.

Stiles noticed movement and nudged his companion for both to then pick up their weapons and slowly pull back the hammer, covering the mechanism with their free hand to muffle the click. Two figures were emerging from the fog, the first of their shapes that could be seen being their black French gaiters and then the blue of their tunics. Both were intent on the stream, neither was looking beyond, the silence of the valley had lulled them into believing that the English were way beyond and gone. Blithely unaware, the two descended to the stream, each with canteens and one having his arm out of his tunic and in a sling. Whilst one filled the canteens, the other washed his wound in the freezing water. Neither had attached any importance to bringing their weapons. They were the worst of Napoleon's conscripts; it was very possible that neither had yet fired a shot in anger, yet here they were, invading Spain, and confronted by a very capable and very dangerous army. The first they knew of their danger was the movement of Stiles

270

and Peters approaching to stand on the far bank of the narrow stream, then they saw the blank muzzles of the muskets pointing at their faces, at which point both froze in horror. They both stood and raised their hands. Without looking at Stiles, Peters spoke.

"We can't be doin' with these as prisoners. What now?"

Stiles looked at both, still with their hands raised.

"These is little more than beardless boys. One's wounded."

He let out a sigh of exasperation and lay down his musket. Using one stone to step across, he then opened the tunic of the astonished wounded Frenchman to examine the wound in his chest, close to his shoulder.

"Bayonet, or sword, more like."

He fished inside his left coat pocket and produced a small package, their "housewife". He found a needle; already threaded, pulled open the boys mouth and stuffed the thread inside, consternation and amazement plain on the boy's face. With the thread now wet he expertly put two stitches into the wound. The boy was too astonished to cry out at the pain, as was the other, who had stood with the canteen spout down, the water pouring unnoticed unto the ground. Job done, Stiles looked along the bank for some moss. He picked a wad, pressed it against the wound, closed the shirt, then took the boy's hand and pressed it to the place. He then looked from one to the other. His words meant nothing but the inclination of his head showed all.

"Now bugger off!"

Both immediately scampered off into the fog as Stiles regained their own bank. Peters looked exasperatedly at him.

"Too bloody soft, you! Too soft by half!"

Stiles was slinging his musket.

"Best get back up. Say we saw some Frenchers what scampered off, but they must now be near."

Peters looked at him again, still amazed at his friend, but his words deep with sarcasm.

"Oh really! You don't think?"

They regained their lines to find the battalion forming up to leave. They immediately told Heaviside and immediately received a stock reply.

"Take heed therefore unto yourselves, and to all the flock. Acts 20, verse 28."

271

Utterly bemused, the two joined their Company, in time to obey O'Hare's order to march away. Lacey was using the fog to obey Paget's order to withdraw. The news from Heaviside that French had been seen, brought but a nod of Lacey's head. Heaviside dropped the subject.

"What's the next town, Sir? Do you know?"

"Constantino, I do believe."

Heaviside saluted and joined the column at the side of his command, immediately matching their step.

"The Lord is good, a strong hold in the day of trouble. Nahum One, verse 7."

It was Deakin, as usual, who answered, as usual with more than a hint of humour.

"Yes Sir. I'm sure all the lads sees it that way, Sir."

All within hearing grinned from within their upturned collars, as much as their frozen faces would allow, as grimly they marched on to await the order to halt and become, once again, the last of the rearguard.

The snow had stopped to reveal a clear, but bitter day. However, for every soul; soldier, woman and child, what mattered most was the fact that they were descending. Before each footfall the ground did not rise, but fell away, each step placing the high, barren, merciless plain behind them. Spirits amongst the followers slowly revived, as did Chaplain Prudoe's concern for the social standing of his wife. During a rest he took her to one side.

"My Dear. I do feel that you are lowering yourself somewhat too far. To actually turn yourself into a beast of burden alongside the wives and what-have-you, does little for the representation that you convey of our rank here. I do think that you should now concern yourself solely with affairs that pertain solely to you and I."

Looking squarely through the restricted opening of her bonnet, Beatrice looked sceptically at the image of her husband before her, it not being one to engender affection nor respect.

"Leviticus. What we are going through, here, now, is the worst I have ever heard of, never mind experienced. I would wish you to know that, were it not for the knowledge, strength and fortitude of these good women, I doubt we would be here. Neither myself, nor you! What I bring to you, to keep you warm and fed, comes from them, blessed by our merciful Lord, who looks over us all."

She paused, but all she saw on him was shock, as much as his bound up face was able to register any change of expression. She continued.

"We are too much in great peril not to share the means we have, that we may all survive the dangers we face."

She pressed home her advantage.

"Please do not mention this again!"

Whether he nodded agreement or not, she did not wait to find out, but went immediately to Mary and Sinead, to see if there was anything that she could suggest or even provide. There was not, not much other than a kind word and the lifting of a collar or the adjustment of a blanket tied crossways over the body of an exhausted figure.

They marched on, into the fog and large snowflakes, whipped by gusts of wind that tortured the fog and caused the heavy flakes to make their own noise as they collided with their fellow travellers beside and before. As usual Nelly and Bridie marched ahead, each harnessed to one pole of the travois, Patrick and Eirin both adding their strength, harnessed close behind, keeping in step. For the two women, the fog added a depression of its own, condemning them to march on, as though on a treadmill, through the filthy slush, trudging towards the edge of the fog, a destination that never came, over a track that remained always the same mocking distance before them. The two glanced at each other, sharing a look that they never gave to the others; a knowing look, no words, but a look which silently spoke that this could not long go on. It was Nelly who turned away first, to look ahead, not to see, but in the direction of something that she had just heard.

"What was that? Sure, was that not the bellow of some animal, a cow or a bull or something?"

Bridie looked ahead, then nodded herself as the sound came again.

"You're right! Didn't I hear it, just then, myself?"

Subconsciously they quickened their pace, surprising Patrick and Eirin behind. The sound became clearer, that of several animals, but the first they saw of whatever it all was, was a tall Redcoat, shako, greatcoat, musket and cross belt all in place, then another, then another. The first turned to greet them, a smile across his dirty bearded face, but behind him there was definitely a team of bullocks, harnessed to a large cart. The two women could not speak from bewilderment, but the soldier had been expecting them and he spoke first.

"Morning Ladies! Some supplies, compliments of the General."

It was Bridie who spoke first.

"Supplies! Supplies you say? What sort? What's there?"

The soldier retained his grin.

"Supplies for the Spanish army; uniforms, food an' all sorts, but they b'ain't here. So the General's grabbed 'em for his, an' left us, the Guards, to eke it out, like, so none get's ruined."

Bridie and Nelly had halted and were now accompanied by a close group of fellow followers. Seeing the crowd of speechless faces, the soldier continued.

"This has been held for the Reserve Division, an' if you're their followers then you're entitled to a share."

His grin widened.

"But you've all got to be good girls and not rush at once."

It was Nelly who reacted, now stood with arms akimbo.

"Just listen to the gombeen!"

She fixed him with a ferocious stare.

"You think we're the sort as'll smash to bits what'll keep us alive! You tell us where to go and we'll take our share, then move on for the next to get theirs."

The Guardsman was taken aback by the belligerent response, but he did not lose his grin, in fact he laughed.

"It starts here, good Mother."

He pointed with his thumb to the cart behind.

"This has boots, Spanish, but they'll serve. Easily onto Corunna."

This time he pointed with his finger to the back of the cart.

"They'm laid out there; small, middle an' large. Take a pair if you has the need."

This they did, looking in wonderment at the pile of footwear remaining in the wagon, then, wide eyed, they wandered on to gather further of the wondrous bounty awaiting their pleasure in each wagon.

The food obtained was immediately cooked on the spot and, as luck would have it, the Reserve Division came marching up, just as the pots were boiling, a combination of pasta, dried beef, beans and Spanish brandy. The 105th, being held back, arrived a little later than the main Reserve Division and, after his men had obtained their full share and more from the wagons, Lacey stood and watched, but mostly listened; he had not heard such cheerful talk and laughter for some time.

He and O'Hare filled their flasks with the fearsome Spanish spirit and toasted each other's good health. Around the messes of the battalion, after the consumption of the hot food, many debated the merits of donning the new Spanish uniforms that the carts contained. Many decided against, but all had opted for the new boots. Sat with Halfway amongst their followers, Deakin could not help but feel a sense of satisfaction grow within him as he watched the children eat their second helping and then came the sound of Nelly chiding her husband Henry, which domestic discourse told Deakin that things must be now set on an upward road to rights, albeit with a very long way to go. Whatever, they were now in better shape than 24 hours earlier and they had extra on the extra. There was still substantial left in the carts, but an extra share of this was obtained by using the silver rescued from the army's treasure. Stomachs were full and haversacks were bulging, and several haversacks that could not be carried were lodged safely on the travois and Mary's mule. 'Money still talks,' he thought, 'even though the chief currency is biscuit and salt pork!' He settled under a blanket next to Bridie's sleeping form, took one last look at the children, then fell instantly asleep himself.

***

Chapter Seven

Each day, that follows another!

"If it snows on the moon, I may as well be there!" Such was the thought that occupied the capable, but uneducated mind of Joe Pike, as he stood and contemplated the utterly bleak vista that surrounded him in all directions. These surroundings were revealed by weak sunlight through cloud somewhat less dense than of late, therefore, at least not shedding snow to add to what already lay, thick and frozen, as far as he could see. 'As far as he could see' was a mind numbing exposure of white or black, with the odd variety of grey, that stared back, cheerless and mocking, all painted at random over a jumble of hills, cliffs, and gullies. Then his mind returned to the task in hand, 'If there are trees on the moon.'

They had been following a track now for two days, a route indiscernible on the snowbound landscape, rather one that, for all intents and purposes, existed solely within the minds of their guides. These were no longer Mangara and his merciless associates, for the nine had been passed on, to the next guerrilla band, this led by as cruel eyed a villain as Pike never wished to meet. At the hand over, the new leader, whom everyone called El Navaja, including the guerrilla himself, had looked with burning hatred upon the nine British soldiers. Byford had no idea what this name meant, but perhaps the huge cut-throat razor stuck in his waistband was a clue. There had been a long conversation between this El Navaja and their Mangara, which eventually ended with a nod from El Navaja and a handshake between both. From studying the gestures between the two and the odd word known by Byford, it seemed that Mangara was enjoining El Navaja to deliver these 'Soldados ingleses' to their army, as a matter of 'honor', both to himself, this being Mangara, and to Spain.

The nine British did their best to part from Mangara with good cheer. There was certainly plenty of such warmth from the Spanish side, handshakes and 'buena suerte' all round, Saunders receiving an extra thwack on the top of his arm, accompanied by 'el buen hombre, ingles'. This parting had taken place the previous morning and, for the rest of that day, throughout the night and the next morning, always at least three of them had kept constant vigil in some form, either on

276

sentry or walking out a flank, where they could keep better watch. Again, Ellis had forcefully ordered his small command, on no account, to unsling their Baker rifles, thus Joe Pike's was now swinging awkwardly down his side as he carried a felled log down to the broken bridge that was holding them all up. El Navaja had seen the downed bridge from some way off and sent one of his men to investigate. Ellis sent Alf Verrity forward with the scout as a gesture of co-operation, but El Navaja merely shrugged his shoulders. The stones of the bridge were scattered around, easily found under but a thin covering of snow, meaning that the demolition had been quite recent. The guerrilla leader spoke briefly, then spat in contempt.

"Caballería francesa."

Ellis nodded agreeably.

"Si. French Dragoons."

However El Navaja ignored him, rather he pointed at the three green uniformed Riflemen, then at the positions he wanted them to take to watch and guard the work they were about to undertake. Four more of his own band were sent up to add to this guard and those remaining set about felling trees to bridge the gap between the truncated arches of the bridge. About nine foot, Pike accurately estimated, being an ex-fencer on a large country estate, before being dismissed because the daughter of the estate was taking too keen an interest in him, this in the form of teaching him to read. That thought also entered his mind as he carried one end of his log down to the bridge, for it to be stood vertical, and then allowed to topple forward over the gap. Tom Miles had been at the other end to carry the log and something still rankled within him, his being a temperament easily riled. Chosen Man Newcombe had grinned as he walked off, carrying his rifle, as the Lights of the 105th joined the guerrillas to fell the necessary timber. The insolent grin was one thing, the words were another.

"Have no fear, soldier boys! We'll be watching over you."

Ellis had quickly discerned Newcombe's contempt for them being "Line Regiment", for it also annoyed him, but he knew enough to keep Miles away from the Rifleman. The last thing he needed was a fight in which either could be injured, or even killed, then he would have an issue on his hands. What Newcombe was capable of he didn't know, but of Tom Miles he knew everything he needed and this dictated that they be kept apart. What he did not realise was to credit himself for Newcombe keeping his tongue in check; Ellis' own iron character was enough to persuade the Rifles Chosen Man to indulge but lightly his liking for heavy sarcasm.

Within an hour the bridge was "bridged" and they marched over the rough collection of logs, good enough for a train of pack mules, but not for a wagon; however, that had to be left for more peaceful times. They now climbed the valley of the stream that ran under their bridge and a half-mile from the top they halted for bread, dried meat and wine that tasted more like vinegar. The soldiers sat in a close group, their rifles to their front, the barrel leaning in the crook of their necks and shoulders as they ate and they ate rapidly, to finish before the guerrillas and then sit waiting. The hand gesture in their direction from El Navaja to continue their march was not long in coming, whilst his band all stood when he did.

The valley widened at the top to show a clear, level ridgetop, now but half a mile distant. As they trudged on further, ever upwards over the powdery snow, Ellis seized John Davey's arm to bring him to a halt.

"John! Hear that?"

Now, with no noise from their own footfalls, Davey listened intently, his brows furrowed, then they widened in recognition.

"Muskets! That's muskets. There must be fighting, over that ridge up yonder, if I'm any judge."

Ellis nodded.

"Not too long now, I'm thinkin', but fighting means French. Where are we, behind them, or up beyond the lads?"

Spirits sank, when a careful approach to the ridge top, revealed the former. They could see a heavy French skirmish line being held by an equally heavy line of Redcoats and Riflemen, but the French were supported by a strong squadron of cavalry; Dragoons, so Davey's perfect eyesight told him. They slithered down the back slope to discuss their next move, but there things took on a very nasty flavour. El Navaja clearly considered that he had carried out his task, that he had brought them to the English army, whereas it did not seem to count that, in this situation, the nine rejoining was impossible. He was stood before them pointing over the ridge and repeating five words.

"Hay su ejército. Puede ir."

Ellis looked at their one source of Spanish.

"What's he saying Byfe?"

"It's more the pointing, and I do know that puede means "go" as in "leave". He wants us to go."

Ellis let out a long sigh whilst looking at El Navaja, who was repeating his words from a mouth almost obscured by beard, but not obscured enough to not reveal large yellow teeth. Ellis soon decided and replied to Byford, but all could hear.

"This cove don't want no more of us. We'll part company and push on alone. We can keep an eye on the lads and try to get up to 'em. Shouldn't now be too hard."

Ellis knew enough Spanish himself to say what was now needed.

"Mucho gracias, Senor El Navaja."

He stood and offered his hand, but the gesture, which he refused, motivated El Navaja to speak further, and more aggressively.

"Mi pago es uno de sus fusiles!"

Ellis knew enough to gather the gist of what El Navaja had said, but he needed confirmation.

"What he say, Byfe?"

"In essence, he wants us to pay him for his services by giving him a Baker."

Ellis immediately began smiling cheerfully at the guerrilla, trying to make light of it by saying 'No, no,' and simultaneously shaking his head and waving his right hand. Davey now spoke up.

"Tell him what you said to that other cove, Byfe."

Byford cleared his throat.

"Ningún fusil. Disparó por ingles"

To back this up, he held his own rifle out of sight and pointed a finger into his own chest.

"Disparo. Bang! Ningun fusil."

El Navaja repeated his own words, growing more angry and, in addition, pointing at Spivey's rifle, which was the newest of the nine they carried, and looked it, being well cared for by its owner. Ellis didn't like what was happening. El Navaja's band were watching and must have heard all, so for El Navaja this was now a matter of face. He couldn't back down in front of his men, but Ellis had had enough. He spoke to his own men first.

"No! To Hell with it! If we meets French I wants nine who can fight, not eight."

He then looked directly at El Navaja.

"No, Senor El Navaja. Perdon, no, perdon, no."

He continued to shake both his hand and his head.

"Byfe. Tell him thanks for his help."

Byford gave his best ingratiating smile to support that of Ellis

"Gracias por su ayuda. Adios."

Ellis then repeated the adios, which brought a change of expression into El Navaja's eyes, which Ellis did not fail to notice. He spoke to his men.

"Walk back. Don't turn your backs, just step back."

The nine did as ordered which drew from El Navaja a deep growl. He reached for the sling of his own musket, but Miles was carrying his rifle cupped in his hand. Before El Navaja could lay a hand on the wood of his own musket as it swung to his front, he was staring down the barrel of Miles' Baker, it cocked and ready with Miles' very malignant right eye at the far end of it. In addition, if El Navaja wanted to enforce the handing over of a Baker, he had placed his men badly and not allowed for the fact that he was dealing with trained soldiers. Although fifteen in number, the guerrillas were grouped together and, with Miles coming to the "present", the others, in less than a second had unslung their own rifles, cocked the hammers, and were at "stand to", their weapons pointing forward, but not yet up at "present". In addition, by instinct, the nine spread out to cover the group of Spaniards and prevent them deploying in any fashion. Ellis continued to slowly step backwards and his men followed, Miles still with his rifle pointing directly at El Navaja's head, which held the guerrilla rigid; him not making any sound, save a growling, nor him making any move. As they walked backwards, Ellis felt the need to repeat Byford's words, it might help.

"Gracias por su ayuda."

Ellis and his men walked backwards, for 50, then 100 yards. The guerrillas made no move and Ellis was satisfied.

"Halt."

They did.

"Keep your rifles in hand. Form two's."

This was quickly done, each man casting anxious looks behind him. Ellis took the lead, giving El Navaja a cheery wave in the hope that it may be returned. It was not. Ellis released a long breath.

"Could this get much bloody worse? Frogs to our front an' those murderin' Devils at our backs!"

He looked ahead for the best route in the hour of daylight left, whilst Davey looked behind.

"Sarn't. They're movin'. Goin' back down the valley."

Ellis turned to look for himself.

"What do you think that means?"

"Bugger all!"

***

A darkening dusk is a time for the gathering of evil spirits and so it proved for the followers of the 105th. Now rich with food and supplies and again marching ahead of their men, who were again some way back in the rearguard, they had made good progress and crossed the bridge at a place called Constantino, exchanging ribald comments with the Engineers there, these busy placing their charges to blow it up, but not too busy to examine newly arrived female company. Now they were all camped in a shallow valley, populated by a scattering of trees. Snow still fell and the wind still blew, but here there was wood for both fires and protection and the capable women soon made lean-to shelters as taught to Bridie by John Davey and soon copied by most others. The fires were cheerful and the children slept, right back in the lean-to where it was warmest.

The first they knew of approaching danger was a scream and a young girl, a teenager, came sprinting back into the light of their campfires. Malevolent shapes followed behind her, bulky, the shapes of men, and one, particularly large, was in the lead. Whilst the Reverend Chaplain Prudoe shrank back into his own shelter, having now zero confidence in his authority regarding deserters and scavengers, Chaplain's Assistant Percival Sedgwicke swallowed hard and walked forward, towards the man who was evidently the leader and a man he recognised.

"What do you want? These are the followers of the 105th, their women and children. They should not be harmed!"

Seth Tiley did not even break stride, but marched up to Sedgwicke and pushed him back, not caring to even speak, but Sedgwicke managed to keep his balance as he slithered back on the ice and remained confronting Tiley, who overtopped him by more than two feet. The practising Christian inside him had come to the fore; charity was a duty, to even such as these.

281

"This is no way to go about this. We can give you a share of what we have and no-one gets hurt and we can all go our own way, under the protection of our Good Lord."

The last words gave Tiley some pause, enough for a sarcastic laugh, before finally pushing Sedgwicke thoroughly aside.

"The 105th. It couldn't be better!"

However, at that point his confidence took a severe knock. From his time in barracks he had come to know, if only from a distance, Mrs. Nelly Nicholls, and it was her voice he heard now, in its most harsh and grating form.

"You come any further into this camp, now, Seth Tiley and it'll be the last thing you do on this Good God's Earth!"

Nelly Nicholls had stepped forward into the light of the nearest campfire and Tiley could see immediately what was giving such strength to her words. She was holding a musket up to her shoulder, cocked and pointed straight at him. With the return of the girl, she had immediately pulled it out from their travois. Tiley gave his best oleaginous smile, but the flickering yellow light of the fire leant it instead a sickening malevolence. His words were ingratiating, but any measure of conciliation they carried was lost in the malice of his tone of voice.

"Come on, now, missus. Words got around that you've been supplied. We only wants but a share. B'ain't we human creatures such as yourselves?"

Nelly lifted the musket higher, then shouted as loud as she could.

"What would a gombeen gobshite like you know about being a human creature? We let you in, you and your shitecaked tripehounds, and you'd take every last bone and crumb."

She paused to let her defiance fully sink in.

"Why don't you take yerselves over to the French? See what you can rob from them, instead of leavin' the likes of wimmin and children to starve and die! Get on over to them, they've as much as us. Take it off them, them as is tryin' to kill our own men and husbands."

Tiley was stood still, the threatening musket was enough to cause that, with the firelight glinting on its grey barrel and brass lock, but he knew that all he had to do was to hold this harridan in place and his men would soon engulf the camp from elsewhere. He smiled again and spoke further in his sickly tone.

"B'ain't that musket gettin' heavy, missus?"

That confident statement also took a bad knock, for Nelly did lower it to waist level and then stride forward to stand barely three yards from him, the musket muzzle but two yards from his copious stomach. Yet his confidence was still high.

"You don't want to point that thing at me, Missus, 'twill only make things 'arder for you an' these others round about."

Nelly Nicholls looked up into his face; the firelight catching the grey of her stretched back hair.

"If it does, you'll be the first to go! 'Tis no good thing to die with a ball through your guts. 'Tis a bad way to go, I'm thinkin'."

Tiley was waiting, confident that his men would soon be in amongst the camp, then he would dart forward and deflect the musket, when the attention of this warrior washerwoman was turned away, but, as he sensed his men moving forward, that hope also faded considerably. Another voice was shouting from the shadows of the camp.

"And you'll not be the only one! And that ye all need to hear, ye gang of blackguard murderers?"

Bridie Deakin was stood off to the side and behind Nelly, a huge Dragoon pistol held out in front before her chest, held in both hands, and she was not alone, all the women had found their own similar weapon, distributed by Jed Deakin and the other husbands, French pistols salvaged off the field of Sahagun. Nelly alone had the strength to lift a musket, but a Dragoon pistol was well within the strength of all the women there, including Eirin Mulcahey. They were as numerous as the number on any skirmish line, all with pistols, all pointing out at the threatening shapes stood at the edge of the firelight. It was a tense standoff. Tiley's band were hungry and desperate, but it meant death for many to attempt what they wished, that being to ransack the camp and take all they could carry.

Chaplain's Assistant Sedgwicke had meanwhile hauled himself to his feet and gone to his own possessions to gather two haversacks, each with biscuit, meat and brandy. He took them towards Seth Tiley, but Nelly Nicholls halted him several paces away from where Tiley was.

"You'll hand over nothin', Parson. Nothin'! Not one scrap! There's plenty of hunger on down the road from here. You gives nothin'; not one bit to this piece of Devil's shite!"

She hoisted the musket up to her shoulder again.

"Your choice, Tiley. You can die here, and some of these gobshites here with you. Those that don't die will get some food from us here, but you'll not be one of them! And that's the God's Honest Truth!"

Seth Tiley weighed up the odds. He was not a brave man; it was the cold calculation of the success of some robbery or similar crime that had kept him alive so far. Neither was losing face of any concern, he'd club into bloody pulp anyone who challenged his position as leader, but the sight of 30 women, perhaps more, all stood with good firearms pointing at him and his men changed the odds from short to long. He walked backwards, watching Nelly all the time, then, when in the darkness, he turned and disappeared into the snow. His men followed, but he led them away from the French, up towards the coast and the next town; Lugo, or so he believed.

The darkness returned. All was as it had been before Tiley's arrival, all bar the women still pointing deadly weapons out into the dark void beyond the light of their campfires. Nelly Nicholls found that she could not move. It took Chaplain's Assistant Sedgwicke to come over to her and prise her fingers from the barrel and the finger guard.

"I'll take that, Mrs. Nicholls. You've been very brave and done very well. We're all very proud of you."

It needed Beatrice Prudoe to come and wrap a blanket around her shoulders before she could walk unaided from the spot, back to her own campfire.

***

The same night was surrounding Ellis and his small command. He would not let them stop but ordered them to march on, even though their progress was more like a stumble through the darkness, through the trees and rocks and up and down the gullies and clefts. Ellis and Davey agreed that making camp by any measure close to El Navaja was too great a risk. They may lose their way in the dark and aim off away from their army, but that risk was as nothing compared to the risk of the guerrillas finding their camp and cutting all their throats. With a growing dawn, Ellis ordered them to climb, to gain height. What they may see may cause them to retrace their steps, but he hoped to see campfires lit for breakfast and their effort was rewarded. They had not strayed too far from their army's route, for below and over to the right could be seen bright dots, growing in number as fires were lit. What they did not know was if the fires were British or French, but at least their task was defined and simple, to get ahead of whoever was now enjoying a hot breakfast. There was further reward; their height was a ridge that extended off in their best direction and so they progressed on, along its length. Spivey suggested that they link arms to prevent slipping on the smooth ice beneath the latest snow, ice formed from snow melted then frozen, and thus they progressed in close company, like some odd form of green and red caterpillar, but their passage remained clear across the pristine whiteness.

Once into the trees Ellis granted a period of rest. A fire was lit inside a dense group of thick trunks and they shared their food, Joe Pike being required to put slightly more into the communal pot, as he had been given extra by Mangara's 'girls'. Ellis looked at the sky through the tangled, but bare branches, above him.

"I never thought I'd want it to snow, but I do now."

His wish would not be answered, at least not in the immediate future. If anything the sky was clearing to give weak sunlight, as an improvement on the merely better daylight of the day before.

"Our tracks are still out there, leading to here. I'll not count on that El Navaja not being one vengeful bastard, enough to take the trouble to follow us all the way."

He sighed and scratched his head, then took his share of the hot food. Davey finished his mouthful.

"He won't want to be too close to the French. He'll know, that what he does to them, they'll do to him, if they catch him."

The others looked at him, as he developed his thought.

"If we march on, keeping as close to the French as we dare, him and his crew will keep off and, if he is that determined, he'll have to waste time worrying about French cavalry. If it's just us and them workin' this through, they'll track us and kill us. These are their mountains, but if he has to worry about the French, on top, that'll be to our good."

It was Newcombe who now spoke.

"Too close to the French, you Redcoats will stand out like a beacon, and the French cavalry will sweep us up!"

Ellis finished his mouthful.

285

"Right, you're right, but we all has an army blanket. It's grey, so from now on we all march with that wrapped round us. No bad thing anyway, not in this cold!"

Newcombe had more to say.

"Our green lets us blend in enough, and 'tis awkward marching, holding a blanket round yourself."

Ellis was aroused. He would indulge either of the other two riflemen, but not this one, so full of himself.

"Then tie the top two corners with string, or make a hole in each corner to go on your buttons. Either will serve! I thought you Rifles was chosen for initiative!"

He leaned forward.

"Each of us wrapped in grey, we'll just look like a group of rocks from a distance. That makes sense to me, what about you, John?"

The ex-poacher nodded and grinned.

"Grey's a good colour if you don't want to be seen. As good as dark green, I'd say, probably better."

Ellis now treated Newcombe to a hard stare.

"So that's my order, Chosen Man!"

Newcombe looked down at his food and Ellis considered the matter settled. With the food consumed, Byford was now rummaging about in his pack and he brought out what looked like a large watch with a flat glass face. He held it out for Ellis.

"My brother gave me this, before I left."

He gave a small laugh.

"He was the only one who would speak to me. It's a compass."

Ellis held it in his hand staring at the swinging needle as Byford continued.

"The silver end of the needle always points to North."

Miles now joined in.

"So how does that 'elp?"

"The route to Corunna is roughly North West. If we get lost or aren't sure, then if we keep that needle pointing off to our front right, we'll be, more likely than not, going in the right direction."

Ellis looked in Byford's direction, puzzled and perplexed, but he nodded all the same.

"Right, every little help. Can I hang onto this?"

"Certainly, Sarn't. I hope it does help."

Ellis nodded his thanks, now plainly grateful, but Miles delivered his usual verdict on such sophisticated contributions from such as Byford.

"Book learned bastard!"

However, he was temporarily appeased when, as they all arranged their blankets over their shoulders, Ellis came to stand beside him.

"This is the first chance I've had to say thanks, for putting a stop on that El Navaja, when he was pulling round his musket."

Miles looked at him, suspiciously and with thinly disguised animosity. There was no love lost between these two.

"Well, all right. But perhaps you'll remember that, next time you finds a dirty button!"

Ellis nodded.

"I will. For the first, but not for the second."

With that, they parted company, for Ellis to lead them out of their cleft, then onto the flat uplands, him regularly consulting Byford's compass.

***

The sounds of conflict were far behind them and it was to the rumble of distant cannon fire that the 105th crossed Constantino bridge, watched by a squad of Engineers standing guard over what looked like a collection of thin black ropes, laid over a piece of wagon tarpaulin. That the 'ropes' led under the bridge told all, for the Engineers had finished their preparations to blow up the bridge; they were fuses. Their Officer, a Captain, approached Lacey. He looked clean, well clothed and well fed, not at all like someone who had endured almost two weeks of hard retreat.

"Excuse me, Sir, but we have been told that a regiment and a battery are yet to come. After you, that is, Sir. Can you confirm, please?"

Lacey dismounted his horse. He was as weary and careworn as he looked, the almost continuous conflict and tension of the past week were taking their toll, but he faced the Captain, returned the salute, and spoke civilly.

"I can. The battery, as you say, for that's what you can hear, are behind us, also the 52nd. I suspect them both to be here with the fall of darkness."

He felt it to be only polite to divulge more.

"That is how it works, you see. We hold them off, with guns and infantry, and then take ourselves off, when night falls. It seems to be working quite well, so what would be useful is if you could find some way of lighting up this place, torches or lanterns or watchfires. Not all may come by the road, you see, Captain. An aiming point would get them to your bridge that much quicker."

"Yes Sir. I'll see to it."

"Your orders are to blow the bridge when the guns and the 52nd are back across?"

"Correct, Sir."

Salutes were exchanged and no more said, so Lacey followed his men to their position 100 yards beyond the bridge. Already they were making camp and lighting their cook fires, which made Lacey realise that his request to the Captain may have been superfluous, but it would do no harm.

Carr, Drake and Shakeshaft simultaneously slumped down onto a low stonewall and watched Morrison achieve a blazing fire in the space of a minute, then to prepare the ingredients for their evening stew. Drake, ever cheerful, watched the pot fill up.

"And what's on the menu this evening, then, Chef de Cuisine Corporal Morrison?"

Morrison permitted himself a smile. He liked Drake, in fact he had a lot of time for all three, but Carr had spent the last two days in a very black mood, still depressed by the loss of 11 good men. Today was the 5th; he had lost his men on the 3rd, but it was Lieutenant Drake that was asking, and Morrison was a cook of some experience.

"Well Sir, I'm using a base of biscuit and ground acorns, with main ingredients of horsemeat and turnips, seasoned with a little pork barrel salt, and what I believe to be sage, Sir. We've all been eating it for the past few days with no ill effects, so I feel confident about continuing. Sir."

Drake slapped his hands onto the dirty cloth of his breeches.

"Excellent! Most splendid! My favourite, "Potage de Corunna". I trust you will be able to repeat the recipe when we entertain our Generals back in England."

Morrison now laughed out loud.

"Yes Sir, mostly Sir, although it may be a struggle to obtain horsemeat, days old."

288

"Oh, I'm sure any old knackers yard could provide the necessary. We'll all enjoy it around the mess table, I feel sure. It'll bring back warm memories. Who'd want to forget such as this?"

Shakeshaft was laughing as was Morrison, but Carr sat with his hands on the pommel of his sword, his forehead resting on his forearms, staring at the snow between his feet. He was making no sound, nor attempting to join in the conversation. His Lieutenants made no effort to include him and so it remained, whilst they talked about their respective home lives, seemingly on another planet, until the stew boiled and Morrison declared it edible. All four shared it, eating in silence and eating quickly. The heat in the food was as important as its nourishment.

Watched by Lacey and O'Hare, the 52nd and their attending battery came across the bridge just after midnight. Their Colonel passed by them and Lacey asked the usual question.

"How many did you lose?"

"Four and four wounded."

"All accounted for?"

"Yes. All from enemy action. No desertions."

Both Lacey and O'Hare touched their hats, as the 52nd's Colonel followed his men off to their campsite. Paget then joined them and the two sprang to attention, but, at Paget's first words, they stood at ease.

"We'll blow the bridge just before dawn. M'sieu won't attempt anything serious before then, and we may gain a few more stragglers, such as those."

He pointed to the bridge approach where three Redcoats were now entering the torchlight, one without a weapon, one without a shako and one without any equipment, bar his musket. They were immediately bawled at by a Sergeant of the 91st, before being kicked further into the camp.

With the 20th standing picket, the Reserve Division slept easy. There were no incidents to disturb their cold slumber, save the crying out of men in their sleep, holding conversations with images in their dreams, of both comrades and combatants, forming pictures of events best not recalled. At the hour before dawn, all were awakened and moved back, out of the way of any flying debris. Paget and his Staff watched the Captain light his fuses and then run back. Within two minutes came the explosion, muffled and second-rate, such that Drake found cause to observe the same to Carr.

"That doesn't sound up to the mark! Not enough to send that bridge down into the depths."

Carr looked at him and hefted up his rifle so that the barrel rested on his shoulder.

"It wasn't. Go around and check weapons and ammunition. We'll be here for the day, with Johnny for company."

Carr was correct. The central arch was still in place. In fact most of the bridge was still in place, bar a section of the downstream parapet, which had been dislodged. Whilst Paget screamed at the distraught Captain, one of his Aides came to Lacey. He was the same that had arrived at Bembribre and this time his salute was punctilious.

"General Paget sends his compliments, Sir, and would you be good enough to deploy your men to hold the bank downstream of the bridge. The battery will crossfire onto the bridge and the 95th will be holding the bank upstream."

"Very good, Captain. I take it we will be here until nightfall?"

"I think you can safely assume that Sir, yes."

As the Captain made off, Lacey looked at the lightening sky.

"Full daylight in half an hour, I'd say Padraigh. Get some men over to the far bank, the Grenadiers. Get them to cut away as much of that thicket as you can. Why give the Johnnies some cover, when we have none?"

Thus, it was a very disgruntled Captain Lord Carravoy who led his Company over the still reeking bridge to clear bushes and brush from the bank that would be opposite them. The men of the 95th soon copied them, but there were no tools. The undergrowth had to be cleared by main strength, several men simply pulling the bush apart. Carravoy stood and watched, his bad mood growing as he watched his men at their difficult task. He spoke to no-one in particular.

"Gardening! Sent over to do damn gardening! Us! Grenadiers, an elite Company!"

His mood worsened to see Ameshurst working beside his men, using his sword, the only blade they had, to hack through the branches sufficient for his men to pull them away and cast them down into the deep river course. He felt even worse, when D'Villiers similarly joined to aid the efforts of his own men. Carravoy's hand went to his own sword hilt, but the sword remained firmly sheathed. With full daylight Lacey called them back over and they took their place to the right of the 105th's firing line, making them the furthest of the British right, furthest from the bridge. Carravoy's mood lightened somewhat when Binns presented him with a mug of hot coffee, the first since gaining new supplies from the Spanish waggontrain. He was still drinking it when the French arrived, but the taste was too good, so he carefully finished the mugful. He had plenty of time, because the first to arrive were cavalry, who halted well out of effective range. Thus began the performance, the first events of the drama, being the Officers studying each other through their telescopes, Paget spending a long time with his. He saw what looked very much like an argument between the cavalry commander, these plainly being Dragoons, and some other ranking Officer. Lacey and O'Hare watched through their own instruments and O'Hare chuckled.

"There's certainly some kind of disharmony over there!"

Lacey laughed his own agreement and then looked over to his left and slightly back, he knew why there was dispute on the far side. In plain sight to the French, were six guns covering the bridge, three each side of the road, so that the six would set up a crossfire that met on the bridge. In addition there were his men and the 95th, in even plainer sight, ready to add their fire. Lacey looked behind to see the last of the 20th marching away; the 52nd and the 91st had already gone on before, so those remaining were now on their own. However, their position was immensely strong and yet, or so thought Lacey, were it not for that bridge still being there, we would also be on our way.

The two sides stood staring at each other for long into the morning. Paget ordered that one member of each mess was to fall out and make a hot drink for his fellows, thus all the men in his firing line were soon stood watching the French whilst drinking tea, many mugs being raised towards the French, but this was long finished when events finally developed on the far bank. The Dragoons turned off the road for marching infantry to then appear. Deakin, stood beside Rushby, saw all and felt relieved when the column halted, but his thoughts were interrupted by Heaviside, stood just in front.

"Colours!"

The two Colours were uncased to hang limp in the still air, and all around, on both banks, was still and quiet in anticipation of what may yet come. Heaviside turned around to check on his Colour Party and felt approval, but there was no quote, merely a nod. All felt oppressed by what seemed likely to ensue, however, few were watching their Captain, least of all Deakin, now speaking out loud to Halfway to his right.

"Be we in line for a repeat? Didn't these damn heartless Devils learn anything from the last time they tried a bridge? That one at Cacabellos or wherever it wer'!"

His friend shared his concerns and spoke his own thoughts

"Seems not. Brave, no question. Stupid, no doubt."

Such eloquence came rarely from Halfway, but Deakin did not notice. O'Hare was running up the firing line giving orders. He soon came to Heaviside.

"Joshua. Hold your men here, in reserve. The Nine, Eight and Seven will support the artillery at the bridge as a firing line. The rest will be to your front in skirmish order, the Lights immediately before you. Send in men as you see fit, when it comes, as I fear it will."

Heaviside saluted as O'Hare ran off, but he immediately returned to watching developments on the French side. Things were already happening; the infantry were advancing forward, in large numbers. Lacey and O'Hare had rejoined each other and stood watching the same scene.

"What do you think, Padraigh?"

"That's a whole Regiment, two battalions. French Regiments have two, sometimes three. So, that's one for the bridge, the second to keep us busy from the far bank."

The accuracy of his prediction was immediately confirmed. The first battalion formed a thin column, only a little wider than the bridge, whilst the second extended out in skirmish order, dividing itself between the riverbanks above and below the bridge. Soon they heard the drums and then the shouts of "Vive l'Empereur!" Lacey's mouth clenched into a grim line before speaking.

"Haven't they learned anything, about us, after what we did to them at Vimeiro."

They both heard the Battery Commander order to load grape, to hold grape for the second discharge, and to have case standing by. O'Hare took another look.

"Right. Nothing else for it, but to get our men ready."

Lacey watched him run off to their three Companies at the bridge, drawing his sword at the same time. Lacey elected to walk to the nearest file of his own men where he could see all and it was not long in coming. He just had time, as they came close enough to see their faces, to marvel at the men in the front ranks walking to certain death. When the head of the column was at the centre of the bridge, a six gun discharge blew apart the leading quarter of the column, men being cast backwards like helpless effigies. As the next ranks came on, O'Hare ordered whole Company volleys to match those of the Rifles above the bridge. This held the French back until the guns reloaded, then a second discharge completed the work, so that there remained less than half the column still advancing. The French soldiers miraculously remaining alive on the bridge tried to fall back, but they were held there by others coming on, so, unable to escape they jumped into the river, a fall of some 12 feet into the deep bed. For a minute the British assault on the column continued, the Company volleys, then the guns. Soon the riverbed was a mass of men who had jumped to escape the appalling weight of fire, whilst above them on the river banks the two sides exchanged skirmishing fire across the deep divide. The fugitives waded or were carried downstream through the icy water, trying to find a way up the steep cliff, but there was none.

Carr's Lights men were lying prone on the bank, which slowed the reloading of their rifles, but they made a much smaller target for the inaccurate French muskets. They had suffered but one wounded so far, whilst they had already downed a dozen French. The French who had escaped off the bridge were soon carried down the river to where they were, almost filling the riverbed and all still vainly scrabbling for a way up. Many of their fellows above, skirmishing on the bank now ceased combat with Carr's men and lowered their muskets down to their comrades and many eagerly seized the butt of a musket to allow themselves to be hauled up. Many French tried to raise up their own wounded, far enough for an arm or a tunic collar to be seized and the man then hauled up to relative safety. French Officers were running around, shouting, but it seemed that none were pointing at their enemy over the gulf between them. Almost 100 Frenchmen were trapped down in the deep watercourse.

The 105th continued to fire. The sight of blue uniforms, their enemy, and instinct, were enough to cause them to fire and reload as fast as possible and soon many French dead were floating away on the river. Fire from the French side had all but ceased and it needed Sergeant Fearnley to approach Carr.

"Sir. This is murder!"

The powerful words drew Carr's gaze to immediately look at his remaining Sergeant, then he looked again at what was happening down in the river and within that one brief look he saw two Frenchmen fall away from the steep bank, hit by the accurate fire from his rifles. The recent loss of his men still festered and so hatred and humanity competed within him, but the sight of so many helpless men eventually brought him to his decision.

"Cease fire!"

This shouted at the top of his voice caused all around to stop firing and Fearnley ran off to carry the order further. Above the bridge the 95th were still exchanging fire bank to bank, but through the bridge arch no French could be seen, the stream had carried all fugitives down to the 105th. What applied to the 95th skirmishing above the bridge, also applied to the Companies of the 105th below Carr's Lights, which included Carravoy's Grenadiers; for them the fight continued. The French opposite them were still maintaining fire, so they did also, but the French ensnared in the river had not floated down beyond Carr's men, who remained inactive and watching as their enemies were pulled up to lie frozen and exhausted amongst their own comrades. Whilst the sounds of conflict continued all around, including the artillery, at that place there existed a small truce. A French Officer, probably a Major, stepped forward to the bank and brought his sword up in salute. Carr acknowledged in the same manner, as did Drake and Shakeshaft. Meanwhile, their own men lay in the snow and awaited orders.

None of Carr's Light Company fired another shot. Desultory skirmishing continued up and down the banks until the light began to fade, but neither side on Carr's front broke the truce. The three Companies of the 105th supporting the artillery at the bridge had long ceased firing, but remained in position in case of another attack, however that was now the remotest of possibilities. Paget came to stand with Lacey and O'Hare and all three looked across the bridge and beyond, the whole now carpeted with French dead. Paget spoke first. French casualties had been appalling, whilst his own had been negligible.

"Perhaps we should make a ploy out of failing to blow bridges?"

Both Lacey and O'Hare detected the irony in Paget's voice, but it was Lacey who replied.

"It's the game we're in, Sir."

Paget nodded.

"Well, we've now held them up for some while, at great cost to them, but Soult's right up to us, with his full army."

He turned to leave.

"Come full dark we'll be away. Make sure your men are ready; it'll be a forced march through the night. Johnny's only been held up for a short while, he's plenty more to replace these."

He waved his hand at the charnel house of the bridge, then walked away and both Lacey and O'Hare punctiliously saluted as Paget walked to the 95th.

Other salutes were being given along the British line. There was still sufficient daylight for each side to see the other and, in particular, the French Commanding Officer, whoever he was, could be seen sitting on his horse, surrounded by his Staff. Some in the 105th were motivated by the fact that they had given the French a very bloody nose, but others, like Deakin, had been appalled at the hopeless and murderous task that he had driven his men forward to attempt. Thus, as the French Commander in the gorgeous uniform studied his opponents, pondering his next move, he was treated through the lens of his telescope to the sight of several two fingered salutes, as had been delivered centuries before by the yeomen ancestors of those in red to the foot soldier ancestors of those in blue.

***

The 105th marched through the night, their guns before and the 95th behind. Troopers of the 15th Hussars bearing lanterns guided them on their way, but all were too weary to attempt any conversation. Dawn saw them pass through the village of Lugo and then three miles further, to the British lines drawn up on the heights beyond, there to collapse with exhaustion, not caring even to cook a meal, but instead to sleep with any shelter, be it a low wall or a leafless tree. Paget passed on the order that they were in reserve of what was now the main army and Lacey spoke his thanks to the Aide de Camp, then took himself off to his tent, the erection of which had to be completed around his sleeping figure. The followers heard of their arrival and came back to sit with their men and be there when they awoke. This coincided with the Noon meal and Deakin and Nicholls were well content to sit, eat and rest within the bosom of their families, all now with more energy and in better spirits than they had seen them in for some time past. Lugo had contained a depot of provisions that would last about four days for Moore's whole force and so it was well fed and clothed followers that they awoke to see, smiling and content, at least for now.

Having eaten, Carr and Drake returned to the main British line at the top of what proved to be a long slope and each surveyed the scene, leaning on a stonewall, before Drake broke the silence.

"Is this to be the long sought after battle, do you think?"

Carr studied the British position. It was immensely strong; the right ended at what looked like an unfordable river, the left finished at a collection of rocky outcrops, whilst the front was covered by low stonewalls at the top of a gentle and very exposed slope. He shook his head.

"That I doubt. This is a good position, but our army is in too much of a shambles, Moore'll not risk a set piece. He'll fight off any advanced French units, but fall back before their main army. If they concentrate here, he'll fall back, he'll have to."

They watched a battery of guns pulled by teams all walking past at leisurely pace, before Carr continued his theme.

"The army's all over the damn place. I took coffee with a Captain of the 20th who told me that Fraser's Division marched off on the wrong road and had got 20 miles before being told to return. They're not yet back and that's not untypical. He'll need more than a day to pull us together. More like three!"

Most Officers of the 105th came to the hilltop to watch for the arrival of the French. This proved not to be the case, but what did arrive were hundreds, even thousands, of Redcoats from all directions; stragglers, deserters, or those simply lost. During the next night this was the main topic of conversation, between Deakin, Halfway and Nicholls. The latter seemed to have gained the most information.

"There's rumour of a battle, so that's bringin' the lads in."

Halfway shook his head.

"Sounds daft to me, to desert, then rejoin, wantin' to fight in a battle. I've seen enough fightin' since we landed in this Spain to last a lifetime."

Deakin looked at his good friend.

"And there'll be more! Of that you can be certain."

He paused.

"And of all those lads comin' back in, what can you say? I've heard it, so've you. Just give us a chance at 'em. There's not a manjack as doesn't think these French are no ways up to our mark and a good fight is one way to get things off your chest, after the past week and all. Take it all out on the Johnnies, like."

With the following day, it did seem that a setpiece battle was building, at first, that is. The French arrived, but as Carr predicted, they arrived in small packets. First two Brigades of cavalry, then what looked like a Division of infantry, but all that day the two sides sat and looked at each other, stood or sat on the exposed slopes. It was full dark when campfires appeared on the French side and Moore allowed his men to stand down, behind a thick picket of Rifles. The next dawn found the British deployed once more to watch the arrival of more French, this time two Divisions of infantry and one Brigade of Cavalry. From their stonewall; which by now was boringly familiar, Carr and Drake observed their arrival. Once more Carr gave his opinion.

"I'd say M'sieu's army is in no much better shape than our own. Soult must have had stragglers and fall outs, just as us, especially when they have to range far and wide to get food from the locals, and there must have been damn all of that. to be found easily."

Drake made no reply. The 105th were in reserve, back from the front line. Strictly speaking they had no business where they were and his appetite had reported in.

"Breakfast!"

Both left without a word, but, with breakfast thankfully inside them, the order came for the 105th to form column, the easier to advance should they be needed. They stood for 30 minutes, then Lacey allowed them to sit. The sounds of battle reached them, beginning with a furious cannonade which finished almost as soon as it began. Shortly after came a conflict involving musketry, but that also lasted but minutes. If there had been a battle it seemed over and done, as the day settled to little more than the sounds of odd engagements between skirmishers, then even that died away. Carr, Drake, and Shakeshaft were sat together, all listening to what was not happening and all were more than pleased to see O'Hare moving along their column, ordering all to draw rations and return to their mess fires. Drake passed judgment.

"Well that's that, then!"

He was almost correct, as the next day proved. Lacey called all Officers to his tent and there, stood outside, he read out an order from Moore.

"A battle is at hand. Your men are to waste no ammunition on skirmishers. Leave that to the Rifles. Withhold your fire, saving ammunition for the supporting columns."

Lacey lowered the piece of paper and regarded his Officers, with a "make of that whatever you like" expression on his face.

"Tell your men."

As they returned to their mess lines, Carravoy spoke to D'Villiers.

"Well, that does not exactly fill me with confidence. I thought we had been re-provisioned at Lugo. We are about to fight a battle, but now it seems we are short of ammunition. What next from this incompetent, no ships, when we get to Corunna?"

D'Villiers grimaced. He had no reply; he feared too much that his Captain may be right. No sooner had they read the order to their men, than another order from Lacey arrived telling all to reform into their reserve column of the day before. Within ten minutes all were stood upon the same piece of frozen snow and slush as the day before, awaiting orders but listening intently for sounds from the main position. Lacey had forbidden his Officers to wander up to the main British line at their leisure, but he nevertheless took a walk up there himself with O'Hare. Across the valley he saw the French holding a position as strong as their own, the blue of their uniforms almost as prominent against the snow as their own red. Lacey folded his arms and slowly exhaled.

"There'll not be a battle, Padraigh. If that's all Soult has, he'll not attack. We're too strong and perhaps this one has learnt what happens when you walk up a slope to try conclusions with firing lines manned by the likes of us. He'll hope to hold us here until his supports come up, which will happen, and in which case Moore would be mad to wait here at Soult's convenience to allow him to build his strength."

O'Hare looked at his Colonel and good friend.

"Could Moore attack?"

"No. If things look too dicey for Soult, if he sees too great a chance that he's about to be beat, he'll just fall back onto his supports and offer battle again, but this time reinforced. And where would that leave Moore, if he pursued? Further from Corunna and then with the bayonets of Soult's whole army right in his back. To begin a set-piece here Moore would be an utter fool!"

He took a deep breath.

"No, we're done here. Pass the word, but not too loud."

The men of the 105th, standing along the edge of the main road, had drawn their own similar conclusions but they used the evidence of several battalions of the main army marching past them in columns, all to rejoin their own followers at their mess lines. The order from Moore to fall back, break camp and prepare to resume retreat had been speedily, if very reluctantly, obeyed. Another order came to Lacey, but this time from Paget. It was brief and told all.

"105th and 95th to form rearguard. 18th, 20th, and 91st, hold in support."

Lacey showed it to O'Hare, who read it, then looked at Lacey.

"The Reserve Division continue in reserve!"

Lacey smiled wryly.

"It would seem so."

For the rest of the day the army prepared for more marching and the preparations included the destruction of 500 cavalry and artillery horses who could no longer go on. For two hours it sounded as though a minor battle was taking place as each was shot, but canny followers were walking to and fro, carrying lumps of horsemeat. Fuel to boil the meat, the better to preserve it, was in no shortage either as the artillery caissons that no longer had horses to pull them were knocked to pieces. Bridie and Nelly had set up what was almost a production line of a whole series of various pots on fires, each boiling its fill of meat, to be wrapped in whatever, to keep it for a few days. Then another order arrived, 'Reserve Division to maintain campfires until past Midnight'. Specifically, this would be the task of the 105th and 95th, spread along the ridge.

With the arrival of this order, Carravoy's Grenadiers were despatched to gather any available fuel, which included the wood stock of the followers boiling their meat. The Grenadier who entered the "kitchen" of Bridie Deakin and Nelly Nicholls to purloin their wood, found himself being called all kinds of a 'gobshite' and had genuine fears for his own safety as Nelly described threatening curves with a huge metal spoon. He left with what he could carry and himself intact, for which he was grateful.

Deakin used his rank to absent himself along with Nicholls and Halfway and to go to a dump of stores that did not seem to be about to be loaded into any wagon. They were guarded, as before, by The Guards. Deakin approached a sentry.

"What's the plan for this lot, mate?"

The sentry swivelled his eyes suspiciously whilst maintaining face front.

"To be destroyed."

"How?"

"No orders yet."

Now he did face Deakin.

"Can't be burnt, no fuel. Just chucked around, would be my guess."

"What's there?"

"Flour and biscuit."

Inwardly Deakin fumed. An army could march on biscuit and flour, if they had a little fat to make dough cakes, but he held his peace. Instead he reached behind his head to scratch his neck whilst at the same time pulling a silver piece from his pocket with his other hand.

"What'll this get us?"

The Guardsman's eyes fixed on the glinting silver guinea. He then looked up and looked all around, especially at the two Guardsmen on his side of the pile.

"Same for those lads?"

Deakin nodded and the Guardsman quickly took the coin.

"Seems like there's a stack there as is soon to fall over. If you could just give us a hand, if you've a spare minute?"

Five minutes later, a bulging harvest sack, full of bags of flour and biscuit, was being carried back to the lines of the 105th by Deakin and Halfway, with another almost as full over the meaty shoulder of Henry Nicholls. Eventually, by a roundabout way, which avoided any tents occupied by Officers, they reached their comrades.

Staff Officers were galloping all over in the gathering dusk, despatching orders for an orderly withdrawal that would see an organised column comprising the whole army, progressed far along the road to Corunna by the time the next day had dawned. In the gathering gloom the battalions of the rearguard spread themselves over the whole British ridge to maintain the deceitful flames of the fires still burning to confuse the French. No one was exactly sure what "past midnight" meant, but all knew that campfires died in the early hours and so that was considered good enough. The followers were to leave with the main army and so Deakin and Nicholls bid their farewells to their families, all now with bulging haversacks and with extra on the travois and on Mary's mule, now well established as the children's pet, often being fed hard biscuits, which it seemed to enjoy, especially those most full of weevils.

Deakin, Halfway, Nicholls, Stiles and Peters took over a lonely campfire just as the Regiment that had kindled it marched away. Then it began to rain, heavily and constantly, with no hint of any cessation. The snow and ice melted in an hour and the fires spluttered and died, for they required constant feeding to stay alight in the heavy rain and the meagre wood supply was quickly exhausted. Soon the 105th were doing no more than squatting before cold, dead fires, listening to the raindrops spitting in the last of the embers. It was a miserable night, but all were now veteran enough to know how to keep reasonably dry, all by now had a piece of wagon tarpaulin and the rain raised the temperature to at least above freezing, often beyond.

When the first streaks of light came in the East, Paget gave the order for his four foot Regiments to close up, ready to march off. The 95th and 105th were to be the last to leave, the 105th grateful that they had been given the road to picket, whilst the 95th were to guard the far right flank, close to the river, which made them the furthest from the road. They would cause them an oblique cross-country trek reach it and then form for their own march. When Lacey saw the 15th Hussars trotting onto the road, he gave the order to "form fours" and the 105th set off. Drake was the last to leave the British position and he stood on the road, just at the point where it would disappear over the ridge for the French, and he gave anyone watching through a telescope a cheery wave, using his shako for extra emphasis.

Thankfully the road had not thawed, it was quite firm beneath their feet, but that was the only bonus of the dawn. All around it was plain that Moore's orderly retreat had turned into a complete fiasco. Columns of Redcoats could be seen way out in the fields both to the left and the right, these being Regiments that should by now be way down the road towards the next town, this being Valmeda, as they had been told. Several times the Reserve Division was halted to allow such wayward battalions to regain the road ahead of them, because for them to join behind the rearguard would incorporate them into the Reserve Division, which would only cause extra confusion. The Officers of the 105th learnt later that day, when they took a brief rest and met some Officers of the 14th of Leith's Division at a provision train, that the stone walls, so useful for defence had created total confusion in the dark, so that many Regiments actually marched in circles. In addition there were again hundreds of deserters and stragglers who had absconded once more when they had learnt that there was to be no battle and others who simply felt an overwhelming need to find some shelter from the rain.

On entering Valmeda, it was pleasing to see that the small town had not been torn apart as had Bembribre and Villafranca. Deakin, Halfway and Gibney noted such as they marched in, Deakin being the first to voice his thoughts to Gibney.

"Pleasant to see a place still as it should be, Sar' Major. Nothing smashed up, even some people still here. Could be a market day!"

Such idle conversation did not sit well with the likes of Cyrus Gibney, but the sight of an undamaged, well kept town had done much to lighten his own mood."

"Aye, that's the right of it. But ah'm thinkin' that's because the sort as'll do damage to such as this, is now off in t'ills, deserted and gone, dead, or dead drunk, most like."

Halfway felt like making a point.

"Now having full haversacks must help though, Sar' Major."

Gibney leaned forward to look at Halfway further in the ranks.

"Aye. 'Appens as that's true, n'all!"

This was said with such finality that nothing more was said, but Gibney's deserters and drunks were not quite in either form of "dead", at least not those close to the army. Hundreds came in from the hills, lost since Lugo, but now guided in by local peasants, who took their payment in biscuit and horsemeat and considered it a good bargain.

That night the rain began again in torrents and orders came from some Brigadiers that the men be allowed to seek shelter wherever they could find it, but nothing of the sort came from Paget. Lacey, also knowing the confusion this could cause, but wanting his men out of the rain, specifically ordered his Officers to find shelter for their men and make a note of where they were. Carr was leading just such a group of his Light Company into a deserted cottage on the outskirts of the town when he met someone that he did not expect to see again, at least not in Spain. Carr was carrying a lantern which he hung on a beam and the light fell further across the floor to reveal a figure sat hunched in the corner, head down, perhaps sleeping, perhaps not, wrapped in a thick horse-blanket which was now standard practice throughout much of the army. Carr walked towards him, feeling in better spirits, even enough to be jocular.

"Evening to you, my good fellow. Is there any objection that you can see to my men sharing this decent shelter with you?"

The head rose up and it was some time before Carr placed what suddenly struck him as familiar. The man was as dirty, dishevelled and careworn as any of them, but it was definitely Tavender. Carr was taken aback, but he was the first to find words and he remembered vividly the condition Tavender had been in when he was taken off the field of Vimeiro.

"Tavender! What a place to find you. How are you? What of your wounds? The last we saw of you, you were none too well."

Tavender looked stonily at him and managed a reply.

"Better, I thank you. A month in Lisbon brought me back to some degree of health, sufficient to rejoin."

Carr found his flask of Spanish brandy and offered it.

"Can't speak for the vintage, but it does the trick, you know, straight to the spot where it'll do some good."

Tavender took the flask and drank. Having swallowed, he did manage a smile of sorts.

"Vintage! More like lamp oil."

He handed back the flask, but Carr felt that sufficient human contact had been made to give him excuse to sit down besides the forlorn figure. This was, after all, a fellow Officer, severely down in his circumstances

"Your care in Lisbon must have been of the best!"

Tavender did not turn his head.

"Nuns! We, that is Officers, were sent to a Convent, and there our care was, as you put it, of the best."

Carr now felt obliged to do what he could, for here was a Captain of Horse, a fellow Officer, squatting alone in a Spanish hovel.

"And now? Is there anything that we can help you with?"

Tavender looked full at him, both his face and tone indulgent, the sentence both carefully spoken and carefully enunciated, but not quite to the point of sarcasm.

"Now, I am infantry. My horse gave up, a week or so past, so that now, I am making my way back, as you, on foot, hoping for passage home."

Carr decided to be practical and ignore the tenor of the reply.

"Well, we can help, I'm sure. I'd invite you to attach yourself to us, but we're rearguard and it's none too pleasant and we, the 105th are, after all, infantry, as you say."

He smiled at the half attempt at humour, but there was no reaction from Tavender, but Carr continued.

"However, I do know some of the 18th. They're with us and I do believe that they've kept a string of spare horses. On top, they've suffered, as have we, and would not say no to a new recruit, I feel sure."

For the first time Tavender's face significantly brightened.

"You think so?"

Carr nodded.

"Why yes, I do, very much so."

He paused and nodded his head.

"I think we should try."

Without looking at Tavender for agreement, Carr looked up to his men, settling in for the night, and saw a Corporal he knew.

"Fergusson. Go and find where the 18th are. I do believe they turned off left, where we turned off right."

Fergusson dropped his blanket, saluted and left. Carr then noticed that his men were lighting a fire in the grate of the room and preparing the ingredients for the pot. There was another that he recognised.

"Thomas!"

The Private sprang to attention.

"This Officer is to have some hot food. I will see that extra arrives for you tomorrow morning to make up your rations. Is that clear?"

"Sir!"

Carr turned to Tavender, whose face remained blank.

"No acorns! That I can guarantee. Filling, but bitter, and they give you shocking wind!"

Again no smile nor any form of similar reaction from Tavender. Carr was beginning to detect a slight feeling of resentment from this Officer whom he was trying to help, but he thrust that aside and pressed on.

"If you are in agreement, I will return in 30 minutes, time enough for me to check on my men. By then you will have eaten and Fergusson will have returned."

A blank stare and a slight nod from Tavender, so Carr left and within 30 minutes he returned. He had found for himself where the 18th were, but Tavender appeared as morose as ever, remaining in the same corner. Carr looked at him in the dim light, but, as before, he pressed on.

"Shall we go? They are not too far."

Tavender pulled his sabre up from alongside his right leg and used it to lever himself up. When he walked it was with a slight limp on that leg, which Carr felt it improper to comment on. Once in the street, they began their journey, Carr having to slow his pace to allow Tavender to keep up, but social requirements dictated that he make some attempt at polite conversation.

"Have you heard from your people at home?"

Tavender looked at him as though he were an idiot and Carr quickly qualified his question.

"I mean, since Astorga. None of us have received anything since."

Tavender nodded.

"Yes. One caught up with me some days before then."

He paused.

"Usual chit chat, about my sister, my niece, my horses, the estate, local politics."

He continued wearily.

"I could go on."

He looked at Carr.

"What else is in letters from home?"

Carr nodded.

"True, but on the other hand tedium is good news, isn't it? That's the charm of a letter from home. It means that all is as it was when we left. That's a comfort. Well I find, at least."

Such sentimentality brought another querulous look from Tavender, but no further conversation. Carr started again.

"Do you know the 18th. They're Hussars. Light horse."

Tavender stared straight ahead.

"I know what Hussars are! I was in the 20th Light Dragoons, therefore I have somewhat more than a distant idea."

Carr sighed and dwelt on his own thought, 'Ah, well, I've given it as good a go as can be expected. The fellow's as much of a stuffshirt as ever!'

Finally he found a reply.

"So you think you'll fit in? If they'll take you?"

Again the querulous look, but there was no answer, nor any more conversation for the few minutes before they came to the door of a large and well lit building. Carr entered followed by Tavender, but Carr immediately recognised an Officer of the 18th that he knew, a Captain Jones. Several times he had ridden back in through Carr's picket line with a small patrol and over those occasions a standing, humorous exchange had been established between them, always beginning in the same form, "Jones. 18th and riding!" "Carr. 105th and walking!"

Carr's face split immediately into a wide grin.

"Captain Jones!"

Jones turned and grinned in reply.

"Carr of the Foot! Not frozen yet?"

Carr ignored the question and turned to Tavender.

"May I introduce Captain Tavender of the 20th Light Dragoons, but at present horseless. His mount gave up some weeks back, so, can you help, that being to get him a horse and perhaps take him into your ranks, at least until we reach Corunna? He was at Vimeiro."

Carr immediately regretted mentioning that cavalry disaster, but Jones appeared not to have noticed or had decided not to mention it, politeness requiring otherwise. He walked forward, offering his hand, which drew a smile out of Tavender as he took it. Jones looked at Carr as though he had just asked a highly foolish question.

"But of course, Carr! Where could there possibly be any doubt? Of course we have a horse and of course we have a place. We're a bit short of good company in the Officer's Mess, for one reason or another. If we had a Mess, of course."

Carr was grinning openly.

"Well, it's your own fault! Perhaps you should start eating your own horses."

Jones rose up in mock umbrage.

"Carr the very idea. I'd rather eat my own boots first, no, rather your boots. If you haven't worn them out yet!"

Carr laughed and conceeded defeat, then turned to Tavender who was now looking marginally more cheery.

"Well, I'll leave you in these capable hands, Captain, and bid you good luck."

The reply was merely a nod, a definite one, but he did not offer his hand. Carr turned to go, but with a backward look at Jones.

"Good luck, Jones!"

"Good luck, Carr!"

The following day was unmercifully clear, for, marching ahead of their men, Lacey and O'Hare saw what convinced them that Moore's army had finally fallen apart. All over the hills, on both sides, and at the furthest range of their telescopes, they could see swarms of Redcoats, all fallen out, deserted, or simply lost. The kindness of some Brigadiers orders in Valmeda to allow their men to get out of the rain had virtually destroyed the cohesion of their battalions and these Redcoats were now little more than looting marauders breaking into cottages and hamlets to steal and pillage anything of value, which went beyond the simple search for food. Twice Lacey stopped to examine such through his telescope and there was now a constant pattern, for any visible building had its own collection of plundering Redcoats. Eventually he could no longer bear to watch, instead he marched alongside his men, as did O'Hare and Simmonds, encouraging and promising better times ahead. However, ensuring that his men were well fed did more. At Lugo, O'Hare had gathered up some horses, less exhausted than most, fed them bags of biscuits to keep them alive and loaded them with provisions about to be destroyed. Regular hand-outs of biscuit as they marched and meat for mealtimes kept his men content and, more importantly, together.

The day finished at a bridge over the river Ladra. It had not been destroyed and so Paget set the 20th to guard it for the night, with the 105th in reserve. They were not disturbed, neither physically, nor in their thoughts, by the sight of any French campfires glinting in the far distance, for there were none, and the point most often made around their campfires through that night was that Moore's ruse of leaving Lugo at night had worked, albeit at the cost of his army's cohesion. The following day they came to a familiarly half demolished bridge over the Mendeo, which had been guarded by the 91st. On their arrival, the 91st formed up and marched on, before the 105th then crossed, to follow them into Betanzos, where, for the first time, they saw the sea.

***

The nine were lying on a bank just within the tree line. They had been there but minutes before Ellis had given judgment.

"This'll do!"

It had been four days since they parted company with El Navaja. The snow had gone but they had still left tracks in the mud, which for the first two days was a constant worry to Ellis and so he and Davey were in constant conversation concerning how they could make themselves difficult to follow, particularly for such as El Navaja, were he still bent upon revenge. However, keeping as close to the French as they dared, they had been unmolested, but then their worries disappeared, because for the following two days there were tracks everywhere from roaming Redcoats. Now, plainly, they had come up on the junction between the British rearguard and the French advanced guard, but their food was exhausted and all were very hungry. They had kept themselves separate from the bands of deserters and marauders that numbered thousands all around them, or rather it was more the other way around. Some attached themselves to the nine, but soon slid off into the night when they found themselves subjected to Ellis' iron discipline, especially if they had no weapon and also, more pertinently, when they found that they had more food than the nine and were required to share. Ellis had utterly refused, even in the face of Miles' protests, to indulge in any form of thievery or pillaging, but nevertheless, there was now almost a second army of detached Redcoats in the hills all around. If El Navaja wanted to avenge himself on British soldiers, he now had plenty to choose from.

Because of these factors, they remained the original nine, but all, from their high vantage point, had decided for themselves that this was their chance and probably their only one. From their position they could see that they were now opposite a large portion of a large village split by a river. They were on the "French" side, but all the buildings were still held by the British, evidenced by Redcoats leaving and entering, although those leaving appeared to be in a much higher military order than those arriving, the latter plainly being the now common battalions of stragglers. They called it the French side because off to their right was the wide deep river, probably tidal and the village contained the only bridge. The British army was formed on the far side and the other black spot in the picture was French cavalry, a substantial body, in plain sight some three quarters of a mile to the left of the village, which they would have to enter and pass through, then cross the bridge to gain the British lines over to their right. Ellis rolled over to look at his companions.

"Weapon inspection in five minutes! Get 'em clean!"

The Riflemen looked at him as if he were mad, but Ellis' 105th companions knew him of old and all brought their rifle around to their front and to give it a clean and a check. Newcombe had had enough, this was absurd.

"You're wasting time! The French are on their way, five minutes could cause us to be cut off again."

Ellis gave him a look that told that this martinet of a Sergeant would brook no argument.

"You listen and listen good. Down there we will, like as not, find ourselves in a fight. I wants nine rifles as works, and I'll tell you this, if yours misfires that'll cost you your stripe, Chosen Man!"

Annoyance spread across Newcombe's face and stayed there, but he was a trained Rifleman and he had finished cleaning and inspecting his own weapon well within the five minutes. After that time eight rifles were held forward for Ellis to look at, hammer back and frizzen plate open. Ellis gave each a brief inspection, for, indeed, he did not want to spend too much time, for Newcombe did have a point and he knew he could trust veteran soldiers to maintain their weapons up to the standard required. However, he and Miles still found time to share a look of intense dislike, before he peered into the priming pan on Miles' rifle.

"Touch hole?"

"Clean!"

"Flint?"

"New!"

With no acknowledgement to Miles, Ellis gave his next order, speaking to them all.

"Load."

All nine immediately found a paper cartridge and bit it open. Within 20 seconds all had a loaded rifle and Ellis led them forward. Joe Pike as usual keeping close to John Davey.

"What's that village called, John?"

"Haven't got a clue, boy. Come on,"

On the slope above the river, the 105th were in their customary position as rearguard, all with a panoramic view of the scene. Carr's Light Company were formed on the left end of the 105th's line and the

95th were further down the slope forming a skirmishing line someway back from the river. Drake and Carr, as usual were exchanging observations on what they saw. The conversation had opened with the usual question from Drake.

"What's this place called, again?"

The answer before had been "Betanzos", but now Carr was passing judgment, ignoring the repeated question.

"There is such a army of stragglers and deserters between us and Johnny that we are practically redundant. They make up more than enough between him and us to occupy his advanced guard. Look, there's another crowd coming over that hill."

Carr pointed unnecessarily to a group of about 200 Redcoats, all running to enter the village before the French cavalry. There was another group behind them that had almost no chance of escape, but Carr was right, there were plenty of their British enemy to occupy the French Light Cavalry, without any need to involve themselves with the far more potent Reserve Division.

Jed Deakin was also stood watching, but he was the first to see, using his sharp country eyes, the nine emerge from the trees to the left and run across the winter dead field towards the village. At first he paid them little attention, dismissing them as yet more "drop outs", there were several other groups around, but something held his attention and so he studied them further. After half a minute he turned to tug at the sleeve of Halfway.

"Toby! That group, thur, running to the village from the left, just comin' out of that hedge. Ours and Rifles mixed, don't they seem familiar?"

Halfway shielded his eyes to look. The group had at least ten minutes running to do and he gave them a careful study.

"They are! Are you thinking what I'm thinking?"

"I am."

Deakin looked around for an Officer, but saw none. The Ensigns did not count.

"I'm goin' down to the Lights to tell Mr. Carr. If Heaviside comes back, tell him that and what we think's goin' on."

With that Deakin left the front rank and ran down the line to the Light Company. Carr and Drake were still out at the front and so Deakin ran up to them, came to the attention and saluted.

"Sir. Beg to report. Sir!"

Carr felt a slight annoyance at their conversation being so summarily interrupted, but this was Colour Sergeant Deakin.

"Yes Sergeant? What do you have for us?"

Deakin pointed.

"Sir. That group there, runnin' for the village from the trees. About nine, Redcoats and Rifles, Sir."

Carr and Drake identified them as they came into clearer view.

"Yes. What of them?"

"Well Sir. We do believe them to be some of yours, Sir, Light Company, what we lost at Cacabellos. That tall one, well, I'll lay money that's Zeke Saunders, them two out front looks like Ellis and John Davey, an' that one in the centre, Sir, movin' like a sneaky weasel, well, that's Tom Miles. They last could be Byford and Pike, Sir."

Drake looked away, choking back his laughter. It was very bad form for any Officer to laugh at any humour made by a member of the "other ranks", but the humourless Carr had found his telescope and was studying the group. He had been since he heard the name "Zeke Saunders", the rest of Deakin's words he had ignored as he attempted to focus his glass. Within seconds he knew that Deakin was right, here were six men of those he thought lost, six of his best men. He turned to Drake.

"I'm going to the Colonel. I want to take us down into the village and get those men out. They're six of those we lost at Cacabellos."

Drake nodded, just before his own telescope arrived at his right eye, and Carr ran off. Lacey was stood behind the Colour Company, but Carr was immediately disappointed.

"No. Remember what happened at Bembribre."

Carr was so animated by the possible recovery of six he had thought lost, that he felt inclined to argue.

"But Sir, could I not take a section down to give them at least early support, at the bridge? The 95th would be in support, Sir, down there. Being there may mean we get them back across. It could make the difference. Sir."

Lacey looked at Carr and thought. Six of his own were in danger, but he vividly remembered how men were cut to pieces by the Cuirassiers as they ran from Bembribre. He fixed Carr with a fierce gaze.

"One Section only and 50 yards only in front of the 95th. Stay close before the bridge, definitely not over it. Is that very clear?"

Carr nodded gladly.

"Yes Sir."

He saluted and was gone. He did not even pause when he returned.

"Number One Section, with me!"

Drake's men immediately followed Carr down the hill.

Ellis, with Davey at his side at a fast trot, was heading straight for the nearest gap in the houses, a narrow alley. Through it, on what must be the main road, could be seen a continual stream of Redcoats, all hurrying through to reach the British line beyond the river. The nine entered the alley, dodged around a scattering of broken furniture and came to the main road. There Ellis was surprised, he expected to see mainly able bodied men; deserters, thieves and marauders, but what surprised him were the number of wounded and injured, and what surprised him more was that the great majority still carried their muskets and had their equipment slung about them. At that moment intense firing broke out at the French end of the village, distant but continuous. He looked to his right and could just see the green uniforms of the Rifles up on the slope and over the river. There were also some Redcoats running down. He looked again at what was before him, which re-affirmed the proportion of sick and wounded, many on crutches, but what he now realised was that there was a total absence of any badges of rank. He looked left to see smoke in the distance and the sounds of musketry intensified. He turned to John Davey.

"Seems there's some good lads down there, putting up a fight."

Davey simply looked back at him, expressionless. Ellis had decided.

"Come on!"

He turned left towards the fighting. Miles looked aghast at what was happening, Pike, Saunders, and Byford were merely bemused, but they followed. Ellis looked behind to see only two Riflemen and he soon realised who was missing.

"Where's Newcombe?"

It was Spivey who shouted back.

"Turned right!"

"Shite-auk!"

Running in the opposite direction to the flow of soldiery, they soon cleared the village to run out onto the bare fields on the long hill slope beyond and, when over the shoulder of the first rise, they saw what was happening. Attacked by French cavalry, the fugitives had gathered into a dense crowd and were holding off the French, but they were making no movement back. They were a good 300 yards beyond Ellis and his few companions. He looked over the short distance to the only road that ran down the slope, where more Redcoats were streaming past, eager to get to the village, which was now at least 400 yards back.

"Follow me!"

He ran to the road, stood in the middle of the flow and bellowed his orders, but first at his small command.

"Middle rank of a square, across the road. I want your rifles ready to give fire."

His pitiful command of seven soldiers looked at him in puzzlement, but Ellis seemed to know what he was doing and so they lined up across the width of the road, rifles at the make ready. Meanwhile Ellis was yelling at all and sundry, those approaching, passing, and even gone.

"Form square, on them!"

He pointed at his small line, almost a caricature of his beligerent self.

"On them! Front rank fix bayonets! Move! Move! You dozy lot of brain dead bastards! Move!"

The sudden appearance of this fiercesome Sergeant, yelling orders, shocked many out of their state of single minded self-preservation and then the drilled in blind obedience, reinforced by fear of the lash, took over. All who could, began to build the square. It included several wounded, but within minutes it was substantial enough to withstand cavalry.

Ellis ran up and down the front of the square; all was not as it should be.

"Call yourselves soldiers. All you kneeling in the front rank, fix bayonets! Back two make ready. Christ! You're in a defensive square, not a choir for a singsong! Sort yourselves out! Face up! Receive cavalry!"

The stunned soldiers fixed bayonets or loaded their weapons, but Ellis, stood out beyond the front rank, was now looking at the

313

beleaguered group further on. He turned to look at his now solid square, with the front rank kneeling, bayonets extended.

"Stand up!"

The front rank did.

"Advance!"

The side nearest to him shambled forward, then the other three sides, suddenly finding themselves detached, followed in the same direction. Ellis knew that he had to take his men forward to draw the cavalry away from the surrounded Redcoats further out and soon it had the desired effect. Ellis' square, advancing to offer battle, was soon noticed, the first sign being what seemed to be an Officer raising his sabre and detaching men from the attack. They rode down upon Ellis, in substantial numbers, but in no set formation. Ellis resumed command.

"Halt!"

The whole shuffled to a standstill.

"Front rank kneel."

The obeying of the order ran around the square to meet in the centre of the rear side, but they were soon ready. Ellis entered the square to take his place in the centre of the middle rank of three. He had to rely on their training, there was no time to issue instructions.

"Middle rank. Present!"

Around him, at least, men were acting as they should, the front rank kneeling, their bayonets forming a protective hedge, middle rank at the "present", third rank at the 'make ready".

"Fire!"

The front of the square exploded with fire, followed by the weapons along the sides, as the cavalry parted and lapped around down both flanks. There were many French dead and riderless horses, but there yet remained scores of cavalry all around, still so many that the other "square" could not be seen. Ellis continued.

"Rear rank. Present."

The rear rank's muskets came down around him.

"Fire!"

Again the explosion and again, when the smoke cleared, they could see the effect. The disorganised French attack had suffered significantly.

"Independent fire!"

All around, his men reloaded and fired at the cavalry milling around before them, but Ellis could only hope that someone in the other

314

square would see what was happening and begin to pull back, now that they were in less peril from so many surrounding cavalry. After about two minutes Ellis was relieved to see exactly that happen, for, like some cumbersome reptile, the mass of Redcoats began to shuffle back, still holding together and still offering resistance to the cavalry all around them. It took almost ten minutes before they reached Ellis and then continued on, to finally reach a point back beyond him. 'There must be at least one good lad in there', he thought, because, when they were 50 yards beyond, they halted. They were going to cover his withdrawal, which he now began.

"Pull back. Slow! Keep together!"

His square began to move, the front rank now pushing at the others, but the noise of the musketry was incessant. Drop outs and marauders they may be, but they were all veteran infantry and could manage at least three reloads a minute, so any French Hussar that came within five yards of either square was blasted out of his saddle. Ellis took his square to 50 yards down the slope, beyond the other, then halted to cover their retreat. Then they pulled back to halt once more, as did Ellis, by which time they were at the buildings of Betanzos where the cavalry could not operate, but their attacks had by now lost all of their ferocity and much of their confidence. Expecting easy prey at first, the French horsemen had been demoralised by the determination of the resistance, which had caused many casualties and, what was worse, it had been dealt out by what had been, at first, a disorganised mob. As the squares neared the buildings, the Hussars held off, hoping for an opportunity at broken infantry, but that never came, for Ellis was not done. The street up through Betanzos was wide, wide enough for a cavalry charge, and so as the squares broke up he ran around seizing some men and pointing at others, allocating them to a pair of thick skirmishing lines. As the first one formed up, he found John Davey.

"John! Take charge here. I'll make another 50 yards back."

As Davey ran to command the line that was already forming across the road, rubbing clean his Chosen Man stripe as best he could as he went, Ellis ran back, yelling at everyone. Many from the two squares, now feeling themselves under orders, had not run back through the town but remained to take part in what came next, which was the forming of the two lines. Davey looked at the French on his front, who, thankfully, now seemed thoroughly intimidated and were holding off, then he looked back at Ellis' line further into the village. It was formed.

"Fall back! Reform beyond."

The line dissolved and they all ran back, Davey looking over his shoulder at the French to see, worryingly, that they were now trotting forward in the hope of targets. He ran on through Ellis' line to become part of his own that was rapidly forming 100 yards behind. It was not long before Ellis and his men did the same and so they progressed through the town. Not one shot more was fired. The French had had enough; the British, somehow, as the Hussars saw it, had formed themselves into too tough a nut to crack.

Beyond the town both lines, now near the bridge, with the Rifles on the far side, and now feeling themselves safe, dissolved to cross, then to climb the slope and join the ranks of their army, now growing on the summit above. Amongst the last remaining in the village, Ellis' five 105th found each other again, but Spivey and Verrity had disappeared, having run on to cross the bridge and join their own comrades of the 95th. The six Lights approached the bridge, but it wasn't long before they heard their names being called, Zeke Saunders first, him being the first recognised because of his height. Ellis stopped and looked over to the right to see arms waving on the far bank. Against orders, but feeling himself safe, Carr had led his men right down to the river, so, as they set foot on the bridge, Ellis turned to Davey, at his side.

"I think that's some of our lads over there."

Davey peered through the crowd and nodded.

"I think you're right, and that's as good a way to go as any!"

His answer was for Ellis to immediately turn right and walk off towards the shouts and waving and it was not long before faces were recognised, most importantly Ellis recognising Carr. Ellis nudged Davey.

"I'd better go first to the Captain!"

Davey nodded as Ellis marched off to present himself before Carr, rifle grounded and held by his left hand, whilst saluting with his right.

"Myself and five men, reporting back. Sir."

Carr was grinning like Ellis had never seen him before.

"Welcome back, Sergeant. You've been missed. What have you been up to?"

"Beggin' your pardon, Sir, but it's quite a story."

Some yards away, Byford and Miles were in the middle of a knot of 105th Light Company climbing back up the hill, Miles using the advantage afforded him by their return for all it was worth.

"Now we b'ain't sayin' not one single word till we've had some grub. Like now, straight away, some biscuit and water from you lot; we'n bleedin' starvin'. Nothin' till we've got somethin' inside. An' a drop of summat won't come amiss, neither!"

It was only when they slumped down amongst their comrades that the five realised just how close to exhaustion they were, so that after merely a few mouthfuls of stew, which they could barely chew, they rolled themselves in their blankets and slept, al bar Joe Pike. He lost no time in running to the mess line and enquiring the whereabouts of the followers and many, recognising him, were eager to direct him to where they knew he was desperate to go. It was the imposing figure of Nelly Nicholls that he recognised first and he increased his speed to get to her, knowing that Mary would be near. Nelly soon noticed his hurrying figure.

"Jesus, Joseph, and Mary, Mother of God!"

She turned to find Bridie, working behind her.

"Bridie. It's Joe! Come back! He's not dead!"

Bridie saw the look on Joe's face and recognised immediately his extreme anxiety. She ran forward to stop him and Nelly followed. Between them they managed to halt his progress, but it was Bridie who began talking loudly and urgently, holding firmly to his arm.

"She's not dead, Joe, she's fine, just fine, but Joe, there's something you've got to know."

She seized his crossbelts and shook and pulled him to make him look at her.

"Joe! She's lost the baby, Joe!"

Whilst shock and dismay entered the face and eyes of Joe Pike, tears welled up in hers, the words coming in a torrent.

"'Twas the cold and the hunger, Joe. She was suffrin', so far down she went that we thought we'd lose her, but we didn't. She couldn't hold the baby, Joe, her body gave up, to keep her alive, her body let go. 'Twas no surprise, Joe, and for the best. It was lose the bairn or lose both. Joe!"

Joe Pike was recovering from the shock as he listened, but he became so sunk in depression that both women thought that he was about to collapse and Nelly saw the need to put her arm under his.

"Joe!"

There was no reaction.

"Joe!"

This time he did manage to pull his face around to hers.

"Joe, but she's fine! Sad, like you, but fine, as are we all. It's not the end of the world, you're young and you'll make more, sure you will. I've got three left, from six, six! Two I lost and one died of fever. It's how it goes, Joe, for the likes of our sort, but the sadness don't last forever. It fades as life goes on, especially with more children."

Bridie now seized his crossbelts again to gain his attention.

"She needs you strong Joe, not some mopin' spineless sop! You're the man, you're to carry it through, for you both!"

Joe Pike seemed to straighten slightly, then he took a deep breath, but said nothing. Nelly still had his arm.

"You're back, safe'n well, and she's up and about, an' copin' like the rest, now the provender's come up. You're both all right and together now, praise be to the Good Lord above."

Bride tugged again at his crossbelts.

"Sure now, isn't that the right of it?"

Joe was expressionless, giving no reaction, but at least he did not shake his head. Bridie was encouraged.

"Come on, let's go find her."

He allowed himself to be led forward, past the cooking fire and the children feeding the mule, to then see Mary sat on a log, mending her coat, sat with Eirin. She glanced up at the arrival of the red coat then screamed as half recognition made her look again to see the full figure of Joe. She was up in an instant to run to him and he dropped his rifle to run forward to her. She sprang into his arms and both fell to their knees, each to hold the face of the other, their tears mingling as their faces pressed together. Both Nelly and Bridie dissolved into weeping themselves and searched in vain to find something to wipe away their own tears, so they gratefully returned to their cooking pots. Finally Joe was able to speak, but the tears still gushed down his face as he searched for the best words to say.

"It's all right. I know, they've told me. It wasn't your fault, none of it. It's all right and I'm back. 'Twas nature, and none of your doing."

She clung to him with all the strength her arms could muster and they clung to each other for almost an hour or more, the arm of one, at least, always around the shoulders of the other and, for Joe, sleep did not come for a long while.

***

The day wore on, in anti-climax after the drama of the morning. The Reserve Battalion went about its business whilst the last Regiments of the main army moved off, but not before Paget had passed his judgment on the 'dregs and the dross', as he viewed them. None who had emerged from Betanzos were to be allowed through the Reserve Division lines until they had been searched, but Carr's presence meant that his six were not included in the round up. The 91st and 20th were allocated to the task and soon there came a large pile of pillaged and stolen objects; brass candlesticks, bent double, bundles of knives of all types, copper saucepans and every type of kitchen utensil, almost all of negligible value. Deakin and Halfway wandered over to watch and they soon noticed the growing pile and went over to examine it. They stood for a minute, then Deakin spoke, his voice weary and resigned.

"Does this pile of stealings here, look familiar, Toby, at all?"

Halfway nodded, the picture of a similar mound in a Lisbon road coming quickly to mind, then both turned and walked away.

The 'dregs and dross' now much lighter in their loads, were allowed onto the main road to march on in the hope that they would then attempt to rejoin their Regiments. As the last received his cuffing from a NCO of some level of rank, Moore had his army drawn up, not to offer battle, more to intimidate, for all that could be seen of the French were the cavalry, in force, these being those that had failed against Ellis' "square" but were now well occupied with rounding up yet further British stragglers, crossing the hills beyond the village.

The night was spent eating because another supply column intended for La Romana had arrived some hours before. All of both soldiers and followers were now supplied with food and the soldiers with ammunition, so that, come the dawn, all of Moore's army, including those which Paget had called the "dregs and dross", and the followers of all Regiments, had moved off with their own haversacks bulging full with flour and salt pork, even some dried fruit. The remaining Reserve Division were left with more stores than they could carry away, also three thousand muskets and five guns. The muskets were not needed and the guns had no horses to pull them, so these were

319

all spiked and Carr's Lights, still watching over Betanzos, were entertained by them being cast down the slope to roll into the river, then to watch parties of the 20th inflicting the same fate on armfuls of muskets which they had to carry down to consign to the water. However, they were unmolested, for Betanzos was deserted of troops of all colours. The French cavalry had pulled back, French infantry had not arrived and the British were more than content to remain on their side of the river.

The only activity on their side of the town, and that was on the outskirts, where the Engineers were preparing to blow the bridge. D'Villiers and Carravoy, sat on an empty musket crate, watching from afar the barrels of gunpowder being manhandled under the bridge. D'Villiers was feeling more cheerful than of late, as were they all, now being well fed, but Carravoy remained morose and surly. However, D'Villiers felt inclined to be positive.

"A pipe of claret to a bottle of Spanish brandy, that that bridge remains standing."

Showing a stony face, Carravoy turned to him.

"Which are you betting?"

D'Villiers retained his good spirits; he saw no reason for gloom.

"I'll bet the claret if you bet the brandy!"

Carravoy exhaled noisily through his nose, his ill temper remaining paramount. He did not answer. Thirty minutes later they heard the explosion and walked back to the ridge top to see.

"That's a pipe of good claret you've lost. You should have taken the bet."

The partially destroyed bridge had to be guarded for the rest of the day and the task was given to the 105th, who were building a reputation as the 'specialist part destroyed bridge watchmen'. Thus they sat the ridge above the river whilst the rest of the Reserve Division marched off. In the growing gloom Lacey's men observed the growing number of French campfires in the hills beyond Betanzos, which meant that French infantry were arriving. Carravoy was still in a foul mood over nothing that he could specify, other than his detestation of General Moore, and so he wandered the lines alone, picking his way between the sentries and the various mess fires. He came across Carr and Drake standing at their most favoured place overlooking the river and he felt much inclined to deliver a barbed remark, whatever the consequences, as he passed by.

"I hear some of your deserters have returned to you, Carr."

Carr swallowed the mouthful of dried apple that he was enjoying.

"Deserters! That's a bit strong, Charles. They'd been cut off, but they fought their way back and came in with all equipment. Hardly desertion!"

However, Carravoy had not paused in his ill-tempered perambulation to argue the point. Having delivered the comment, he wandered on, face grim, hands clasped behind his back, walking further into the growing dark. Drake, now with a mouth empty of dried something or other, watched him go.

"What's wrong with him?"

Carr, just prior to pouring some raisins into his own mouth, gave a short reply.

"When is there never?"

The 105th were assembled long before dawn and marched off, leaving a pile of supplies for the oncoming French, it had been impossible to destroy it all, but only after they had piled as much of it as they could upon themselves. Some had been thrown into the river, but eventually the barrels and sacks began to pile up from the stream bed and threatened to give the French a way across, so the effort was abandoned and a mound now remained at the top of the ridge. Drake, ever the humourist, would not let the opportunity pass by and had chalked a message on the side of a barrel.

"Les compliments de la boulangerie de petit gâteau de Pied 105. Vous voir dans Corunna."

Carr looked at it as they prepared to march off.

"And that says what?"

"Compliments of the 105th Foot biscuit bakery. See you in Corunna."

Carr nodded and his mouth twisted into a wry grin.

"Tres droll, I'm sure, although I doubt M'sieu will appreciate Navy biscuit, full of weevils, but it'll spare some local family being robbed of their winter food, assuming if it keeps them here, and not out foraging."

Well fed, dry and enjoying better weather, the 105th kept up a good pace, however, they were halted to form the rearguard for Paget's Division, which he had drawn up on the far side of the river Mero. They were not to cross but to maintain a watch backwards from a good watching position, but the sight of the French campfires in the absolute furthest distance set Lacey's mind at rest that they were on the wrong side of the river. Then more supplies came up from Corunna as the 105th made their camp and, as they gorged on salt pork, practically a joint between two for their evening meal, the same as midday, Deakin and Halfway, sitting with Stiles and Peters, passed the time in contented conversation. Halfway made the most relevant comment.

"Much more of this and we'll be puttin' on weight!"

Deakin sniggered as he swallowed and took a drink of wine.

"An' a bottle of this each, an' we'll be rollin'. Things has come around, that's for sure."

However, what came out of the gloom was their Captain, who waved them back onto their log as they rose in his presence.

"Eat, that thou mayest have strength, when thou goest on thy way. One Samuel, 28, verse 22."

As usual it was Deakin who replied.

"Yes Sir. I'd say that about sums it up, Sir."

Heaviside nodded lugubriously, as though considering the merits of such a response to so well known a Biblical text. Then he lifted his head, suddenly cheerful.

"How are our two Ensigns?"

Deakin put on a cheerful look himself.

"Fine, Sir. Seem to be holding up well, Sir, now we'n through the worst and all better fed."

Heaviside nodded, slowly and thoughtfully.

"Pleased to hear it. Unto whomsoever much is given, of him shall be much required. Luke 12, verse 48."

Deakin looked at Halfway, both wholly clueless, before giving his stock reply.

"Yes, Sir. I'm sure they both sees it that way, Sir."

Heaviside passed on, to exchange words and impart quotes to the rest of his company and consume, with thanks, the food and tea that he was often offered. However, the harmony of the night was somewhat marred when O'Hare came around and entered the mess lines of the Colour Company. Heaviside was now spreading the same warning elsewhere, as it was his men that were to stand sentry for this night. Seeing O'Hare, the four did spring to attention, but O'Hare soon put them at ease.

"Ah now, boys, are youse getting some of this good eating inside of you, now?"

Peters thought of a reply first.

"Yes Sir, and very tasty, b'sides very welcome!"

O'Hare nodded.

"Ah good, that's good."

He paused.

"Now, we're hearing that there's some desperate characters not too far and hereabouts, them as are the worst and'll not rejoin. All have been supplied now, including the fall outs and whatnot, so there's food around aplenty, whilst they have none. So double the guard. There's some out there as'll think nothing about sneaking in, slitting a throat and making off with a full haversack."

Deakin was instantly worried.

"The followers, Sir?"

O'Hare waved his hand in dismissal.

"In Corunna by now, but a day's march away, and well looked after. It's us you should pay a mind to."

All four saluted as he walked off, then Deakin made his arrangements.

"Alf! You'n' I first."

He picked up his musket and subconsciously fixed the bayonet.

"Then you two."

He led Stiles out onto the sentry line, found the sentry on his left and kept him well in sight as he extended right, whilst Stiles did the same. For an hour he stood in silence, changing the position of his musket approximately every five minutes, mostly for something to do. He took off his shako, wiped his face with the cloth knotted around his head, then replaced all, then he chased a piece of pork lodged between his teeth. Finally he took a drink of water mixed with brandy and some honey, this luxury found in La Romana's hoard. This was augmented when Parson arrived, having undertaken to tour of the sentries with a bucket of spirits, which, when Deakin drank his portion, he concluded tasted like nothing on earth, but it did warm his insides. Parson had probably mixed up Spanish brandy with Navy rum, Deakin told himself, wanting to give good measure rather than one or the other watered down. He felt filthy and knew that he was, the tails of his greatcoat, for example, were stiff with mud, but that he took to his advantage, they would not blow up in the wind! Thus, all was at peace in the mind of the goodman Deakin. The fact that he was on campaign, barely five miles from an enemy that would unhesitatingly try kill him, given the chance, did not enter his mind. He felt sure that the worst was over and they had come through. The tragedy that had befallen Mary came to mind, but to such as him, from his place amongst the common herd that the 'other ranks' were drawn from, such a heartbreak was far from uncommon across all layers of society, but particularly his. He found a biscuit and chewed contentedly, thinking of Bridie and the children and was grateful that they were safe. He made the biscuit last until Halfway arrived, then he returned to their fire, covered himself in his blanket and fell fast asleep.

Nonesuch contentment dwelt within the mind of Seth Tiley, sat in a cold, ruined house that had been picked clean long before they arrived. He was hungry as were his remaining men. He walked to the door and saw campfires, which he believed to be French, then he stared in the direction of some others, which were in the far, far, distance over to his left, which he believed to be British. He was in a quandary as to what to do. The black space before him would soon be filled by the French, or so he believed. But what to do now? Two of his men had spoken of rejoining and, to end such talk, he had clubbed both almost senseless, but this violence, his answer for every problem, had merely motivated them to make off into the darkness, accompanied by three others. He was left with but four, all as hungry and desperate as he, but there now being so few comprising his band, limited their capacity to rob and pillage, and he did not trust them to not make off at the first opportunity. He remembered Lacey's threat to hang him at his first offence. Well, he had assaulted his guard in order to escape, that gave a hanging sentence, and, on top, there was his attempt on the followers that had been stopped by that hag Nicholls. What to do? He leaned against the doorway. What to do?

***

Chapter Eight

Corunna.

The sea was not unfamiliar, unsurprisingly, but usually to them all it was a source of some anxiety, because to be carried upon it always meant a period of storms, seasickness, and ever-present danger. However, as they woke in their camp in the growing light of the next dawn, with the spires of Corunna becoming more distinct and the ocean beyond and beside, there was not one man in the Reserve Division who did not look again with fondness upon the cold grey expanse that stretched over the horizon, as had the main army a day's march before them, for beyond was England, home and safety. In addition the weather, although January, was Iberian January, and they were now off the cold, high plains. Were they in England they would be remarking to each other on this very early Spring weather.

To rejoin the Reserve Division, the 105th at last reached the Mero and descended into the valley, hiding the sight of Corunna, which most men had sent a grateful and relieved glance towards over each of the past hours. Experience told them that a valley meant a river and a river meant a bridge and a bridge begged the question regarding whether or not it would be destroyed. Down in the valley was a small hamlet, built around this crossing point of the river, it being plainly on an estuary, for, as the 105th marched across, they could see the mid-brown, scum laden waters surging inland with the building tide. There were no civilians, the word of one army being followed by another carried its own warning of probable conflict, in fact almost a certainty at such a strategic point as was their home. The bridge was about to be destroyed and Paget, sat on his horse at the far side of the village, waited until they were across and then sent the 105th far up onto the valley's backslope, into reserve. Paget now positioned his men, the 95th he deployed far upstream, to a point furthest from the bridge, thus extending his right beyond the 52nd, who were positioned immediately upstream from the bridge. The village itself, now known as El Burgo, was to be held by the 91st, whilst the 20th were downstream, watching the tide come in.

Lacey and O'Hare, knowing they were in reserve, allowed their men to rest, make their messfires and eat. Both sat themselves down on a rock outcrop, waiting for their servants to bring them some tea and whatever could be prepared in quick time. Both watched Paget awaiting the blowing of the bridge, both knowing the question which would be very much uppermost in his mind. The answer came within five minutes with a terrific explosion that sent lumps of masonry hurtling into the air, several that came from houses alongside the bridge itself, for, when the smoke and dust cleared, there was nothing left of the bridge nor its neighbouring buildings. Instead, there was a levelled space of 20 yards radius, centred on where the bridge once was. With the noise, billowing smoke and dust, and lumps of rock descending disconcertingly close, Paget and his Staff were having difficulty controlling their horses, but, with these calmed, Paget and his Aides walked off to their own place of rest and repast, but not before an icy look from Paget had banished the self congratulary grins and smiles of the Engineers, all now emerging from the extremes of the village.

Sat behind a stonewall which gave shelter from the sea breeze, the messes that habitually cleaved together were all sat watching their fires and awaiting the boiling of their food in their communal stew pot. All save one member, which caused Davey to remark to Miles.

"Something's wrong with Joe."

Miles looked at the solitary figure, sat alone someway down the slope. He had not divested himself of any of his kit and his Baker was upright and sloped back into the angle of his shoulder and neck. He looked as though he had been ordered to stand ready, ready to be sent into action at a moment's notice, sitting staring at the empty road over the valley that they themselves had come down not an hour before and would also be the route for the oncoming French. He seemed to be intensely anticipating their arrival. Miles turned his head to look at Davey, but neither spoke; both knew well enough the cause of Joe Pike's state of mind. Eventually the cooking pot boiled and soon the ingredients contained therein was pronounced ready by Tom Miles, after he had stirred in four biscuits. Miles spooned out Davey's portion, then concerned himself with Joe Pike.

"Joe! Food's here! I needs yer pannikin."

Pike reached behind for the flat metal dish that was lodged between the straps of his pack and threw it back to Miles, to land spinning at his feet. Miles looked again at Davey, each sharing a worried expression with the other, but Miles spooned out Joe's portion, spread extra salt on top and then placed a piece of bread on all, bread but one day old. He thought of calling Joe back to fetch it, but human

326

feeling for once got the better of him and he rose to carry it the few yards down to Pike. It was not possible for any kind of look from Tom Miles to lighten the mood of anyone, but he did wait until Joe looked up at him before releasing the dish. There was a look in Pike's eyes that he had not seen before, one of both rage and despair, but, as Pike finally looked down at the well-stocked dish, Miles spoke, as cheerfully as he could.

"'Tis fish! Sent up from Corunna, with peas and potatoes, but watch out for the bones."

Pike nodded and, as he turned towards his fire, Miles clapped his grimy hand onto Pike's shoulder and gave him one single shake. This did elicit a response.

"Thanks Tom."

"You're welcome, boy."

However, Miles did not return immediately to the fire, but halted to study Pike's hunched figure.

"Get that inside you, Joe, then get some rest. 'Tis us as'll be down at the river come tomorrow, or even tonight, with Frenchers on t'other side, I shouldn't wonder. You'll not be much cop sightin' that Baker with no sleep what's cleared yer 'ead."

Pike turned his attentions from his food to look back at Miles, but said nothing. It was Miles who continued.

"'Sright! Get yer 'ead down for a few hours."

Pike nodded once, then returned to the eating of his stew. Another explosion, further upstream, made them all look up. Another bridge had been destroyed, but that dwelt but fleetingly in their thoughts, their food was too important.

In the event their sleep was undisturbed, save for some being given the duty of standing sentry on the churning waters of the incoming tide, or the rushing melt water from the hills around at the time of the ebb. It was full dawn when the French did arrive, cavalry first, halting immediately on the valley skyline at the obvious sight of the Reserve Division drawn up on the opposing slopes on the far side of the river. A Commander of some form arrived soon after, marked by the amazing uniforms of both himself and his entourage. Then the cavalry were sent off up the valley whilst the Commander remained, at first a lone-silhouetted figure, but that for mere minutes before infantry arrived and were immediately dispersed into skirmish order to descend the valley. With his battalion still in reserve, Lacey and O'Hare watched all from their vantage point and, as the French infantry disappeared into the trees on the far side, Lacey stated the obvious.

"Here we go!"

He had barely time to draw breath before the firing began, first within the buildings of El Burgo, held by the 91st, then the conflict spread upstream to the 52nd. The crackling sound of the incessant musketry continued throughout the morning, to little effect from both sides. This was evidenced by the few killed that were dragged back onto the grass behind and the few wounded that made their way back to their surgeons; some supported, some making their own way, but it seemed to be the 52nd that was suffering most, but hardly in any real discomfort. The exchange continued all throughout the morning and at Noon, Paget rode along the rear of his men, these all well positioned within the trees or behind walls, or within the buildings of El Burgo. The 20th below and closest to the sea, remained unmolested. Inspection done, Paget spurred his horse at the slope up to the 105th. Lacey, O'Hare and Simmonds rose to meet him.

"Colonel! Your Light Company are armed with Bakers, are they not?"

"Yes Sir. Correct."

"Right. The 95th are giving better than they are receiving, whilst, for the 52nd, it's just a damn waste of ammunition, they're too far away for a musket to be of much use but I'd say that the French were getting the best of it. That's where M'sieu feels he may get over; in fact, if I've got it right, they're gathering up bundles of any kind of fabric and furniture, to throw in at low tide and get over."

At this O'Hare focused his glass on three farms on the valley slope to see mattresses, sheets, blankets and furniture of all kinds being dragged out. The same could be seen at the French held buildings of El Burgo. All the while, Paget continued.

"These are handy fellows, these French, a cut above, I'd say. The 52nd will handle any serious assault of that nature, I feel certain, but a little discouragement of any such notion as attempting a crossing will not come amiss. So! Get your Lights down there and see if you can't spread a little serious disaffection!"

With that he was off, cantering rapidly down the slope to check on the 20th, still sat idle. O'Hare set off to his left.

"I'll see to it."

Carr was sat on some tarpaulin, with Drake and Shakeshaft, idly watching the unchanging events below, but all three rose at the approach of O'Hare, who wasted no time.

"The General wants you down there."

He pointed to the 52nd.

"To support the 52nd against a possible assault. Your riflefire could provide the required discouragement. Take care of any that look too eager and keep them back."

O'Hare looked at each of the three faces.

"Take care! This is just pointless, annoying, bickering over a river that cannot be crossed. Tell your men to get into good cover and choose their targets carefully, always starting with Officers and NCOs. Just pick off a few to keep them worried. I don't want to leave any dead behind; none, at least as few as possible."

All three saluted and Drake and Shakeshaft ran in opposite directions to instruct their Sections, leaving O'Hare alone with Carr.

"Conserve your men, Henry. There's going to be a battle, a full set piece. Only a clear defeat of the French will get us home. We have to knock him back, right back, to buy time. As we speak, our ships haven't arrived!"

O'Hare saw the alarm in Carr's face.

"But, for the moment, keep that to yourself."

He paused.

"Now, down you go!"

Carr saluted and followed his men down the steep slope. Drake led his One Section, containing the file of Davey, Miles and Pike, down to the river. First they had to run through the growing lines of 52nd wounded, then the three entered a clump of trees and undergrowth and immediately they heard the buzz and hum of musket balls from the French opposite. Drake could not be seen and so Davey looked for someone in command and found a Sergeant, hunched over and looking worried, at the point of lowering his smoking musket.

"We've been sent down here, Sarn't, 'cos we've got Bakers. Where do you want us?"

The NCO did not withdraw his attention from the far bank for one instant, ceaselessly studying the movements of the French opposite, particularly looking for any that were making him their target. The reply was terse, as the Sergeant subconsciously reloaded.

"Just find a position and pick off what you can. These is tasty lads, seems they've got a more accurate musket than is common amongst most Frogs."

Davey nodded and returned to Miles and Pike, the former flattened to the rear of a thick tree, but Joe Pike had gone well forward, almost to the river edge, and was already firing his rifle, having jammed his sword bayonet into the wood to act as a rest to steady his aim. Davey didn't like what he saw, Pike was too forward and too exposed for his liking. He was the first he addressed.

"Joe! Keep in. Careful shots. All as is needed is just to pick a few off, just to keep 'em back. No risks, now. These has got some gun as is better than the average. So take care."

The reply was a brief look up and a nod, whilst he reloaded from the cover of his tree. Davey then looked at Miles, who was peeking out from behind the trunk of his own tree, studying the French opposite.

"'Tis them tassel swingin' bastards what we saw off at Maida, John. I've been wonderin' when they'd show up. But there's others there besides, them with all the red!"

Davey took a look for himself. There were indeed the yellow decorated shakoes and shoulder epaulettes that he remembered from Sicily, with the distinctive tassel beside the shako, but there was another type of French skirmisher present, distinguished by a red highlighted shako and a distinctive red collar.

"Now, just who are you, all sparked up in red?"

The question was spoken out loud and, immediately answered by Carr speaking from behind him.

"Tirailleurs. French sharpshooters, just like us. The others, our old friends, are Voltiguers; assault troops, they've just got a musket, but those in red have a hunting gun. Give them priority."

Carr left immediately to pass on the instruction elsewhere, leaving Davey to pass it on to Miles and Pike.

"Tom. Joe. Did you hear that? The ones with red, get them first, they've a better musket!"

Miles nodded, before sighting from behind his tree trunk, to send a musket ball across the river. He, as his training dictated and his natural cunning confirmed, was firing obliquely across the river, which meant that he could remain concealed behind the tree, out of the view of anyone opposite. Joe Pike, on the other hand, was choosing his

330

targets from anyone he saw, both opposite and further along the bank. To sight on a target directly opposite meant leaving the security of his tree, wholly contrary to their training and it was simply reckless. Davey's worries increased, added to by the fact that Joe was much further forward than them both and the buzz of French musket balls was too thick to make him risk going forward to join him. All that any French sharpshooter opposite had to do was wait for Pike to emerge. It was wild and careless, which Davey realised and he was now annoyed.

"Joe! Fer Chrissakes! Keep behind that tree, fire up or down but not straight over. Keep back in, or they'll bloody well get you!"

The only reaction from Joe was to look up as he reloaded, then look back down to tip some powder into the firing pan. Davey despaired. He looked over the river and there indeed, but 50 yards back from the bank was a Tirailleur, kneeling with his musket trained all too likely on the point where Joe would emerge, him now halfway through reloading. Davey shifted his loaded musket to his left shoulder, not the natural side, but he had no choice. He rested the end of the barrel on a very fortuitous branch, sighted using his left eye and fired. All was immediately obscured by smoke, but not enough to prevent Davey from seeing Pike bring up his own rifle, then take the time to look for a target downstream, but plainly he had failed to see one and would not wait for one to arrive, so he emerged further out, to sight and then fire at an enemy directly opposite. Davey gave vent to a huge sigh of relief as Pike stepped back into cover to reload, then he looked directly across himself. Of the Tirailleur there was nothing to be seen, but there were many other French, of both varieties. Davey looked across at Miles, now reloaded and patiently waiting for a target upstream.

"Tom! Joe's got to come back. Out there he'll get himself killed, an' it don't seem to bother him, one way or t'other! Stay loaded, an' wait for me."

Davey reloaded within 20 seconds and, still left-handed, trained his rifle across the river. By now Pike was again ready, but Davey was angry and determined.

"Joe! Listen!"

No reaction.

"Pike! Listen! Damn well listen!"

The expletive did cause Pike to look back to Davey.

"You're comin' back! You hear? Back to us, and that's an order, and if you disobeys and still lives, I'll bloody well get you flogged! See if I don't!"

John Davey, although superior in rank to both Miles and Pike, had never before given either what could be called a direct order. This first time did give Pike some cause to stop. He looked back to see as angry a face on John Davey as he'd ever seen and from that face he received specific instructions.

"Find a target upstream. Upstream! After we fire, then you fire, then you runs back, here, to us."

He paused and was relieved to see Pike still looking at him.

"Clear?"

Pike nodded. He reached around to pull out his bayonet and dropped it into its scabbard, then he hefted his Rifle to his shoulder and sighted at a target that was, much to the relief of Davey, as he had ordered. Davey then switched his attention to Miles.

"Tom! Find a target."

The answer came quickly back.

"Got one."

Davey sighted his own weapon on the nearest threatening Tirailleur, directly opposite. He needed to be further exposed himself than he wished and a bullet buzzed past his head.

"Fire!"

Both fired, followed by Pike. He sprinted back the ten yards to Davey, but Davey was still maddened. As Pike ran back, Davey tripped him, then dragged him into cover. He immediately grabbed his cross belts and hauled him closer, to look directly into his face.

"Now you listen! I knows that you wants to kill Frenchers and I knows why, but, sure as God's in His Heaven, they'll kill you, if you goes about it like some no-knowin' half-baked newcome fresher!"

Davey stared harder into the astonished eyes.

"You'm lucky you'm still alive, and where would Mary be then? Well?"

No answer as a bullet hummed above them, but the outburst had calmed Davey. Somewhat.

"We wants to kill Frenchers, too. Any surprise?"

He released Pike's cross belts.

"Mary's of our own. She and your child matters to us 'n'all, y'know. We wants to pay 'em out, same as you, but still to stay alive. So's we can do a few more."

He let the words sink in, but the faint humour lightened both their faces. Davey's tone dropped to one of giving advice, rather than admonishment.

"But you carry on like that, you'll get killed and perhaps one of us on top, tryin' to get you out of it!"

At that point Ellis arrived, none too pleased with what he saw; two good men lying behind a log, neither sighting across the river, nor reloading, for both their muskets were on the ground.

"And what the bloody Hell's this? You two on for a handy kip?"

Davey held up his hand, which halted Ellis' forthcoming tirade; just as well, for he was winding up for plenty more.

"Sorry Sarn't. Joe here found himself a bit far forward and we had to cover him coming back. We're alright now!"

Ellis was only marginally mollified.

"Well then, I suppose we can now say that all is just fine and dandy for you both."

He raised his voice to a shout.

"So now perhaps you'd like to join in what the rest of us is all about doin'! Start by pickin' up them bundooks!"

With that Ellis clutched his own rifle across his chest and ran on. Davey looked at Pike.

"There! That's the other kind of trouble your lame brained ways can get us into!"

Somewhat chastened, Pike recovered both weapons. He gave Davey his, then began the process of reloading, whilst Davey looked carefully at him for a full half minute. Pike's eyes were narrowed and focused on the task in hand, but his mouth was set in a grim line, with several more furrowing his brow, and it was plain to Davey that here was no longer the trusting, unknowing boy of the past month. Pike completed his reload and pulled back the cocking hammer with relish, then rose to move forward, but Davey placed a hand on his forearm.

"Wait for me. We'll ease forward together, and fire from prone if needed. No one's getting' killed here, not for bugger all use. Just to answer their fire is all as is needed."

Davey quickly reloaded and they both crawled forward. Davey allowed Pike to take a tree on his left, whilst he found one that enabled him to fire obliquely downstream, across Pike's line of fire. In this way he kept a close eye on Pike's actions and was relieved to see him obeying his training and not leave the cover of the thick tree trunk. Feeling more confident that Pike's moment of madness was over, he took time to examine what was happening opposite. There were now many blue uniforms prone on the ground, with some writhing in agony and some dragging themselves back into cover. Their Bakers had made a difference and the French skirmishers were now much more wary. There were far fewer moving forward to close the range across the river, in fact most were unseen, hiding behind cover. Davey concluded that if he, himself, saw no point in taking risks if there was to be no major assault, why shouldn't the French opposite think similar?

Night fell, but there was no abandonment of their side of the river by the British. Most withdrew to their mess lines, leaving only a thick picket line, composed of the remaining Companies of the 105th and the 20th, who had not been in action at all. The moon was full and appeared regularly from behind broken cloud and, whilst Deakin and Halfway, now in the picket line, kept watch from behind a thick fallen branch, Stiles, Peters, Nicholls and Sergeant Hill, snoozed in the dip behind. Being friends for almost two decades, Deakin and Halfway had no need of conversation between themselves; besides, in the dark, the French could be mere yards away, therefore both kept silent watch on the bank opposite. The stream, now much lower, was no longer soundless as it ran over and amongst the stones and tree roots of its riverbed, instead it now tumbled and gurgled on its way to the sea but a half-mile distant.

Suddenly, in the moonlight, Halfway brought his musket up to the ready, resting on the log.

"Jed! Look yonder."

Deakin looked at Halfway, who was pointing with one finger across the rough bark. He looked and saw a figure, plainly a French soldier from his white breeches and white cross belts, approaching the far bank.

"Wait!"

Both studied and awaited developments, the first being to see a clutch of canteens, the second being the soldier lowering himself down the bank to submerge them all into the water. Halfway was affronted.

"Cheeky bugger. Come down 'ere, under our guns like that! 'E's takin' one bloody great chance! Who's he think we are, a bunch of wet recruits?"

Deakin settled back behind the log.

"Ah, leave'n' he! He just come down for some water. Can't be much up there on top, not with us holding the river. You shoot him and you'll start up a firefight and lads'll just get killed or maimed, and to what end? I just wants to get back to Corunna and get on out, an' I don't care how many live Frenchers I leaves behind. That's for some other day."

He settled his arms across his chest.

"Just watch'n. Make sure 'tis only water 'ee's come for, an' that 'ee's on his own."

A minute passed, then Deakin sensed Halfway rising up above the log, but it was not to use his musket, but to make the point he wanted to earlier.

"Now say thanks, you lucky French sod!"

Deakin looked at him in the half dark.

"What was that for?"

"Well, I wasn't goin' to let'n think that he could fill all they canteens without no let nor hindrance from such as us!"

Chuckling softly, Deakin rose to look over the log, but of the "watergatherer" there was no sign, but one canteen was floating down the river, perhaps dropped in surprise at hearing Halfway's admonishment. Both settled to watching, listening to the river and watching the patches of moonlight track across to give strange shapes of shadow to the trees and shrubs opposite. However, their next disturbance came from behind, in the shape of their Company Commander, Captain Heaviside.

"Anything to report?"

As the Senior, it was Deakin who answered.

"Nothing Sir, bar one lad of theirs what come down for some water. We let him get on, Sir, saw no point in startin' things off again. Sir."

The reply from under the silhouette of his shako was immediate and supportive, and a quote that Deakin well understood.

"Blessed are the merciful, for they shall receive mercy. Matthew 5, verse 7."

"We can all only hope so, Sir."

335

With the dawn, the 52nd, the 95th and their own Light Company came back down to relieve them, in anticipation of a renewal of hostilities, but not one shot was exchanged, although blue uniforms could be seen moving amongst the bushes and trees set far back. Those of the 105th who had kept watch through the night, retired for food and rest, whilst their own Light Company and the two front line battalions, maintained a watch. At mid morning the bugles blew "recall" and "form up"; their shrill notes that echoed around the valley being given added urgency by Paget's Staff riding hither and thither, pushing all into a hurry, until the whole Reserve Division was ready to march off, with the 105th in the lead. Carr's men had to run to arrive on time to join their column as they started off and as he passed him he inquired of Simmonds.

"Why the urgency, Sir?"

Simmonds halted his own hurrying.

"Frog cavalry have got over the river inland. Place called Celas, seven miles inland. So, it's quick march for us back to Corunna."

Carr turned to salute the Battalion's Second Major and, as he did so, he faced across the valley, where he could see hordes of French infantry pouring down the slope to get to the river and, more importantly for them, the site of the bridge, now merely two forlorn buttresses.

***

Relations were not good between Beatrice Prudoe and her Chaplain husband. As their situation had improved, in terms of both health and strength, he had looked with covetous eyes at the mule which was now being led on by Mary and Eirin. The patient animal bore on his back the two smallest children, these being Kevin and Sinead Mulcahy, and he also pulled the travois that carried rations and bedding. He, that being Pablo, having been kept alive and defended, was now a firm favourite with all the children and was also spoken to as though a family member by Bridie and Nelly.

Nevertheless, as the spires and cathedral towers of Corunna revealed their finer detail, Chaplain Prudoe had allowed his thoughts to dwell more and more on his status as a King's Officer, and a spiritual one besides. He considered it to be lower than his dignity to continue walking, therefore he must ride, albeit on a mule. True, it was of lower status than a horse, but superior to a donkey. Thus, on the evening of the last day before entering the city, he had gone to the tethered mule and boldly untied him and led him off, with no word at all to the women, merely a terse sentence to Eirin, who felt in no place to argue nor even say a word. Prudoe was a Commissioned Officer, but one down from God.

"I have a requirement of this animal."

However, minutes later Pablo was being led back by his wife, Beatrice. She returned the animal into the charge of Mary, with but one sentence from herself.

"I'm sure that you have a continued need for Pablo here, therefore I am returning him into your charge."

All expected a heated exchange to ensue from around the Prudoe campfire, but there was none. Instead, come the following morning, Pablo, continuously being patted and rubbed by both Mary and Eirin and continuously chewing biscuits, was being led sedately towards the town gate. Of the Chaplain there was nothing to be seen, whilst of his wife, there was much more, she was walking with Bridie and Nelly, talking about everything from sewing to the sea voyage home. Chaplain's Assistant Sedgwicke was with the children, teaching them their ABC and their numbers one to twenty. It was as pleasant and carefree a scene as it was possible to behold, despite their approaching the lowering walls and battlements of the vital port and city. However, in one instant it was shattered by a colossal explosion coming from somewhere over on their left, inland. The only creature that did not start in alarm was Pablo, whilst the devout Catholics in the followers feverishly genuflected, but there was no following silence, for this was filled with the sound of falling glass, the explosion had shattered every window in Corunna. Bridie looked beyond the walls to the now gaping windows.

"Jesus, Parson! What on earth could that be all about?"

Sedgwicke was now enough of a soldier to realize what a retreat and an evacuation meant; the wholesale destruction of stores and munitions. He pointed to the huge cloud of smoke, ascending and extending over on their left.

"Gunpowder, Mrs. Deakin. Being destroyed, all in one go it would seem."

What this destruction also meant, in addition, was confirmed when they entered the city gate, to be confronted with yet another threat to Pablo, because a Provost Corporal came forward to seize his bridle.

"All animals that cannot be taken aboard the ships is to be destroyed, and those that we are taking are only horses. So, this one's for the chop!"

He was immediately surrounded by howling children, but, more potently, also outraged women, which included the most formidable that he had ever encountered, these being Nelly Nicholls and Bridie Deakin. Both were fighting him for the bridle, whilst the mule himself, seemingly well aware of the fate described for him, was pulling back, but nothing matched the vehemence of Nelly Nicholls.

"Ye'll not take that mule, ye'll not, ye gobshite! He's took us through snow and ice for the past week nor more and that means he doesn't get kilt just because we've reached journey's end. You'll not take him, if it means we has pull out our own guns to keep youse murderin' Devil hands off, an' we all gets flogged for it! He stays with us. He's needed to take us to our camp, which'll be behind our men, that's the one hundred and fifth."

Surrounded, and so verbally assaulted, the Provost paused in his attempts to pull the mule away, then Eirin spoke up, being quite prepared to lie and also give a try to the effect of her very pretty face.

"He's not army! We found him wanderin', so we used him and he's given good service, so he has."

She fixed his eyes with hers, being large Viking blue.

"So, there y'are! He's not British, he's Spanish, and when we've finished with him, we'll let him go, or give him to some local farmer. To do him in would be a surely sinful cruelty."

The Provost gave in. Plainly these followers still had some use for the animal, which was his excuse, but perhaps a long study of Eirin relaxed his grip on the bridle and allowed Mary to pull the animal's head away.

"All right, if you've still a need, but keep him out of sight. Out of sight! Those are the orders, to destroy any horse not fit for the passage home."

Eirin had more to say and the Provost had no objection to studying her some more as he listened.

"Well, he is fit and he's not a horse. On top, he's ours, we found him. But, sure, we're grateful for your indulgence, 'tis a kind man that y'are."

With that she rose up and kissed him, which left him standing in shock as the mule was finally pulled away and his new state also spared him the force of an evil look from Nelly Nicholls. Finally, now recovered, the Provost had another thought.

"Missus!"

Nelly turned, fully ready for another confrontation, but it was to scc a much more affable look from the Provost.

"You'd best not stay inside the city if you want to keep your mule. "Sides, there's to be a battle, the ships aren't here yet. You'd do best to take yourselves over to that ridge there and get behind a place called Elvina. Your men will end up there, or close."

Nelly was calmed somewhat.

"Well, now, we thank you for that, your Honour. That's a kind thought."

However, Eirin was enjoying the moment.

"And if you pass by our mess fire, you're welcome for a cup of tea!"

Nelly was immediately appalled at such forward conduct and moved quickly in Eirin's direction, hand raised to cuff the back of her head.

"Get away, ye shameless hussy! Have ye no sense of what's proper at all? Just wait till your Uncle Jed hears!"

But Eirin had sprung sideways, to also duck away from the additional threat of her Mother, leaving the Provost chuckling as he moved away himself.

***

"What on earth was that?"

The colossal explosion had easily reached the ears of Lieutenant Royston D'Villiers marching besides his Company Captain, Lord Charles Carravoy, who looked disdainfully to the horizon and the pall of smoke, before bothering to find the breath to reply. He felt he knew only too well.

"Moore, I expect. Blowing up what may be useful, if we're to fight, as rumour would have it."

D'Villiers looked apprehensive.

"Fight a battle! But what shape are we in? Losses have been huge since Astorga, and the men look like half starved scarecrows, in fact that's exactly what they are. Does he really intend to fight a battle with such? Too huge a risk, surely?"

Carravoy now turned his disdainful look in the direction of his subordinate.

"Well, as usual, our noble Commander is at the mercy of events, rather than directing them. There are no ships, and that's more than rumour, that's fact. I bumped into Lucius Tavender yesterday. Remember him from Taunton? Well, he's now attached to the 18th and has a clear understanding that our embarkation will be mightily delayed. There are as yet, assuming there ever will be, no ships. So; a battle is inevitable."

D'Villiers looked shocked and studied the ground passing beneath his feet. He had no reply, but his Captain had further thoughts.

"Or!"

He paused.

"Or, he negotiates with the French. He strikes a deal which allows us to leave."

D'Villiers suddenly had a thought of his own.

"You mean a Cintra for us!"

Carravoy gave vent to a contemptuous release of breath.

"Cintra! You think the French would be foolish enough to give us what we gave them?"

He waved his hands for emphasis.

"We'll be stripped of our arms, swords, cannon and the rest of it, then allowed to troop home with our tails between our legs."

D'Villiers straightened up.

"Right, then we fight! We've never been beaten and we've given a very good account of ourselves throughout the retreat. We'll smack M'sieu on the chin to stand him off, then embark for home, ourselves and our arms intact."

Carravoy's reply dripped with sarcasm.

"You got that from Moore, did you?"

D'Villiers stood his ground.

"We deal or we fight! It's one or the other and I don't mind trying conclusions with this lot that've been on our tail for the past two weeks. From what I've seen our men are as good as theirs and better. The men want to fight! Now that must count for something. The men are all back in the column now, no stragglers anywhere, or at least none that I've seen. That must count for something as well."

Carravoy scowled.

"Let's see! Let's see what we look like when we actually come to form up, on a ridge or wherever he chooses. Then we'll make a judgment, when we get to take a look at Johny, opposite, and then what remains of us!"

D'Villiers was still in the argument.

"That could be a bit late. You talk as though you think we're bound to lose!"

"Well, what faith can one have? Today is Friday the 13th! Unlucky for which of us is yet in the hands of fate, but my money's on us being prisoners before the end of next week!"

D'Villiers actually smiled.

"I'll take you up on that. 100 guineas?"

He grinned to himself, self satisfied at the stand he had made in the face of the strongly opinionated Carravoy and there was no reply.

"So, how's your French?"

"Good enough to negotiate an exchange, and I'd advise you to brush up on the necessary phraseology yourself!"

They marched on in silence, both following Lacey and O'Hare, who were walking ahead on foot. It was well into the afternoon when both Senior Officers were stopped by an Aide de Camp, who engaged them in a short conversation, before pointing upwards to a hill and handing them a sealed letter. Marching up to stand just behind; Carravoy, D'Villiers and Ameshurst overheard all, mostly from Lacey, after he had broken the seal.

"We are leaving the Reserve Division, and are now back with Bentinck's Brigade. Still with the ….."

He raised the letter.

"4th and 42nd."

O'Hare asked the question which was on all their minds.

"And what sort of shape are they in?"

Lacey nodded. The question was one he posed himself.

"Well, we'll soon find out when we get to this place …"

He again studied the letter.

"Elvina."

He looked at O'Hare.

"Up there, did he say?"

"He did, Sir."

"Very well. Onward and upward and call up the Colour Company. If there's to be a battle, I do think that being led to our

341

position by our Colours would be the appropriate thing, especially as we are marching up to join such as the 4th and the 42nd. Don't you agree?"

O'Hare grinned hugely.

"I do, Sir. I agree very much."

Both looked at each other and grinned, as O'Hare continued.

"Yes, I think a bit of swank and show as we march up, won't come amiss. We are, after all, the fighting One-Oh-Five! It's right, is it not, that they should be reassured by the sight of us, more, I am of the opinion, than we are by the sight of them?"

Lacey caught the uplifting mood, adopting a theatrical, heroic pose.

"If The Guards can do it, so can we!"

He looked back at his three Grenadier Officers.

"You would agree?"

Wide grins came back from both D'Villiers and Ameshurst and a loud "Yes Sir!" but from Carravoy, a quizzical stare. However, Lacey was about the business described.

"Sar' Major!"

Gibney came running up, to come to the attention and salute.

"Sir."

"We're forming the men up. Lord Mayor's Parade! We're rejoining General Bentinck and we will rejoin with some style. Let them all know who's come back to the Brigade. No bunch of Detachments any longer, as if we ever were. See that things are up to the mark!"

Gibney fizzed a salute and hurried off, to immediately find fault with all and sundry in the leading Company, the Grenadiers. O'Hare marched down the other side, having the same effect, but with different methods, simply saying that they would soon be marching before their General and two other 'Royal' regiments. They must not let the Colonel down. Soon there was a drummer placed between every Company and all was now ready.

"Colours to the fore!"

Under Gibney's command and supervision, Number Three Company swung out of the column and, when they were at the head, all followed on, with Gibney screaming about arms and feet above the sound of each step. O'Hare stood back a little and watched his men march past. They did look appalling, with their uniforms more likely to fall off than to stay on, but there was nothing wrong with the swing of their arms nor the beat of their step. Nor the cleanliness of their weapons, he noticed! Satisfied, he allowed all to pass, including Carr's Light Company at the rear, then he joined on, immediately picking up the step and the rhythm of the march.

On the heights, they were greeted by the same Aide de Camp, who again pointed out Elvina, now very evident by its prominent church with a high front containing twin bells. However, he directed them over to the right, where they would drop down slightly onto the opposite side of the ridge to overlook Corunna. There they would immediately see Bentinck's other two regiments, so he said. As the Aide turned away, Lacey spoke his thanks, then he led his men on and, as they crossed over the ridge and the ground fell before them, then they saw for themselves the high walls and battlements of Corunna, these defending the whole isthmus on which the town was built. What he did notice, which depressed his thoughts somewhat, was the empty harbour, empty save for the fishing vessels of the last few brave, or desperate, souls who owned and worked them.

The rear of the ridge was almost fully covered by the camping grounds of the majority of Moore's Regiments and, for a moment, Lacey felt that he was bound to lose his way. Then he saw a figure he recognised from two weeks back, two weeks that had seemed an age; Bentinck was sat on his horse, accompanied by two Aides, waiting for them, which was very decent and noble of him, so Lacey thought. Bentinck was positioned between two encampments, which Lacey reasoned to be those of the 4th and 42nd, so he led his men on, through the gap, to the first available space. Leaving O'Hare to lead them there, he left the head of the column and took himself over to his Brigadier, then he came to the attention and saluted. For the first time since they had known him, Bentinck was without his irked "Headmaster" expression and actually looked pleased with what he saw.

"Evening Lacey!"

"Sir."

"Your men look in good spirits. You must have had a good retreat!"

Both men laughed, as Lacey answered.

"Well, I've known better, Sir, and that was in the Americas, where the weather was, on the whole, kinder."

Bentinck nodded.

"That I don't doubt."

He eased forward in his saddle.

"I'm pleased to have you back, and Paget is sorry to lose you!"

"That's very kind of him. We did our best."

"Just so! His trust in you was total. His exact words, Lacey."

"Again, Sir, that's very kind of him."

Bentinck grinned some more.

"Now, I've a welcome back gift. Bentson here will show you."

Bentson saluted, as did Lacey, for the conversation was evidently at an end, and the Aide led Lacey to what could only be described as an arms dump, for there were hundreds of brand new muskets, stacked in piles of ten, with new bayonet scabbards and bayonets hanging from their muzzles. Behind these were stacks of wooden crates, the very familiar cartridge crates, also with piles of soldier's cartridge boxes with their straps draped beside them. Upon the cartridge crates were smaller boxes that Lacey recognised as flints. Bentson pointed and grinned.

"You are to help yourself, Sir. Corunna is stuffed full of such, and anything that cannot be used is to be destroyed."

He paused, whilst Lacey feasted his eyes of the display of brand new equipment.

"Welcome back, Sir!"

Lacey returned the salute.

"Thank you. These are most welcome."

Lacey knew that his men had taken scrupulous care of their weapons, but, nevertheless, they were worn and rusty and showed the effects of the retreat. He looked for the nearest soldier of the 105th, who were all, by now, busy making their camp. The soldier was evidently on a mission to obtain water.

"Fetch Sergeant Major Gibney."

The soldier dropped his bucket and ran off. Soon Gibney appeared and brought himself to the attention.

"Sir!"

"Sar' Major. I want every man with a new musket, before lights out. When you get to Captain Carr tell him that his men have the choice, to keep their Baker or get a new musket. I want this done quickly, the French are not that far behind."

Gibney rounded an elaborate salute and left. By the time the sun was touching the horizon, every man had a new cartridge box and every

344

man who wanted one had a new musket and bayonet, but these included none of the Light Company where Miles pronounced judgement.

"Bugger that! When I sights on some Frencher as is opposite, I needs to know that I'll put'n down, so's 'ee don't get off a shot back, when I bc doin' a reload."

***

Chaplain's Assistant Percival Sedgwicke was awake long before dawn, partly because on hard ground he was a habitually poor sleeper, unless utterly exhausted, but mostly on this occasion because of the loud snoring coming from the tent of Chaplain Prudoe and his good wife, the sound emanating wholly from the former, Sedgwicke concluded. Immediately they had made camp, the Reverend had strode on into Corunna, rescued a horse from assassination, loaded upon her back a tent that he had requisitioned from a military warehouse and brought it back for Sedgwicke and four other soldiers to erect on his behalf. However, yet another sound roused Sedgwicke from his bed, the sound of explosions as unclaimed stores were destroyed before the battle, which all in the army now assumed was inevitable.

It was early, too early, to consider making breakfast for his Superior, which he usually took alone, for his wife habitually now took hers with Bridie Deakin and Nelly Nicholls. Therefore, somewhat at a loss, Sedgwicke dressed himself, grimacing as he again dragged on the clothing stiff with sweat and mud and God knew what else, then he sparked up his tinderbox and made a fire. Some tea for himself was reviving and, with that inside him, he considered what to do next and his thoughts soon alighted on the idea of joining the campfire of Nelly and Bridie. An extra loud snore from the tent confirmed that Prudoe yet remained in deep sleep and so he took himself the few yards up the slope to find Nelly and Bridie already awake and bustling around after their children. It was Nelly who noticed him first and bade him welcome.

"Ah Parson, a fine good morning to youse. You'll take some tea?"

Sedgwicke nodded. 'Tea' with Bridie and Nelly always included a dough cake, of a quality that he could never match and so he sat on a horse blanket and eagerly accepted both. As he ate and drank it was

Bridie who spoke the subject of the previous night's conversation between herself and her friend.

"Parson, darlin', do you think our men have come up yet?"

Sedgwicke was, after all, the graduate of an ecumenical college and therefore not bereft of good sense nor short on logical thought, therefore his impression, gained from what he had seen and from the sheer passage of time, drew him to the positive conclusion.

"Yes, Mrs. Deakin. I'd say that by now they were."

Bride Deakin screwed up her mouth in a sort of "I don't like to ask this, but......"

"Well, Parson, you don't think, what with you wearin' the uniform and all, and can go places we can't, you could take a go at findin' of their whereabouts?"

She paused.

"Do you think?"

Nelly had listened to all and was nodding vigorously.

"Yes Parson. There's goin' to be fightin' soon, even today, God forbid, and so we'd surely like to see our men, before it all starts off, and, well, there's a whole army spread over these hills and the Lord knows where ours could be."

Sedgwicke put down his cup and sighed.

"I could and I will, dear ladies. But I may be some time and, therefore, the Reverend will need his breakfast and I may not be here to provide it. So, if you could take care of that, I will try to find our Regiment. It is my duty to be with them during a battle in any case; therefore it is my duty to find them. Please to tell the Reverend that, when he awakes."

He arose from the blanket.

"You will see to the Reverend, will you not? That's important to me."

Both women nodded, but it was Bridie who answered.

"Don't you give any kind of concern to that, now, Parson darlin'. He'll get a share of the next lot we bakes up."

Seeing both women still nodding, Sedgwicke began to climb the ridge. The two women watched him go, but this time it was Nelly who spoke.

"The Reverend'll get his share, all right, but it'll be Patrick as takes it to him. I don't like the way he looks at the girls, especially Eirin, not since their fallin' out, between him and her! What with his

346

Missus up here and him alone in his nice tent, I'm not for takin' any chances, so we'll get Patrick to poke his breakfast inside and scoot on back."

In saying that, Nelly Nicholls was being somewhat unfair, but the conduct of the "Man of the Cloth" Prudoe had been whole unimpressive over the appalling hardships of the retreat and so anything that they could level against him, however faint, was worth saying, as they saw it. Bridie passed judgment.

"Yes, but only after the rest are fed. That'll do for him. Squint!"

Meanwhile, Sedgwicke was using the church of Elvina, it's bell tower just visible, as his aiming point. He remembered what the Provost had said and so concluded that Elvina was as good a place to start as any. Most Regiments now had a tent for their Commanding Officer with the Regimental Colours prominent outside, so he headed for one such, but soon turned aside to aim for another, for the first was guarded by soldiers in kilts and too often, so his experience now told him, they could only speak Highland Gaelic or English with such an accent as to render it unintelligible for him. The second camp was undoubtedly English, they were not wearing kilts, but their facings were not of the green he required, these were dark green; nevertheless he approached a Corporal, who was turned sideways to him and much taller.

"Excuse me, Corporal."

The NCO turned, to make no reply, other than to study the unimpressive object before him, but he spat out his own question before Sedgwicke could say a word.

"And what might you be?"

Looking up, Sedgwicke could only give the most correct of answers.

"Chaplain's Assistant Sedgwicke. 105th Foot, Corporal."

The NCO took another good look.

"Well, Chaplain's Assistant, I notes that you 'as no weapon, nor kit, nor nuthin'. There's goin' to be a battle, hereabouts, soon, an' you may be required to join in."

With Sedgwicke stood rigid before him, the NCO turned to a Chosen Man.

"Fraser! Get this "soldier" what he needs to join a firin' line. Full kit, box, bayonet, bundook and backpack. There's plenty spare over by Four Company."

The alliteration and the way the words flowed from one to the other showed that the NCO had said this hundreds of times before. He continued to stand regarding Sedgwicke for some minutes, astonishment plain on his face, until Fraser returned, with all the necessary draped either over his arms or over his shoulders. The NCO now moved the affair on.

"Right. Let's turn this Chaplain's Assistant into some kind of a soldier."

With great ceremony, the NCO took each item from Fraser and passed it to Sedgwicke who adorned himself with each, until he was passed the heavy musket.

"Right. Here we go!"

He took a step back.

"Order arms!"

Sedgwicke remembered enough of his training from almost a year ago to place the musket alongside his right leg, butt on the ground.

"Shoulder arms!"

Sedgwicke heaved the heavy "Brown Bess" up to slope against his left shoulder.

"Present arms!"

The musket was swung off and held directly out, the metal of the barrel pressed against Sedgwicke's nose."

"Shoulder arms!"

Sedgwicke returned the musket to his left shoulder. Partially satisfied, the NCO thought he should now allow Sedgwicke his say, it may possibly be important.

"Now then, Chaplain's Assistant. What exactly was you lookin' for?"

Sedgwicke was now more than a little angry at the events of the past five minutes and spoke directly up into the Corporal's face.

"I am looking for my Regiment, the One Hundred and Fifth. I was told that they are close to Elvina. I am trying to find them so that their followers can join them for one last time before the forthcoming battle."

He deliberately paused.

"Corporal!"

The poignancy and significance of Sedgwicke's request was not lost on the NCO, but he did not show it.

"Now just how am I supposed to know where some Regiment not my own may be?"

Sedgwicke took a deep breath.

"They have facings like my own. A bright green, at least as was. I'm hoping that you may have noticed such, either passing you by, or as you went around, about your own duties."

Again the pause.

"Corporal."

Fraser had been listening and watching all the while.

"Dan. What about that lot as came across here yesterday evenin'? All marchin' like it was Sunday. B'ain't they now in the same Brigade as us?"

The NCO turned in his direction, but did not look at him.

"Did they have bright green? Do you think they're what this cove's lookin' for?"

The NCO looked over at the sympathetic Fraser, who continued.

"Well. Have we a better idea? They went down over the ridge aways. Beyond our Grenadiers."

Dan, now identified as such, nodded and returned to Sedgwicke, before pointing.

"That way! That's the best we can do, the best place to start."

Sedgwicke began to move off, but "Corporal Dan" was not finished.

"As you were!"

Sedgwicke halted and "Dan" studied him.

"I wants to see you marchin'. Marchin'! Proper soldierlike!"

He stepped back further.

"Now!"

A pause.

"Attention!"

Sedgwicke straightened up, musket still over his left shoulder and now feeling heavy.

"By the left, forward march!"

Sedgwicke set off, left foot first, right arm up to shoulder height and in this manner he progressed through the camp. Once through, he'd had enough, and anyway felt safely beyond the jurisdiction of Corporal Dan, so he brought the musket to the vertical, thrust his right arm through the sling and swung the heavy weapon around to his back. Once on the back slope of the ridge, he saw another tent and two cased

Colours erect outside it. Progressing down, he saw what he wanted, emerald green facings, no longer bright, but green on all the uniforms around. Better than that, he saw a figure he knew well and went straight to him. The figure spoke first.

"Hello Old Parson. See they've decked you out for proper soldiering; full and thorough."

Jed Deakin's was the first friendly face he had seen so far on his journey.

"So, what's on? What's your purpose?"

Sedgwicke eased the straps of the heavy kit.

"I've been sent up to find you, by Bridie and Nelly, so that they can come and join you, before the forthcoming battle."

He now looked alternately concerned and puzzled.

"But now I've got to find my way back to them, from here!"

Deakin laughed.

"Well now, Old Parson, I'm sure we could manage that, now, between us, if we both goes about it together."

He turned Sedgwicke around and led him back the way he had come.

Within 30 minutes, the followers were threading their way through the camp of the 105th, finding their men and sharing kisses and hugs, even between the pair Henry and Nelly. As usual the mess fires of Davey and Deakin were not far apart and so the women of both were sat close together, tending their fires and sending cooking aromas around the hunched figures that were far superior to anything that had wafted over them during the past week and more.

The rest of the day was spent in peace and domesticity. Water was laboriously lugged up from a stream that ran down the back of the ridge into the river Monelos and the necessary errand gave cause to a continuous procession to and from the 105th's camp. Men tried to clean themselves and their clothing, for they were filthy, as were their clothes and all knew the result of dirt entering a wound, either from foul skin or dirty clothes. Despite the cold, throughout the morning, the camp sported hundreds of naked men, whilst their family, if they were there, did their best to wash their clothes into some degree of hygiene. If not, they did it themselves and no one paid much attention to the immodesty, barrack life had inured all to such concerns. Bar one, this being Beatrice Prudoe, who felt the requirements of her position dictated her absence and so, having received water from Sedgwicke, confined herself to her tent and performed her own ablutions, both for herself and, with some reluctance, also for her husband.

Mary was setting out to obtain water for herself, but Joe knew the weight of the task she was undertaking and sprang up to accompany her. He took both buckets and smiled.

"Come on."

She smiled back, the happiness in her face at their being re-united plain and shining for all to see, but it was not there to be seen upon Joe. He spoke but once on their journey down to the river.

"I'll carry both buckets. Water's heavy."

She talked, inconsequentially, but Joe merely looked blankly ahead, but on the way back, he again spoke but once.

"Do you remember? This is how we first met, back on Sicily, I carried water for you, back down the hill."

She grinned, much encouraged, and threaded both her arms around one of his.

"Yes, I remember."

She teased him.

"And you couldn't say a word, all the way down. Not even when we got back to our camp and Mother asked you to stay for supper!"

He smiled and nodded, but spoke no more. Even when she washed his back and then his clothes, whilst he washed himself further, little was said, other than what was needed to finish the cleaning. Now, as clean as was possible on a cold Spanish hillside, Joe Pike rolled himself in his blanket and slept, this as good a way as any, he thought, to spend the time before the evening meal. He was not alone, for all expected to be sent up to a battle line, with the ending of the daylight. Mary sat and studied the prone figure, anxiety growing within her. As he stirred, muttering in his sleep, his plainly not being either deep or restful, she covered him with a horse blanket and then stood, to walk to where Bridie was sewing a hole in Jed Deakin's greatcoat and Nelly was checking Henry's tunic.

"Bridie. It's Joe, I'm worried."

Tears quickly filled her eyes and her voice trembled, as did her chin.

"He's so different, so down in the mouth. And so quiet! He can't bring himself to say ten cheery words. He's changed so much! I'm worried."

351

She sat beside them, wringing her hands together in her lap, to finally thump both back against her chest. Both women looked at her, but it was Nelly who spoke first.

"Well, don't forget now, he's been through a lot, as have we all. And he never did say too much, not like that scut Tom Miles. Now that's one in a never endin' argument!"

There was no change in Mary and so Bridie took it up and tried a different approach, this time more serious, drawing on perhaps deeper thinking and her own experience.

"Losing a child like that gets to different men in different ways. Joe loves you with all his heart and soul, that's obvious, and all I can say is, that in some way, he blames himself for what happened. To you! I'd say he feels protective of you, like the good husband that he is and 'tis a guilt that's weighin' heavy on him. Guilt that he wasn't there, nor could stop it happenin' in the first place."

She paused to look at Mary.

"That's the best way I can see it, and the best I can say."

Nelly had been listening.

"Bridie's right. 'Twas his child too. Not only could he not protect you, but no more the bairn. 'Tis heavy on his mind."

Mary looked at both.

"He might do something silly!"

Nelly continued.

"He might. I doubt not that he's holdin' a powerful hatred of them opposite."

Both women noted the alarm that came into Mary's face, but it was Bridie who spoke next.

"I'll tell you what. I'll get Jed to have a word with John and Tom. He'll tell them to keep an eye on him and not let him get too far into trouble, at least not too far as can be expected in a battle!"

Mary immediately began to cry and Bridie ran to her younger sister.

"There now, don't fret yourself. John and Tom will take care of him. John's as calm and sensible man as you'll find on this Good God's earth and the Devil himself takes care of Tom Miles! Stickin' close with them two will see him out of it."

She put down the coat.

"Now, I'm away to find Jed."

Within ten minutes, Jed Deakin was sat with Miles and Davey, with Saunders, Bailey and Byford in audience. Deakin looked squarely at Miles and Davey.

"Joe Pike. What's different? The women think he's a bit mad!"

Miles' brows came together.

"How'd you mean, different?"

Deakin came straight back.

"Sayin'! Doin'! What's different?"

Davey looked at Miles, then at Deakin.

"He's down, right enough, and can't get off enough bullets at the Frogs. He's both wound up and miserable. Both, is how I sees it. He can't kill enough Frenchers, perhaps seein' that as a way out. We saw it at El Burgo, the last time we closed with 'em, t'other day."

Deakin looked at Miles, who nodded agreement. He had heard all he needed.

"Right. In the comin' set to, I wants you both to keep an eye on 'im. A close eye. This ain't worth takin' risks for. I don't know what our Generals want, but I know what I wants and that's for us all to see home again. Moore may want to deal out what they calls a 'crushin' defeat' but that gets too many good lads killed. All's needed is to knock 'em back some, to give us time to get in the boats and bugger off. So!"

He looked at both.

"No risks! Follow orders, yes, but hold 'em off enough, just enough, so's they pulls back an' you stays alive; an' so does Joe."

He paused and his look even intensified.

"Joe gets home an' so do you!"

This time he looked directly at Davey.

"Your Molly's waitin', an 'Tilly, an' your own newborn son!"

No more was said. The words of Jed Deakin carried as much weight as an Officer, probably more so, as his pronouncements were sourced from deep within the fund of knowledge and working culture that governed the lives of all such as Deakin, Davey and Miles. Deakin rose and walked off to sleep at the side of Bridie and her children. Miles and Davey nodded and exchanged knowing looks that shared their mutual concern, before claiming some sleep for themselves.

The evening was closing and all the Officers of the 105th sat by their campfires, many looking at the far from empty stewpots and wondering if they could eat yet more. They were all now clean, for the same reasons as had motivated the men, but theirs had been achieved in a far more modest manner. A section of the river Monelos had been section off, banned to "other ranks" and hundreds of Officers from Moore's army had stood knee deep in the icy water, sharing the soap that was now available and cleaning themselves of the two week retreat. Meanwhile, significantly upstream, their servants had scrubbed and beaten their clothes clean, both sets of uniforms, for each was equally foul. Now they wore one of them. Whilst both had dried before a blazing fire they themselves had sat wrapped in their greatcoats before gratefully dressing again in clothes, damp in some parts and fiercely hot in others but, above all, clean. Now they sat in the dying light of day, the flames from the campfire replacing the fading light of the sun, discussing the forthcoming battle. The conversation between the Officers of the 105th Light Company was not atypical, in their case begun by Drake.

"I haven't seen many cannon!"

Shakeshaft halted his spoon during its journey from his bowl to his mouth.

"Nor I."

Each looked at Carr, for both judgement and comfort, but they received neither when he spoke.

"The ships have arrived and the guns are being loaded as we speak. I was talking to an Artillery Officer down at the river and what he said was, that we've no horses to pull the heaviest up here. There are nine up here, round and about, but that's all."

That vital number was also being relayed to Carravoy at much the same time, by D'Villiers.

"Nine! That's not even two batteries! Did he not read any account of Vimeiro, where our guns shot them to pieces before they even came within range of us? And so our guns are now safe aboard the transports that arrived this afternoon. And he gave those priority!"

D'Villiers remained quiet in the face of the outburst.

"Does he want to save his guns, or to save his army? The guns, it would seem! An army with no guns gets beat, so his army ends up lost, but he's saved his precious guns."

He rose and stalked off, leaving D'Villiers to poke the fire. However, the conversation was being progressed further around the fire of the Commanding Officers of the 105th. O'Hare making the next point.

"I'm not keen on this ridge. The top is too shallow. Wellesley had the right of it, when he put us on that sharp ridge at Vimeiro. We held back behind it and came over at just the right moment. If Soult brings forward guns that we can't answer and we're out on the slope, well, I'll say no more!"

Lacey took another mouthful of his brandy. O'Hare was right, Wellesley had used faultless tactics at Vimeiro and the French had been practically annihilated, but he had some sympathy with Moore.

"What was it that Roman General said when challenged by another, him saying 'If you're a great General, come and fight'?"

O'Hare knew and gave the answer.

He replied, "If you're so great a General, make me fight against my will."

Lacey nodded.

"Moore'd rather not fight. He'd rather file onto the ships and go home. He has to defend the port and this is the ground he has to use. No choice! Wellesley chose Vimeiro very carefully, a luxury that Moore doesn't have. Soult will be up tomorrow, with an army larger than our own. He has to be knocked back, or at least intimidated enough into leaving us alone. We could not hold off an army larger than ourselves and walk onto ships at the same time. We have to persuade M'sieu to be content that he has kicked us out. Let him call that a victory, and welcome, as long as we've fought him so a standstill."

Lacey swallowed from his own glass.

"But no guns!"

Lacey nodded.

"He's counting on the men to hold the ridge. If we fight them off, the job's done, as I've described, but like you, I'd feel better with a couple of batteries either side."

This ended the conversation between these two, but at the Light Company the topic had turned to a lighter subject, as Drake asked of Carr.

"Have you written?"

Carr released all the breath in his body in a heavy sigh.

"Ah, well, yes and sort of no!"

On this particular subject, Drake spoke as the superior.

"And that means?"

"Well, it means that I kept up a sort of journal, whenever I could, but I needed some of the pages to keep me warm, they're good inside your coat you know, to keep out the cold, and perhaps some have got a bit rumpled and one or two got lost."

Drake looked exasperated at his Captain, but Carr did his best to be reassuring.

"But I've read what's left and it doesn't seem to make any difference! It's all a bit of a ramble, you see."

Drake started laughing, as did Shakeshaft, and soon all three were sharing a thoroughly good hoot. When back in control, Drake looked at Shakeshaft.

"Anyone waiting for you, back home, Richard?"

Shakeshaft looked at the ground, slightly embarrassed.

"No! No, not really. No-one."

Drake looked at Carr.

"Well, we'll have to put that right."

As Carr nodded, both looked at their Junior Lieutenant, and it was Drake who spoke, of how.

"When we get home, you will be a veteran hero of, now let's see, a heavy skirmish, two, no three battles, including this one coming, the occupation of a foreign capital and a fighting retreat. If that doesn't get the girls hanging on your every word, then I'm no judge. We'll have to order a few up, back in the Dear Old."

However, Shakeshaft was not convinced, in fact, he appeared even more embarrassed.

"Well, I'm afraid things tend to get a little awkward when I'm in female company."

Drake came straight back.

"Stuff! Having come thorough this lot, you'll have no trouble whatsoever holding your own with a bunch of girls! And besides ......"

He looked at Carr.

"...... when it comes to being a bit of a dunce with the ladies, you're far from the worst in the room!"

The last word on the dying day came from Captain Joshua Heaviside as he knelt on quiet ground, out before his men, using the last of the light to intone his evening prayer before his Bible, his Cross and the two pictures of his wife and children. He had no need of written words, the words had been composed and re-composed several times throughout the day. He would have preferred silence, but there was none, not that the noise came from behind him, rather from in front, this being the rumble of French guns crossing the wooden bridge thrown across the river Mero at El Burgo and, bourn on the wind, the sound of French marching songs as Soult's Divisions arrived on the hills opposite.

***

With the dawn of the following day, most of the Officers in Moore's army were eating their breakfast "on the hoof", with their servants running back and forth with coffee and rolls. Almost all had taken themselves up to the forward slopes to observe the French and there they were, Soult's Divisions already in place and more arriving from the crossing at El Burgo, down in the valley behind the French right. There was little conversation, most circumspectly ate their breakfast and watched the French arriving, whilst occasionally looking along the line of their own position. Drake, ever the optimist, spoke favourably.

"This doesn't look too bad, you know. I'm told that we're on something called the Monte Mero, so at least the Spanish recognise it as some kind of height, and that's a long climb for Johnny and a tough job, with us waiting at the top."

He looked left and pointed, almost excitedly.

"And isn't that a pair of guns I see, just over?"

Carr slowly turned his head to look himself and indeed there were a pair of guns, well back as yet, however, of small calibre, he noted, merely six pounders. However, he saw no cause to dampen Drake's mood.

"You're right! They're not going to roll over us, not before they've had the fight of their lives. Nothing's changed, they'll come on in their way and we'll meet them in ours. If they're to push us off this hill, it'll cost, probably more than Soult wants to pay."

Carr allowed a silence and then spoke to finish, with no need to state the topic of which he spoke.

"Perhaps today, in the afternoon. If not, certainly tomorrow."

With that he took his leave and walked back to their camp. Drake had noted the usual change in his Captain when in sight of the enemy, his tone was now grimly soldierlike, dour, determined and spoke of his proven capabilities. He and Shakeshaft took one last look and followed, but they were passed by General Moore himself, riding up to take a look in full daylight.

It seemed to Drake that Moore must have taken a more troubled view of what he saw, for, within the hour he was ordered away from his coffeepot. The order came for Bentinck's three battalions to form their firing lines and advance forward to the ground that they were to hold. The order included the word "immediate" and so there was little time for farewells back with the followers. Bridie and Nelly embraced Jed and Henry and kissed briefly, while Mary clung to Joe for all she was worth, until the very last; when Joe had to physically detach himself from her fierce embrace. Her face showed an utter wretchedness, which caused him to cudgel something to say from within his own distracted thoughts. He smiled down at her tenderly.

"Now don't I always come back. I'm with the best two soldiers in the Regiment, which makes me lucky. I've got the luck that brought me you! I'll be fine."

Such tender words brought a profound change in Mary, this was the Joe that she knew, but then he was gone, soon to be lost in the multitude of Redcoats running to take their positions. Lacey stood in the centre, waiting for his line to form and watching Bentinck ride a short course up and down before them. Soon the 105th were ready and also the 42nd on their left and the 4th on their right. Bentinck looked around and, seemingly satisfied, he motioned his brigade forward. Lacey turned to O'Hare and nodded. O'Hare took a deep breath.

"By the centre, advance!"

The whole line moved forward, but O'Hare had more to say, which brought a smile to more than one serious face, as they heard the broad Irish accent.

"Make it smart, now boys. Johnny'll be watchin' through his glasses. Let him know who he's up against! Keep the step, now. Make him think you're no one he'd want to ask to the dance. Sure, he'll be sayin' those boys over there look just too damn sprightly to be partners for any kind of a jig or reel!"

Elvina church grew before them, the village was fully in their front. When 300 yards above it, Bentinck called a halt and rode back to Lacey. He commented on what they could all see.

"He's up there, Lacey, strongest opposite us, and ….."

He pointed across the valley."

"…..are those guns? A whole battery or more?"

Lacey brought up his own telescope and a short adjustment of the sections brought the French cannon into focus, such that he could count them and he did so.

"Ten, Sir."

Bentinck nodded.

"He's making us the soft spot!"

Other telescopes were now focused on the French ranks opposite, including that of Captain Lord Charles Carravoy.

"Guns! Ten in all."

He slammed his telescope shut.

"Johnny got his guns up to this height, why haven't we? Nonsense about no horses! Any number of men would have dragged our guns up to here last night, volunteered, never mind ordered."

He stalked off to no-where in particular, swearing and cursing to no-one in particular, but specifically about one person in particular. Meanwhile Lacey and Bentinck were exchanging opinions. The subject, although unspoken, was obviously their enemy opposite.

"What's he about?"

Lacey found the answer from over on their left, on the French right, from where more French Brigades were marching up from the main road.

"He's still building, Sir. I suspect that he suffered as much over the past weeks as us. His units are still coming up, as you can see, and his stragglers are still arriving. I'd say not today, Sir."

Bentinck gave Lacey a serious look, then studied events opposite, through his own glass, which he trained both right and left and back again. Finally he closed it and turned to Lacey.

"Elvina, down there. I don't propose to give it as a present. So, two Light Companies in there, now, to fortify the place. Yours and the 4th. You'd agree?"

"Agreed, Sir."

Lacey turned to O'Hare.

"You'll see to it?"

Bentinck pulled his horse around viciously and cantered away. The "Headmaster" had returned. Left alone Lacey and O'Hare consulted, each looking down the slope to the village. Lacey spoke first.

"It could be worse, Padraigh. We've the village before us and I like those stone walls, between us and it. It could be worse!"

However, O'Hare had concerns of his own.

359

"To defend here, Sir, yes, but what if we have to retake the place? And above the village puts us bang in the centre!"

The two exchanged knowing looks, then O'Hare hurried away to soon reach Carr. The direction involved was indicated by a simple inclination of his head.

"Henry. Get yours down there and fortify the place. You're on the left, the Lights of the 4th will be on the right. You know the drill, block the roads and alleys leading up, make a strongpoint out of any likely buildings, loophole the walls, make them easy to set on fire, that sort of thing.

He paused.

"No tools, I'm afraid. Just do the best you can."

Anther pause, for an important afterthought.

"And don't forget to establish a quick way out!"

Carr saluted and signalled for his two Lieutenants as O'Hare ran off and they ran over.

"We're to get down there......."

He pointed.

"....... and strengthen it."

He walked forward, down the road that led directly to Elvina.

"Bring your men down, I'm going on to take a look."

Carr entered the deserted village, the atmosphere in the innocent dwellings eerie and fearful. In each home, the evocative signs of family life, pathetic and helpless in the face of what was to come, were all around, now discarded and forlorn; a wheelbarrow with a rake at the backdoor, a coat on a hook, boots, now dry, placed toes up on the grate. He passed on, to a building he chose, the lowest building on the left side of the village, a low, block of a dwelling, with walls three feet thick. This would be a strongpoint in the beginning; he could enfilade both left and right from there. Some of Drake's men joined him and he gave his orders, the first being to block the windows, save a narrow firing slit. He then left, to supervise what was happening elsewhere.

Now content with the activities of his men, Carr took himself to the very top extremity of the village and had further thoughts. There was a lone and substantial farmhouse as the highest building of all. In that position it was protected by the rest of the village from cannon fire from across the valley and it had a field of fire all round. Impressed, he took himself further up the hill, where he had seen O'Hare. The two exchanged salutes, before Carr pointed back to the building.

"Sir, that farmhouse nearest us. If held, it would break up any French attack directly up through Elvina and even one on this side of it. I am of the opinion that we should plan to keep it, even though driven out from the rest of the village."

O'Hare took a long look. The lowest windows were five feet above the ground, difficult to enter, and the floor above had twice as many windows, but the door he could see was wide, with two panels, not one.

He nodded.

"Agreed. But that door is your weak point. Don't use it as a firing point, just pack in behind it, with whatever's there."

He looked from Carr to the building.

"What's on the other side, where they'll come from?

"A kind of back porch and windows above."

O'Hare nodded and waved him away. Carr saluted and trotted off, to organise yet another working party.

Davey, Miles, Pike, Saunders, Byford and Len Bailey, found themselves ordered to build a wall to block a main alley that led up between two solid buildings. The only materials they had were from the garden walls behind the two houses and so it became their laborious task to carry the stone, from the back to the front. Davey immediately took charge.

"We can get it higher if we apply a bit of technique. I've done a bit of dry stone walling, even won a prize. So, you bring them up and I'll do the placing."

Miles immediately took umbrage.

"So now you'm sayin' that you was some kind of champion wall builder, along with bein' champion poacher, ratcatcher, ditchdigger and God knows what else?"

Davey answered the challenge with a direct look.

"That's right! Yes! I happens to actually know a thing or two!"

But Miles had changed the subject, he was looking at the Church at the top of their alley.

"You reckon there's any Communion wine in there?"

At that moment the giant Saunders passed him, carrying an armful of stones.

"Bollocks Communion wine! Get this wall built. Get some stone down."

Miles immediately rose into temper.

"Alright! Alright! I'm just sayin', like."

However, within 30 minutes even Miles had to acknowledge that the wall was looking like a very useful barrier to anyone approaching on the other side and was building nicely as the five brought the stone and Davey did the placing. One hour more and the six foot width of the alley was thoroughly blocked with a wall five feet high, but now Miles made his argumentative contribution.

"Nice wall! But one cannonball and the whole lot's flying back through the alley, into our faces. We need small stones, earth would be better, over on t'other side to suck up the ball, or at least send it up.."

He looked challengingly at Davey.

"This time brought by all of us!"

Davey stared back.

"All right! All right! Don't give yourself a seizure. I hear you, and I'm not saying you'm wrong!"

Leaving Miles to nod in triumph, he led the way back and all began what proved to be the most laborious of the whole task. They began with straw mattresses, but then it became a question of carrying the loose material down to the wall in whatever could be found, from buckets to cooking pots, even blankets, so that slowly a slope before the wall built up. Miles tipped a bucket of painfully gathered soil and rubble over the front and immediately noticed movement further down the slope, then a shape he recognised. He turned to call back Davey, who was carrying his empty bucket back up the hill for another gathering.

"John!"

Davey ran back as Miles pointed.

"One of they tassel swingers is down there. Two fifty yards I'd say, nosin' about."

Davey looked, then saw for himself the French Voltiguer rise from a dip in the slope, to then take a careful look at what the British were doing up in the village. Davey had seen enough and ran into the building to return with his rifle. He carefully loaded it, wrapping the ball in the piece of thin leather used for extra accuracy. He checked the flint, set the sights and raised it over the wall. Miles looked at him quizzically.

"What's the point John? That's well over two hundred yards. Gettin' on for three!"

Davey sighted along the barrel.

"I'm not arguin', but I'm going to show that noseyparker that he's come as close as is worth it, if he wants to keep his head on his shoulders."

Davey sucked in a breath and held it, to take his final sight. Miles looked at him, then over the wall. The Voltiguer was stood at almost his full height, barely bothering to crouch. Then Davey fired, the wind blowing the smoke off sideways, so that Miles could see the soldier instinctively duck, then drop to his knees. The ball must have passed so close that he heard the buzz of it, alarmingly close. He stood, lifted his musket in acknowledgment of the fact that he had crossed "the line", then retired backwards down the hill. Davey nodded and waved his shako in return, well satisfied. He began to reload, but within a minute Drake arrived, anxious and not a little annoyed.

"Who fired? Why?"

Davey came to the attention.

"I did Sir. There was a Frencher, one of them that has the tassel, creepin' up to see what we're about, Sir. I thought he needed discouragin', Sir."

Drake looked at Miles, then over the wall.

"Did you hit him?"

"No, Sir. But we frightened him off, Sir."

At this point, Drake noticed the wall and was doubly mollified.

"My word, men. Good job! This wouldn't disgrace a country estate!"

Miles now felt the mood light enough to incorporate a comment from him.

"Yes Sir. Davey here fancies himself as some kind of champion wall constructor, Sir."

Drake smiled slightly.

"And with some justification."

He looked over the wall.

"And now you're making some kind of glacis."

He looked at the six.

"Well done! Carry on. Well done!"

All six Lights came to the attention and saluted as Drake took himself back up the alley and rejoined a very different and languid group above the village, just in time to hear yet more gunfire, heavier, but more distant. They looked across to their left, this gathering being the six Officers of the two Light Companies, who had been engaged in earnest, if relaxed, conversation, whilst their men, equally earnestly, if in no way as relaxed, did their best to fortify the village. It was Captain Bertram Trevelyan of the 4th Light Company, who spoke the question that all had asked of themselves.

"Is this it, kicking off, do you think?"

All, almost as one, produced their telescopes and trained them over to the sound of the guns and the growing billows of smoke. It was at the far end of the valley to them, on the extreme British left, that the action was taking place, on the slopes above the village of Palavea that they had themselves marched through, but two days before. Through their glasses they could see French battalions descending from their heights, whilst a lone British one ascended up from the very floor of the valley, on the French side. Trevelyan was the first to lower his telescope, then raise it again.

"What on earth's going on? One lone battalion off on a jaunt. Attacking some guns, perhaps?"

Carr folded his glass and put it away. He wanted to see no more.

"Whatever, whoever's leading them hasn't a cat in Hell's chance. Those are guns we can hear, but not ours, and just look at what's coming down the hill against him, whoever he is, and to the depths of Hades with whoever ordered him to do it!"

He continued watching, unaided by any telescope, the movements of clear blocks of blue and the line of red telling their own story. Within minutes the red line was not longer an ordered formation, but a broken horde pouring back to their own lines. At this point Carr did re-use his telescope to see many prone red figures on the ground and several being led away as prisoners. He said nothing, but Colour Sergeant Deakin, higher up the hill, did speak an opinion, but at least only to himself and to his companion Tobias Halfway.

"Another damn Officer with more hankerin' for glory than brains to make it happen. T'ain't a disease as only afflicts the French, more's the pity."

Halfway nodded, then spoke his own agreement.

"That'll be 50 good lads gone, as should've had a better chance of seein' theirselves out of here."

Deakin looked away and leant his forehead against the leather casing of his Colour, After several more hours, it was a relief when, come the late afternoon with dusk falling, they received the order to pull back to the mess line that they had advanced from in the morning.

***

Chaplain's Assistant Percival Sedgwicke was now in a state of high agitation, which had been growing all through the afternoon, but was now becoming unbearable. He was a deeply religious man, believing in Heaven and Hell and in Salvation only through Jesus Christ, with the same certainty that he believed the sun would rise the next day, but he was now practically distraught at the thought that men may soon die without the opportunity to commune with their Maker. His Chaplain, with the evident onset of a battle which may well have begun that very day, should that morning have conducted a Service for all who felt the need to attend and set their Spiritual needs at rest before becoming a part of what was likely to be a very bloody affair. However, relations between the two had become so coldly formal that Sedgwicke felt that it would do more harm than good to make the suggestion.

With the dying of the day, Sedgwicke's worst fears were surely confirmed, for it was clear that Prudoe had not appeared to conduct any kind of evening ceremony and Sedgwicke was now enough of a soldier to know that the men could be called up to their line way before dawn, when it would then be too late. He sat for a further half hour, distressed and wringing his hands, before he decided that it was his Christian duty, even if not his place nor position, to indeed ask the question or to conduct the affair himself, now justifiably assuming that the possibility of Prudoe performing the duty was very remote. Something had to be offered, in some form, by someone.

Taking a deep breath and summoning all his resolve he strode down the back slope to Prudoe's tent. As was customary he knocked on the tent pole as a form of obeisance to politeness and expected to hear a reply, either from Chaplain Prudoe, or his wife, but he heard neither. Alarmed, he pulled back the flap to find a weak light burning in a tiny lantern and he immediately noticed the substantial figure of Chaplain Prudoe lying on a crude cot, with his back to the entrance. He re-summoned his resolve.

"Sir!"

There was no answer. He spoke louder.

"Sir!"

The figure rolled towards him, but not enough so that he could see his face.

"Dammit, Sedgwicke, what do you want?"

Stunned by such profane language, albeit spoken slurred and wearily, Sedgwicke had to regather himself, but he asked himself the question whether his Chaplain was drunk. He sniffed the air, but detected nothing other than the smell of a large man; rank and unhygienic. Neither glass nor bottle could be seen. Having at least obtained some kind of response, Sedgwicke continued.

"Sir, almost certainly there will be a battle tomorrow and it is customary to hold some form of Service. For the men, Sir. They are all of our Faith and expect it, Sir. The day has almost ended."

The figure had rolled back to its original position, but it did speak.

"So you want me to?"

The shape shifted on the bed, like a beached whale shrugging to gain further comfort.

"Conduct a service, that is?"

Sedgwicke was again taken aback by the ridiculous question, partly because of "who else" and partly because it carried the notion that he, Sedgwicke, had some power of command. He searched for an answer and found one, at least partly diplomatic.

"I felt the need to come and remind you, Sir."

The reply could only be described as a sob, more than one, between deep sighs, but eventually scathing words did come and at full volume.

"The men can go to the Devil, Sedgwicke, and so can you!"

Stunned, Sedgwicke took a step back, then stopped. There was to be no Service conducted by Prudoe, which left but one alternative, for it to be conducted by himself. His mind now fully made up, he turned to the chest that contained Prudoe's vestments and opened it. He considered asking permission to borrow, but in the light of Prudoe's last utterance it seemed pointless. The full range of vestments for an Anglican Priest were there, but Sedgwicke knew that for one, he was unfrocked, for another he was Low Church, and finally, he was but a ranker, not even an NCO. He first pulled out the Dalmatic, the robe which would have covered him beyond his knees, but wearing that would be a display way beyond his station, so he dropped it behind him. The next was the simple green Cope that would hang around his neck to mid thigh. This he accepted and he stopped there, after hanging it around his neck. It was sparse and simple, yet symbolic enough to show that he was the Preacher accepting the role, albeit a Lay Preacher. It would convey some significance, but of what nature and to what level Sedgwicke could not calculate, but he had other concerns. He rose and left the tent, then went straight to his own belongings for his Bible, then to someone that he knew would be supportive, this being Colour Sergeant Jedediah Deakin, found in the act of polishing the shaft of the King's Colour.

"Hello, Old Parson. What brings you over here, and done up in your Regimentals?"

"There needs to be a Service."

Deakin nodded.

"I was wonderin' about that myself."

Sedgwicke's blunt answers fully conveyed his uncertainty.

"Our Chaplain is indisposed."

Deakin drew back his head.

"Drunk, that means!"

"No. I think he is in a fit of severe depression."

Deakin released a sharp breath.

"Depressed! About what?"

What he had left unsaid, was that most in any Regiment thought the Office of Chaplain to be the least onerous in the army, a judgment only redeemed by how assiduously and conscientiously they carried out their Office, but Sedgwicke now answered.

"I don't know, but I have taken it upon myself to conduct one. I was hoping for your help."

Deakin regarded him very seriously and not unsympathetically in the dim light, but he paused before answering. He knew that Sedgwicke was going way beyond his rank, but it was a Service from him, or it was nothing.

"Alright, Old Parson. A few drums and The Colours should do it. Where?"

Sedgwicke looked to either side, then behind him.

"Just behind us, here."

He pointed over his shoulder, using a cultivated index finger, rather than an unrefined thumb. Then Deakin stood up, retaining The Colour in his hand, and shook Halfway awake.

"Toby! Parson here wants to hold a Service for the lads. He needs a few drums stacked up and the other Colour. Where's Harry?"

Halfway sat up and reached for his shako.

"Just over."

Halfway did point with his thumb, in the direction behind him. Deakin issued his orders.

"Right. You find a few Drummer Boys and I'll find Harry."

The three went their separate ways, Deakin to find Harry Bennet, the Colour Sergeant of the Regimental Colour, Halfway to get some drums and Sedgwicke to go and stand where he wanted the "Altar" to be. Soon figures began to appear out of the growing gloom, Deakin with Harry Bennet, each carrying a Colour, then Halfway with a group of Drummerboys and seen to be cuffing one for some cheeky remark. One of the drummers carried a lantern on a tall spike. When the drums were stacked, three on four, Deakin spoke the next move.

"We'd better spread the word."

This was spoken to Bennet and Halfway and so, leaving the Colours, still encased, leaning against the drums, the three departed to "spread the word", leaving Sedgwicke to assemble his materials as he saw fit. For a moment Sedgwicke looked at the brown leather cases containing the precious squares of cloth that were now solely in his charge. He first uncased the Regimental Colour, that of emerald green. The cloth was heavy in his hand, but he knew full well its significance and could easily read the battle-honour "Maida" embroidered upon it. He blessed it, then laid it over the drums, before doing the same with the King's Colour, the Union flag. Then he studied his Bible in the light of the lantern to make a hurried choice of Psalm and Reading and a Hymn, before turning around, to wait, standing in front of the drums supporting the draped flags.

He was surprised to see men already there, stood ordered and waiting with more joining. Soon there was a wide half-circle stood before him, but he waited for what he judged to be two more minutes. Whether or not more joined he had little idea, for the growing darkness hid all beyond the front ranks. Several Officers had joined and were amazed to see Sedgwicke where they should see Prudoe, but none felt justified to put a stop to what was highly irregular, but very well attended by the men, plainly well regarded and obviously meeting a need within them. Sedgwicke took a deep breath and began. At his first words, shakoes were removed. If the Service were conducted by an Officer it would have been an order, but Sedgwicke had forgotten.

"Dear Lord, we ask you to look kindly upon our congregation here, gathered in Your Name in a place which will soon become a field of strife and conflict. All have arrived here to gain your peace and comfort in the name of your only son, Jesus Christ. We beg that you look kindly upon us as unworthy sinners who seek absolution through the way as taught by our Dear Lord, Jesus Christ. We gather here at the end of your Holy Day, knowing that many here will not see the sun set on the day you give that follows this. We ask that you hear our Prayers, and keep us safe, save that of knowing that we place ourselves in your merciful hands, obedient to thy will, and that if our time has come, then you receive us with forgiveness for sins past. In the name of Thy Son, Jesus Christ, Amen."

All spoke the Amen and it came as a deep sound, indicating the numbers attending. He paused, then he began singing his chosen hymn "Light Shining out of Darkness" It was a popular hymn, chosen by Sedgwicke as one that almost all would know and could join in, and, after Sedgwicke's lone voice had cast the first few words insignificantly into the night, the assembled male voices swelled across the hilltop. With the dying of the last note, he spoke the 23rd Psalm; with his Bible closed, for he knew the words too well, and many joined in, including Carr, Drake, Shakeshaft and Ameshurst, stood to one side. With silence re-established, Sedgwicke opened his Bible for the Reading, but he had not spoken one word before a hand came across him and took the book.

"I'll read that."

It was Heaviside and, in the light of the lantern, he instantly recognised the chapters and verses that were appropriate.

"The reading is taken from the Books of Hebrews, 10, verse 39, and 11, verses 1 to 7."

He gathered himself.

"We are not of those who shrink back and are destroyed, but of those who have Faith and keep their Souls."

He continued to the end, and all heard the fitting words, spoken deep and heartfelt by this most devout of men. Then he wordlessly handed the Bible back to Sedgwicke, before returning to his place. Sedgwicke had meanwhile, recovered his own self-control and now spoke again to finish the ceremony; speaking loudly and with deep conviction, using words that all knew.

"Our Father......"

All took up the familiar Lord's Prayer, spoken by all, with a depth of care and feeling such that those who had not attended, but could still hear, rose to their feet, some to join in and some merely to stand, to feel part of the familiar commune that had been a cornerstone of their lives since earliest childhood. Some also to reflect that, on the following day, they would be but specks on a field of battle, their fate determined by pure chance or, so they hoped, the merciful hand of the God that they looked to for hope and salvation through times almost as desperate as this one to come.

***

Chapter Nine

From Noon till Sunset

Joe Pike woke with reveille, the uplifting notes swinging across a hillside still dark, and very winter bare; a wide upland, cold and bleak, cheerless bar his comrades now seen and about their own small affairs. Usually, when in the presence of the enemy, he had found himself being shaken awake in some makeshift hole under a log or beside a hedge, to then take his turn at sentry, but this time his Captain had pleaded the case of the Light Company, namely that they would be holding Elvina all day and should therefore be excused the duty of night sentry, so for this reason they had been pulled back to the messlines. He opened his eyes to see a sky grey and dark, but it was sometime before his mind recovered the thought that today there would almost certainly be a battle and then came the subsequent notion that he may not survive it. He tried to move but could not, Mary was half across him, in the exact same position as she had been in when they had both lain down to sleep.

She sighed softly as he gently detached himself by rolling away, out of the blankets that covered them both, to replace them carefully over Mary and put on his greatcoat against the cold. John Davey had already established a good fire and was mounting their pot over the cheerful flames to heat water for tea. Noticing the erect figure, Davey looked up and made a careful examination of him, but saw nothing of any concern, neither in his face nor posture.

"Morning Joe!"

Joe Pike nodded and ran his hand over his hair.

"John."

He looked at the pot of water.

"How long till tea?"

Davey smiled up at him

"Five minutes, perhaps more."

Joe Pike smiled.

"I'll have a bit of a wash."

He took off his upper clothes, feeling the cold, but he splashed water over his face and body, then brushed off what water he could and re-dressed. By now Mary had sat up and was looking carefully at him, much as John Davey had, but perhaps with more anxiety. He returned her look, half smiling, remaining so, until he saw the anxiety fall from her face to be replaced by a look that had become very special to him, a look that he saw no-where else, one that spoke of his importance to her, that he was vital within her world, but mostly that she loved him very deeply. It was at that moment that he felt something well up within him and the anger and despair that he had been feeling for so many days now became challenged, but to what extent he did not dwell on. At that moment he felt more quietened within himself than he had for days and he now fully remembered what John Davey had said when he had pulled him down by the banks of the Mero and the truth of his words rose powerfully within his mind. His job was to stay alive, get on a ship and get home. There was nothing to be gained by losing his life here. He smiled warmly at her.

"Who's doing the breakfast?"

Her smile was incandescent as she sprung out of the blankets.

"Me!"

As she assembled fuel for the larger fire, John Davey motioned him over.

"I need your mug. Here's your tea."

Pike fetched his mug from his pack and handed it over, for Davey to pour in a measure of the dark liquid. There was even sugar. He drank half, then handed the rest to Mary. All over the ridge and close around were similar scenes of domesticity, men washing, drinking and eating their first meal of the day which had been prepared by either their wives or their messmates, the shadows thrown from their fires becoming indistinct with the growing light of the day. There was little talk and even less laughter; today was a day for which they may not see a sunset.

However, despite the lack of martial order over the ridge, the army was guarded against a surprise French attack. Far down the hill pickets had been posted, far in advance of Elvina, such that both French and British had been in easy sight of each other, but a truce had been established and held throughout the night, even to the point, in some cases, of an exchange of spirits and tobacco. Captain Charles Carravoy was at that moment in Elvina, watching Ameshurst return from visiting his Grenadier pickets far down the slope.

"Anything?"

Ameshurst shook his head.

"Nothing! And theirs have gone back up the hill."

Carravoy's mouth set in a grim line. Maintaining pickets had caused a very disturbed night for all in his Company and the thought of breakfast did nothing to cheer his sullen mood

"Right. It's dawn. Bring them back to here."

'Here' were the outskirts of lower Elvina, that side presented to the French, and through the night his men had made use of the empty houses and newly built walls to gain what sleep they could. As Ameshurst climbed a barricade, Carravoy asked his final question.

"Any sounds?"

Ameshurst shook his head.

"None. They're all assembled would be my guess, that being an educated one, I would hope."

"Right. Leave Ridgway in charge and get some food."

With that, he left Ameshurst and began his climb back to the summit of the ridge, but Ameshurst did not follow. Instead he checked first that his men had lit their own messfires and he himself led a party of his Grenadiers back up to their lines, to ensure that all received their rations, even ordering extra, when it was plain that there would be a large surplus left over. Only then did he return to the messfire now being used for cooking by his own servant and, once sat at his ease, he was one of the few that noticed Moore riding forward with his staff to view the French in the now full daylight. Whilst drinking his coffee and eating his bread and bacon he watched his Commanding Officer study the French and then ride over to the left. Ameshurst then poured himself some more coffee; he was determined on a good breakfast. That done, he personally checked his brace of pistols then strolled over to the lines of the 15th Hussars to persuade the armourer to put an edge on his sabre.

Within half an hour, drums were beating and all Regiments were assembling in their battle positions, the 105th again directly above Elvin and again with the 4th on their right and the 42nd to their left. With this, Carr led his men further down to again garrison the village, thus relieving Carravoy's Grenadiers, who now ascended the ridge to take their place on the far right of the 105th's line. Prior to their departure down to the village, the three Light Company Officers had gravely shook hands with Lacey and O'Hare, all done with expressions of 'Good luck' ringing out all around.

With all in place there was nothing to do but wait, but Carr, choosing to be the last one walking down, made a final check of the farmhouse that was to be their final refuge in Elvina. He was content;

the door facing back to the British was utterly blocked by a pile of stone and the porch at the front was well defended by the ground floor windows being now reduced to firing slits. A heavy table and a heavier chest could reinforce the door when needed and both were ready to hand, with the required firing slits through the wood of both doors. The porch would be a 'killing zone' for anyone attempting its door. Satisfied he descended to the lowest extremity of the village, down to the wall built by Davey and his helpers, and there he rested his arms on the top and studied events across the valley. He had a good view of the French lines, he was at the same height as they and, it being January, there were no leaves on any of the trees to impair his view, all seeming to stand trapped, helpless, and sorrowful, waiting for the cannon fire that would blow them apart. Most of the opposite slope was covered in blue uniforms, but there was little movement, bar the odd horseman galloping across in both directions. It was now well into the morning and he knew that if the French were going to attack and use all the day, then by now they should be advancing up the slope. He turned to his men behind; they did not include Drake but they did include Byford.

"What's the French for breakfast?"

"Petit déjeuner, Sir."

Carr returned to his studying of the French lines.

"Hmmm, not so petit it would seem. M'sieu is having a good feed. Perhaps he feels this affair can be settled easily in the remains of the day, post "un grand petit déjeuner". Well ......"

He turned to his men.

"....... we'll see.

His judgment proved correct. All through the morning Moore's army stood out on the bare hillside. The wind blew but at least there was no rain and many who could see them turned to look longingly at their evacuation transports now filling the harbour. The six Officers of the Light Companies of the 4th and the 105th naturally gravitated together, this after numerous inspections of their various preparations and of their men manning them. By mid-morning Moore's men had been standing their positions for over three hours and so he allowed a hot drink and biscuits to be ferried forward by whomever could be spared. O'Hare took himself over to the right, beyond even the formed up 4th, to look into the valley of the Monelos and there he saw Paget's rearguard "Reserve Division" actually filing back into the city. He pointed this out to the Junior Major of the 4th, still in his place for battle.

"Has he decided to evacuate? Paget's filing back."

The reply was a long shake of the Major's head and a long shrug of his shoulders. He was as bemused as anyone at the failure of the French to attack and what Moore should now do.

O'Hare returned in time to see the distribution of the rum ration, which brought cheer to everyone and the clanking of the rum orderly eventually lumbering down their own alleyway did much to brighten up the state of Davey, Miles, Pike, Saunders, Bailey and Byford. As the Orderly doled out the fearsome spirit, thumb noticeably well out of the measuring cup, he and Miles shared their usual litany.

"Good luck to you and the lads then Tom. Come the finish, eh?"

"Come the finish, mate. God look over us all!"

The Officers of the Elvina garrison were all leaning against a front wall, enjoying a more prolonged period of weak sunshine. Little was said beyond the usual "off topic" subjects of tailors, boot makers, types of saddle, sword makers, gun makers and breeds of horse, but even these stock topics did not exercise their minds for too long as they all studied the French lines opposite. Carr said little, but spent most of the time with his telescope pressed to one eye or the other, changing, for the eyepiece made each eye hot and uncomfortable from perpetual use. He had been incessantly studying the happenings at the French battery opposite and it was his first words which opened the battle.

"Here we go!"

He had seen the French gunners apply the lintstocks to the vents of their guns and a second later they heard the harsh drone of the balls overhead, then the reports of the guns. He looked at his watch.

"Fifteen minutes past Noon. They've decided to skip lunch!"

**

Up at the main line, Ensign Rushby was on his third sketch, having handed the King's Colour to Deakin, when the cannonballs arrived, taking out a file from somewhere over on the right. There was a brief scream and then silence. They had heard other balls passing overhead and what happened with them was unknown but the 105th had suffered their first casualties. Without a word Deakin pulled the leather case off The Colour.

Lacey had felt horribly exposed on the slope all morning and now he felt doubly so. He turned and shouted, urgently.

"Lie down! Lie down!"

The men behind him, the Second Company, immediately obeyed and this spread all along the line. Rushby and Neape, responsible for The Colours and knowing that they should never touch the ground, draped them over themselves and their Colour Sergeants and so, that done, whilst prone on the cold wet earth, Rushby gave himself the chance to regather his sketches and stuff them and his pencil inside his tunic. Then another volley from the French guns hissed over, seemingly but inches above. The six Officers in Elvina had immediately run to their stations, Carr and Drake to Davey's wall, where Carr calculated the height of the balls passing above, or at least made a judgement of such from their sound. He again pressed his telescope to his eye.

"M'sieu's got this wrong. He should be softening us up down here, not the lines above, and that's going to take some time, a lot more than he's giving himself. Look, he's sending his infantry against us, but his cannonade above."

He passed the glass to Drake, who quickly refocused to see what he estimated to be at least 600 Light troops, probably both Voltiguers and Tirailleurs in the bottom of the valley, with two dense columns descending the French slope behind. He handed back the glass.

"That's a lot, softening up or not!"

Carr nodded. He knew that two Light Companies, his and that of the 4th, amounted to not much more than 140 men. Heavy odds against.

"Right. Get around your men, you and Richard. Set sights for 200 yards. Officers first. I'll try for a better look from somewhere."

As Carr ran back up the alley, Drake turned to the six behind him.

"You heard that?"

Several "Sirs" came back as an answer.

"The wall's all yours."

The six came forward as Drake disappeared into the house next to them. Within a minute rifle barrels were appearing through the slits of what remained of its windows. Davey stood beside Miles and neither much liked the way the French were extending their line off to both sides, to try to enter the village from the sides and outflank the defenders at the front, but their job was what was before them and soon it began. Carr had ordered white marker posts to be hammered into the ground at 50 yard intervals coming back from 200 yards and, as the first were engulfed by the oncoming French skirmishers, all along their front the rifles barked their defiance, causing French Officers and men to spin or slump to the ground. As the range reduced, the French casualties became even more severe from their accurate fire, such that the remaining Officers halted their men and bade them return fire from a kneeling position. Musket balls began to smack against the stonework and there was a yelp from inside the house to their left, but at 100 yards, the rifles were exacting an even heavier toll, even on their kneeling assailants.

Carr's mind was elsewhere, that being what was happening before the Lights of the 4th, over to his right. Here, faced with only inaccurate musket fire, the assaulting French were making better progress, progress that he did not like, for it was, after all, over 300 against 70 odd. Looking over from a vantage point not too exposed, he saw that the French were but 50 yards from the 4th's position and massing for an assault, which was all too likely to succeed and then cut off him and his men. It was too soon for heroic 'last stands'. He turned to see the Bugler he had been given, the Regiment's Private Bates.

"Sound fall back."

As the notes sounded out, echoing around the narrow spaces, Carr took one last look to his own front. His opponents were stopped, all kneeling or even lying down to return fire. He left his position to run back and up through the village and there see Drake's section piling into the farmhouse, whilst Shakeshaft's continued on up to one of the stonewalls. Carr chose the farmhouse and was the last in, then he ran around each floor, checking that the table and chest now barricaded the door and that all else was in place. He took in the sight of his men, all stood ready at a firing position, then he found Ellis.

"Casualties?"

"Two Sir, but just wounded. They'm still in the fight."

Carr then took himself to an upper window and waited. Close by, Davey looked over at Pike.

"You all right?"

Pike nodded.

"Yes John."

Davey gave him a stern look.

"Through this and then home! Right?"

"Yes John."

The replies were flat and vacant, then Davey saw him raise his rifle through the slit and look with evil intent as he sighted on the first French to appear. Davey let it go, they were, after all, in a battle.

Lacey had seen his own Lights abandon the lower village and he now knew that he had only to wait for what would inevitably come next, but Shakeshaft was running up the slope towards him. Shakeshaft came to the attention, but before he began speaking Lacey stopped him. He was much out of breath.

"It's all right, Richard. Take your time and recover yourself."

Being young, this soon came about and Shakeshaft reported.

"The French skirmishers will soon be through the village, Sir, followed by two columns, one either side of it. They will attack around both sides of the village, or so it appears."

As more cannonballs droned overhead, Lacey began to think, but first he dismissed Shakeshaft.

"Thank you, Richard. Now hold your wall to support Captain Carr in the farmhouse."

Lacey followed Shakeshaft's run down to his men to see the first French Light troops emerge from both the central road and the alleyways to be immediately engaged by his own Light Infantry at close range. Suddenly being the subject of sustained musketry, the French emerged no further, but went into cover themselves to return the horribly accurate fire as best they could. Lacey thought further, "Two columns, one each side. One for the 42nd and one for the 4th. We may have to split into two wings, to support against both."

He turned to find a messenger.

"Warn Mr. O'Hare that his wing may have to support the 4th and mine the 42nd."

As the soldier ran off came a further thought, 'We should now stand.'

"Stand up!"

His men behind him obeyed and all along the line his men regained their feet and his Colours rose up high, most prominent being the bright green Regimental Colour, bright against the overcast January sky. As brave a sight as the defiant line was, they immediately they began to take casualties from the cannon fire. Lacey endured this for five minutes, then went to the centre.

"Lower The Colours. They are too much of an aiming point and probably causing us harm. Order arms position."

Both Ensigns allowed the shafts to fall until the butt of each met the ground, then Lacey smiled at both Ensigns.

"Time enough to wave them again in Johnny's face, eh?"

Rushby managed a weak grin, whilst what came from Neape would be more at home on a mortuary slab, but Lacey's attention was drawn away by an eruption of firing on his left, which included the pair of cannon placed there. The 42nd were engaging the column on their side. The wind took away enough smoke for a much relieved Lacey to see that the kilted line was holding. No Scotsman was taking any backward step.

Carr's farm was a madhouse of noise, his men loading and firing at their maximum rate. He ran to the side overlooking the ground before the 105th, to see few targets there, but he noticed the smoke from the support firing of Shakeshaft's section at the stonewalls. He then ran to the front to arrive just in time to hear a comment from Miles, made to no-one in particular, but commenting on the appearance of their French opponents.

"These Frogs looks as bloody shithouse as we do! Not a bit like they sprightly buggers at Vimmy Ro!"

He noticed Carr the moment he had finished, as he brought his rifle back for a reload.

"Oh! Beg pardon, Sir."

Carr gave him a stern look, whilst Davey chuckled, then Carr took a look outside. The French had gathered themselves for an assault, but it was already faltering, most returning again to cover, leaving many prone on the ground. They were not eager to press forward, perhaps content with having cleared the village, or perhaps knowing of what was coming up behind to support them. Nevertheless, for him and his men, so far, so good.

"Well done, men. We're holding. Well done."

Miles, now reloaded, had his face close to the firing slit.

"I hope so, Sir, but I'm hearin' them drums and their shout of "Old Trousers". An' them with the red seems to have disappeared off, Sir."

Above them, Lacey was at present equally satisfied, but in his case apprehension was growing. It was not helped by the return of Bentinck.

"Lacey! Baird has been wounded. Lost an arm, so it's said, which makes me now Divisional Commander. And perhaps you, acting Brigadier!"

He controlled his horse as a cannonball ploughed up the ground to their right.

"I'm off to see Manningham over on the left and get things sorted out. Hold here."

With that he was gone, riding over to arrive behind the 42nd, who were still engaged in a ferocious firefight with their own column. Lacey knew well enough Bentinck's errand, to see Brigadier Manningham, Commander of the next Brigade towards the centre. He swallowed hard and look around. Nothing was coming out of the village, in fact it was now being re-contested, because he saw his own Grenadiers running down to attack the French on that side, presumably ordered back in by O'Hare. Where the 4th's Light were, he had no idea, but he looked up at his own men, who were standing their ground, albeit taking casualties, as he could see all too frequently.

In the line, stood alongside The Colour, Jed Deakin could not bring himself to look round. He had heard the sound of the cannonball, but beside him, where Alf Stiles and Sam Peters had been, there was now an empty space. The men next over closed up to him.

Carravoy took a careful look through the upstairs window of the house they had just run into, looking both down into the street and also into the windows of the house opposite. There had been next to no resistance so far and he saw nothing now, bar one French head at the window, perhaps dead! He gripped his sword and checked the pistol in his left hand.

"Come on!"

With D'Villiers, he led his men across the alleyway and into that same house on the other side. The few French in there bolted out the back, he had no need even to discharge his pistol, but a similar look through the back window into the next house beyond revealed a different picture; the French had retreated to now hold that en masse. Carravoy turned to his men.

"This is as far as we go! Get some furniture up to these windows."

The word "furniture" could hardly be applied to the crude benches and table that sat forlorn on the rush floor, but they were brought up and used to cover the windows, although poorly. As his men took up their firing positions, almost immediately one slumped down to the floor, his head at an odd angle showing that a ball had passed out through the vertebrae of his neck. Carravoy swallowed hard, he knew his orders as given by O'Hare, these being to keep the French sharpshooters away from the cover of the stone walls between the village and their line. Therefore, thus far into the buildings, he was content that he had carried out O'Hare's wishes, but not at all content with the order to leave the line and enter the village in the first place.

"Hold them there, men. Listen for orders."

His men were already filling the room with the noise and smoke of their firing, but all had heard the last part. Some wondered what it meant, but Carravoy knew full well. He felt himself to be in a desperately perilous position and resolved to pull back at the earliest excuse. Before being ordered down, from his position on the right of the main line, he had seen a French column alongside the village, inexorably advancing. It was worryingly large, as big as each that they had faced at Vimeiro and if it came much further they would be cut off. He went to a side window to see if the column could be seen and he was dismayed to find that it could not, there were buildings in the way. His fears increased.

On the ridge above Carravoy, O'Hare was of the same mind, but analytically calm as he viewed the oncoming column. It was now only half the width he expected, which drew him to the odd conclusion that the French were sending half of it up through the village! He shook his head, how could the half in the village hold their cohesion, or stay in line with the half outside, when negotiating alleys and back yards? Then he realised, that was why that on the outside was so slow! He drew his telescope to see that the men in the column were almost marking time, but they were yet closer, now only 300 yards down from him and level with the centre of Elvina and still advancing, albeit slowly. His eyes never left their front ranks until a French Officer appeared from the buildings to stand before them, to raise his sword and wave them forward. Their pace increased and O'Hare spoke to himself 'Happy now, are you? Well, let's see.' He turned to his Bugler.

"Sound recall."

The notes blew shrill and clear, enough for Carravoy to hear, but his first reaction was to curse, they had not been in the village but fifteen minutes. He spoke angrily to himself.

"What good has this done?"

Then much louder.

"Out! Back to the line."

In a bundle of bodies he and his men exited the building using any likely means, both doors and windows. Once clear of the village he was relieved to see that Ameshurst was, without orders, holding the first wall further up to give them cover. He looked back to see French Light troops creeping out from between the buildings, whilst their supporting column inexorably advanced forward. He heard a yell then a curse behind and to the right, one of his men was hit in the leg, but two raised him up and dragged him back to the first wall as musket balls hummed all about. Once they were at the wall and not blocking the view of Ameshurst's men, he heard Ameshurst order 'Fire', then they were over for all his Grenadiers to fall back and join the end of their main line. Carravoy came within earshot of O'Hare but he lipread, more than heard, O'Hare's 'Well done!'. However, O'Hare himself was now worried, no-one had yet reacted to the oncoming column.

Lacey saw his Grenadiers run back and drew his own conclusions, the French column on that side was now up close. He looked over at the 42nd on the left, to see that they had been pushed back a small distance, but now his own 105th had their own problems. He ran across the front of his men and even as he did so two files, four men, were felled from the incessant cannon fire, but his mind was elsewhere, forming his plan. He found a Subaltern of the Colour Company, who happened to be Lionel Farquharson. Out of breath himself, Lacey gave him the order.

"Get down to Lieutenant Shakeshaft. Tell him to pull back to the last wall and hold there. Keep back their skirmishers for as long as he can."

Farquharson ran forward and down as Lacey continued his journey over. Soon, to his relief he saw O'Hare, still erect and very much alive. He was even more relieved to see their General Commanding Officer, no less, arriving at the same time through the gap between his 105th and the 4th. Moore was composed and calm.

"Report Lacey."

"Sir! The 42nd are being pushed back and I've come over to see what should happen here, Sir. You can see for yourself, we've got our own column."

Moore nodded.

"The 4th must fight on two fronts, one wing to face down into the Monelos valley. There's another French column coming up from there. The remaining wing will support you. With them I want you to hold that column. With the 4th positioned, I'll go and rally the 42nd, they'll be your flank on that side, but you must hold above the village. That column, Lacey, is yours!"

With that he spurred on his horse and rode over to the 4th, on their right, the beginning of whose files could just be seen down the slope of the Monelos valley. Lacey looked at O'Hare and quickly gave his orders.

"First, there'll be a thick screen of sharpshooters. Battalion volley for them, then half company volleys for the main column."

"Yes Sir, but one extra thing. With the column, these walls will break their formation. We should close to under 50 and hit them as they try to come over, when they're in disorder. For the sharpshooters, I'll tempt them forward and up to us, so's they can't use the walls for cover. Then our volley."

Lacey nodded.

"You handle your wing, and I'll do mine,"

He then ran off. O'Hare did not even look to see if Lacey had gone but looked at the advancing column now under 250 yards away and its Tirailleurs even closer. There was only one order to give.

"Time to meet our guests, boys. Advance!"

With his sword over his right shoulder, O'Hare led his men forward. He looked across to his left, at his line stretching into the centre. All were in step, all with their muskets neatly at "Shoulder Arms."

Lacey had regained the centre of his Regiment, the Colour Company.

"Raise The Colours!"

Rushby and Neape lifted the heavy flags and set the butts of each shaft into the leather holder. Lacey looked across to see O'Hare's wing already advancing down, so he drew his sword, held it high and then swung it forward. His line advanced behind him.

Carr was busy with a spare rifle firing from the side of the house. Whilst looking through his firing slit, he saw what he had been expecting, the first files of the column emerging from the houses of the village, but in some disorder. He ran to the front to see little other than what he took to be Voltiguers, merely a few dodging about.

"Every other man! Side wall! There's a column advancing up. Give them all the fire you can give, Officers and NCO's first."

He ran back himself to the side wall and took careful aim himself at a leading Officer.

O'Hare had halted his men at 100 yards from the uppermost stonewall. He had seen what Lacey had predicted, this being a now reinforced screen of Tirailleurs, which had gathered before the column. O'Hare knew that he was right, that they must not give these elite troops the option of using the wall as cover which they would do at a range of 50 yards. At 100 yards distance they would be forced to come over the wall to make their fire effective. As the first of them did so and advanced on, O'Hare gave his order.

"Make ready!"

Beginning with the Grenadiers behind him, muskets were raised to the vertical, which order ran all along the line. Men who had been quivering for almost two hours in anticipation of being hit by a cannonball, now felt some relief. A French column, up close, would actually give some cover, as the French gunners would not risk hitting their own men. They held their muskets vertical, butts almost level with their armpits, muskets raised so high that the flintlocks were before their faces. They could smell the gunpowder in the firing pans, but the steel of the barrel felt cold. Fingers flexed nervously around the wood and mouths became suddenly dry, many wishing that they had taken a last swallow of water, but O'Hare strolling about quelled the terrors of many. He now walked calmly back to the ranks of the Grenadiers picking up a musket as he did so, then he spoke, almost conversationally.

"It's you're turn now, Grenadiers!"

Just as he came to their ranks, just down the slope he saw the 4th take position beyond his Grenadiers, a welcome reinforcement. Moore had carried out his promise. Then O'Hare gave his next order.

"Full volley!"

He turned and placed himself into a gap, then spoke to the soldier on his right.

"Sure, would you be minding if I had a share of your cartridge box?"

For an answer the Grenadier grinned and pulled back the flap to reveal the rows of cartridges. As O'Hare loaded his weapon; he waited for a range of 50 yards and was pleased to see some of the French Light

384

troops, as they struggled over the awkward stone walls, being picked off by Shakeshaft's Section firing obliquely across. There, Farquharson had at last found Shakeshaft.

"The Colonel wants you behind the last wall and to hold there. He said something about keeping the skirmishers back. For as long as you can."

Shakeshaft nodded.

"Very good."

Then he raised his voice above the din, so that his men could hear.

"Fall back. Fall back. To the top wall."

As his men obeyed, he looked at the fresh-faced Farquharson, a face that had not yet felt a razor.

"You stay with us. Pick that up."

He pointed to a Baker Rifle, now orphaned, but with the bayonet strap and cartridge box beside it.

"You can load a musket?"

Farquharson nodded dumbly.

"That's not so different. Sergeant Fearnley will give you instructions."

Farquharson picked up all three items.

"Now. Let's hurry."

Behind the last wall, Farquharson, having laboriously loaded the Baker, was now training it over and Fearnley gave his last instructions.

"You've got a back sight and a foresight, Sir. Get both lined up with your target and then pull the trigger. Slowly."

Fearnley adjusted the sight for 50 yards, then Farquharson sighted on a French Voltiguer, but, as he pulled the trigger, the sound of his own discharge was drowned by the roar of full volley, from higher up the ridge. Fearnley tapped him on the shoulder.

"Time to be off, Sir. Back to the Colonel. If we stay here, we'll mask his fire."

Farquharson, suddenly thoroughly enjoying himself, hefted his rifle and sprinted back with the others.

O'Hare had waited for but 40 yards range, before hitting the Tirailleurs who had come up to disturb his line. Now almost none remained. He walked forward to place himself at the head of his men and raised his sword.

"March on, boys!"

O'Hare was taking his men to within 50 yards of the uppermost wall.

Lacey looked over at the sound of O'Hare's volley, to see almost no Tirailleurs and O'Hare's wing advancing, level with his own. He looked forward to see his own opposing French sharpshooters, now revealed by Shakeshaft's Lights running off to the side. These French were also scaling the wall to close the range. He lifted his sword high and his men halted.

"Make ready!"

Muskets were fired from amongst the Tirailleurs, ragged and desultory, but they were too far away for their fire to be destructive. They advanced closer; they had to if they were to damage Lacey's line significantly.

"Full volley! Present!"

Every musket came down to level. He waited until the French were almost all over the wall and the nearest under 50 yards. The Tirailleurs were still firing and a scream came from behind him, followed by another.

"Fire!"

The noise of the simultaneous discharge of half his battalion, close to 400 muskets, was ear shattering. Lacey let the smoke clear and, as had happened with O'Hare, almost every Tirailleur had been downed, the few survivors now scrambling back across the wall, to run the gauntlet of Carr's fire from the farmhouse, but they were fairly safe. Those in the farm were giving all they had to the exposed flank of the column, now fully emerged and formed above the houses. It was the only firing that Lacey heard as he led his men forward once more to the 50 yards he wanted from the nearest wall. He looked across to O'Hare to see that now the whole Battalion was formed together and waiting, muskets now reloaded and at the "Make Ready", barrels high in the air.

Cannonballs buzzed above them, but they were now more protected by the village and the proximity of the French column. He walked over to the Colour Company at the centre of his line and took a deep breath.

"Half Company volleys!"

He heard the order being passed down the line at either side by his Officers. He half turned to look behind him.

"You're first, Heaviside!"

He heard the acknowledgement come from behind him, the order to "Lock on" then "Make Ready" and then he stood waiting, but not for long. Soon he saw the shakoes of the leading ranks of the French column approach the wall, would they stop and use the wall for cover? The brief moment of anxiety soon passed as the first rank scaled the wall, a column was a shock tactic, not one for attrition. He retreated to stand beside Neape, as Heaviside gave the order, "Present." Lacey watched as the first two French ranks jumped down on his side before the wall, with the third now climbing it.

"Now, Heaviside."

"Fire!"

The first rank of the Colour Company delivered their volley to be followed but a second later by the front ranks of the two Companies left and right. For a moment all was smoke, then it thinned for Lacey to see the effect of their continual fire. It was as though the French were marching into some kind of mincer, for as they mounted the wall they were blown back over, whilst on Lacey's side, were left none but dead and wounded. The noise was appalling, as intimidating as the incessant fire. His timing had been perfect and, with few French Officers now remaining to order a halt, the column simply kept on coming until those following, seeing the heaps of dead at the wall and also thoroughly intimidated by the continuous noise, at last halted. At that moment, Lacey heard a voice that he had come to recognise, that of Moore.

"Go on, Lacey, up to the walls. They won't stand. Over and clear the village. The 42nd are with you!"

Lacey took a deep breath.

"Maintaining fire! Advance!"

Lacey was asking his men to load as they walked, but he was confident that it would make little difference to the impact of their fire, nor the cohesion of their volleys. He walked forward himself, the Regimental Colour at his elbow, and his men followed him, still firing and advancing down to contest the wall itself. Faced with such a potent and inexorable advance as a long line of Redcoats, the French fell back even further, scrambling back over the lower walls between them and the village. At the first wall, Lacey halted.

"Fix bayonets."

Within 10 seconds every musket had its bayonet.

"Over and amongst 'em, boys!"

He led his men forward, up and onto the wall, treading on the pile of bodies, but for a moment he stood alone on the wall urging his men on.

"Charge, boys, Charge! Hurrah for the King! Hurrah for the One-Oh-Five!".

Once over, being more elderly than most around him, he was soon overtaken, but his men had gone over the wall like a red wave, yet they did not get close enough to bloody one bayonet. Already halted and confused, assailed both from the front and from the side from the farmhouse, the men in the column turned and ran, leaving their dead and wounded stretched on the winter grass between the rows of stonewalls, all the way back to the houses of Elvina.

Lacey's Captains led their Companies on, running down through the empty alleyways and roads, for the French did not attempt a stand at the top of the village, nor even attempt a delay. On the left, Shakeshaft led his men down to accompany the advance, but he did not neglect a cheery wave to the Lights in the farmhouse as they passed them by. Neither did Saunders, Bennet nor Byford, but their salute was more two fingered, especially when they heard a deep insult, delivered by the unmistakable voice of Miles, shouted through an open window.

The whole of the 105th, bar Carr's Section, ran in fierce pursuit down through the alleys and through the gardens, pushing the French on and out. Some veteran French, encouraged by a brave Officer, tried to make a stand at usable walls and houses, but those that tried were surrounded and killed or captured after fierce hand-to-hand fighting. What faced both Ameshurst and Heaviside was not untypical, a line of French across a road or an alley, but the 105th were too close on the heels of the French to give them time to become steady. Backed by their men, both these Officers, with swords drawn, crashed into the half formed line to bayonet those who stood and then chase on after those who fled. Between the narrow streets and alleys of ancient stone the fight was quick, fierce and brutal.

Soon they reached the Church, then the last houses and last walls, some newly built by Lacey's own men, and eventually this point was reached by him, keeping up as best he could. He came to the wall built by Davey, then he stopped and looked over, but what he saw shook him deeply. Some of his men, led by Major Simmonds, had gone beyond the village, either from losing their bearings or over eagerness, but they were down the slope and out in the open field with the ejected

French now quickly reforming on their own unbroken troops who were advancing up the slope to support the first attack. Worse, the sounds of the ongoing conflict between the 42nd and their column were now increasingly intense. That French column was still engaged on Lacey's left, and it sounded further back up the slope in relation to him, meaning that the 42nd had halted earlier that his men. On top of that, the French gunners now saw red uniforms amongst the buildings and had lowered their aim and the first hit was on the church bell tower. These guns could now be seen by Lacey and he judged them as being at almost point blank range. He immediately gave his orders.

"Hold here! Take positions!. Support our men out there."

In great contrast to the near desperation building at the bottom of the village, inside the farmhouse, all were peacefully cleaning their rifles, except Joe Pike, now examined by Davey. He was totting up a row of pencil marks, clear on the plaster of the wall. Davey was immediately incensed, he knew what it meant. He went to Pike and spun him around, to point at the marks.

"And what the Hell is that all about? How many will be enough, eh? How many?"

He was stopped by Ellis.

"Never mind your cosy conversations. Get back up to the line and get some more cartridges. Take nine men, two boxes between three."

He indicated Pike.

"This one's tough enough to run the middle."

Davey tapped Miles on the back and six others, then they jumped out of a back window, now with its boards removed. They ran up the hill to the last wall, hearing the renewed cannon fire, but once over the wall Davey seized one of Pike's shoulder tabs and pulled him around, As Pike spun, Davey pointed at the utter carnage at the wall, its stones now more blood red than grey. French bodies were piled three, four high, some still alive, groaning in their last agonies.

"Does that count as enough? Take a bloody good look at that!"

He saw the change on Pike's face.

"Will that sort you out?"

He spoke no more, but pushed Pike up the hill to follow the others. They found the stacks of cartridge boxes behind where the 105th had previously stood their line, that place now marked by a row of 105th dead. Without seeking any permission, each three of the nine took two boxes, the one man in the middle holding the rope handles one in each hand. Then they started back down.

For Lacey and his men, the situation had now, indeed, become desperate. They had no formation and many were still back in the houses plundering dead bodies, but the French attack on the village had been immediately renewed and in greater force. Lacey knew that there could be no formation firing, he had to leave it to his men to load and fire as fast as they could and hold off the French in any way available, but, in despair, he knew that he had little more than half his men trying to hold back this new attack. On top, they were outnumbered perhaps 6 to 1. Lacey found a Corporal.

"Get around the village and tell others to pass it on. Fall back, now. Get out and back up!"

The Corporal saluted and ran up the alleyway, then Lacey looked over the wall. There were several red-coated bodies lying out on the grass and one seemed to be Simmonds. Worse, the French Officers had sent forward their front ranks at the charge; no more the steady pace of the column, because now, knowing that they were faced with but a few Redcoats in the buildings, these French were now running forward in skirmish order. Lacey had about a dozen men around him.

"Back. Fall back!"

Doing his best to keep up, he and his men sprinted up the alleyway. Soon he was overtaken, but the last two of his men held at a corner, their muskets pointing down the alleyway and, as he passed them, they fired. Gasping for breath he pulled them backwards to again ascend the slope.

"Not here. Too soon. Further up."

He resumed his upward run, hearing French shouts behind him.

Meanwhile, back in the farmhouse, Carr was talking to Ellis.

"Casualties?"

"Two, Sir. Dead. A few more wounds but they're still able."

At that moment a soldier turned away from his rifle slit.

"Sir. You'd better take a look at this!"

After exchanging a quick quizzical look with Ellis, Carr ran to the same rifle slit. What he saw brought an immediate order.

"Stand to!"

As his men rushed back to their positions, Carr tried to make some sense of what he could see. Soldiers of the 105th, in disorganised groups, were pouring out of the village, to run further up and over the stonewalls. There could only be one conclusion, that the French had quickly renewed their attack and were pushing his Regiment back and out. He took himself over to the right to see Heaviside organising a rearguard at the first wall. Carr spoke to himself, 'Well done, Joshua', but many 105th were still emerging from between the buildings. Some kind of crude order was being restored and the men, at least, were not just running up and away, instead, once over the first wall they were looking for orders, many simply manning the wall to make some kind of a line. Carr turned to his men.

"I want your most accurate fire. Take careful aim and don't risk hitting our own. Take out the first French you see. Give our lads a better chance."

He took up a rifle himself and slid it through the firing slit to take careful aim at the corner of a building alongside a main alleyway. Redcoats were still emerging, then he saw his Colonel, running as though through treacle. A bluecoat was close behind, then Carr fired.

Lacey thought his heart was coming out of his chest, then he felt a hand seize the strap of his sword scabbard. He felt the tug back, then nothing, other than the sound of a bullet hitting what sounded like a skull. There came a cry from the slope above.

"The Colonel! Help the Colonel."

Suddenly he was surrounded by redcoats, some with bayonets pointing back in challenge, some training their muskets back to the French, some simply helping him. He was practically carried back to the wall and then lifted over, to fall in a heap on the other side, but there were plenty of hands there to help him stand back up. He looked into the face of a soldier, it blackened and filthy, with specs of gunpowder stuck to his chin by the sweat.

"Are you all right, Sir?"

He reached around, down to his left side, to his canteen.

"Here Sir. Have a swig of this. 'Tis good stuff, I traded it last night."

Lacey grinned between gasps, then he lifted the canteen to his lips to taste the good brandy. He grinned openly, this returned by the soldier.

"We can't stay here too long, Sir. Johnny's lookin' lively."

Lacey hauled himself upright from resting on his knees, but the soldier was still speaking.

"You get back on up, Sir. We'll hold 'em here, then fall back ourselves."

Lacey nodded and turned to begin his journey, leaving Heaviside bellowing a whole series of uplifting exhortations from The Bible. Soon this was drowned by continuous musketry, his men had formed a solid line and were holding the first wall. Now somewhat recovered, he took himself up to the second wall, to find it well manned, but there was no Officer. Then Private Bates came running up, bugle in his hand.

"Sir! Found you, Sir."

Lacey took a deep breath.

"Well done, Bates."

Lacey looked along the wall, all his men there had their muskets held over the top, almost all had detached their bayonets to make reloading easier. Lacey took another deep breath.

"Make ready!"

The muskets were all lifted.

"Sound recall."

Bates did so and Heaviside's men fell back from the first wall and across the space to scale that held by Lacey and then, led by Heaviside, they all ran on further to form at the next wall up. The French came over the first wall and into view. They were firing, but many were falling from the rifle fire from the farmhouse, which was duly noted by Lacey. He waited until they were thickly gathered on his side.

"Fire!"

All was smoke.

"Fall back."

They did so immediately and ran back to Heaviside's wall, all to send one more volley into the smoke drifting past them. With that, Lacey ordered all back to their position on the ridge but what he saw there did nothing to reassure him, for all was a confusion of men looking for their Company Captains. With the French coming on, he doubted that his men could provide any kind of organised resistance against them and one look down told him that a solid French column was already half way up the slope between him and Elvina.

***

392

Davey's munition party were running down through the maze of gates that threaded through the jumble of walls, when Davey saw his Regiment in full retreat running back up the slope. Two seconds more he saw the blue of their pursuers and after two more he saw the French reaching the front of the farmhouse in dense numbers. No crowd of skirmishers this, but a whole column.

"Run! Quick!"

They ran the last yards, heaved the cartridge boxes in through a window then propelled themselves through to follow. Within a minute the boards were back up, but the French were all around, hammering at the back door and attacking the front. Every man inside was loading and firing at a frantic pace, but Ellis noticed the nine, not immediately manning a firing slit, stood in some confusion.

"You! B'ain't this your fight, nor summat? Get to a window, or I'll do you for desertion, see if I don't."

The nine immediately obeyed and joined where they could. Davey looked at Pike. His pencil was no-where to be seen.

Lacey hurried to his men gathering on the ridgetop, but the confusion had not in any way diminished and a rough guess showed that now there were many of his Regiment missing. Bates was still at his side.

"Sound form up."

However, as Bates blew the notes, an Aide from Moore galloped up.

"Sir! A message from General Moore."

His horse performed a circle as the French cannonfire, now resumed, whistled above.

"Pull your men back and reform. The General is bringing up the Division Reserve to relieve you. Please pull your men back to clear the way."

Much relieved, Lacey nodded and the Aide rode off, then Lacey pushed his way through the melee of men, taking Bates with him.

"Sound fall back."

Leaving Bates to sound off again, Lacey walked to where he could see The Colours. He was utterly tired and weary, but he knew his role, that being to rally his men and that began with the Colour Party. He went immediately to Rushby and Neape.

"Fall back, 300 yards and stand up The Colours. The men will form on you. Hurry, get there first."

As the two young men trotted back, their Colours over their shoulders, Deakin remained to look at his Colonel, his face very anxious.

"Sir. Are you all right, Sir?"

Lacey nodded, then, extraordinarily, he patted Deakin on the arm.

"Simply my age, Colour Sergeant. Just feeling my age."

Deakin nodded, grinning, but the rest of his face showed that he was not reassured.

"Yes, Sir. There's a lot of us feelin' a dose of that, Sir."

Lacey smiled back and gently pushed Deakin, his Colour Sergeant, on after the two Ensigns. Lacey took the time to walk after his men and, by the time he got there, they were in some order. O'Hare came running over.

"Are we in reserve? Could we be called upon again?"

Lacey placed his hand on O'Hare's shoulder, more for his own support than to deliver any comfort.

"I fear so, yes. We must assume so."

He took a deep breath.

"Have you seen that new column?"

O'Hare nodded, then asked a question of his own.

"Have you seen Simmonds?"

"No, not back here. I doubt he got out of Elvina."

He looked fully at O'Hare, but their looks exchanged more between them than Lacey's simple order, which came next.

"Ammunition!"

As O'Hare ran off, Lacey walked on to take a look at his men, perhaps they would encourage him as much as he could them.

On the left of the reforming line, stood Richard Shakeshaft and his Section of the Light Company. He looked at his men; many carried wounds, some disabling, but all were stood their place and all were drinking or eating. They had had nothing hot since leaving their messfires before dawn, bar Moore's tea and hard biscuits. Saunders, Bennett and Byford were on the far left of their Section, which placed them on the far left of their whole Battalion. They heard a sound from behind them, from behind their left shoulders, then turned to see something which even they could not fail to judge as a wholly impressive sight. It was two battalions in line, two deep and all in step, one battalion behind the other in support. Each line two men deep, they stretched across the ridgeline for the width of 400 men and the Colours for both Battalions, although reluctant to spread out fully in the faint wind, were clear enough and special enough to warrant curiousity; there were four and all crimson.

Saunders was nearest as the ranks passed by, these being the Grenadier Company of the new arrivals, in their place on the right of the passing Battalions.

"Who're you?"

A growl came from a huge Grenadier Corporal, with distinctive sideburns.

"First Battalion, First Foot Guards."

Saunders was not quietened.

"And them behind?"

The same growl.

"Second Battalion!"

The lines marched on, drawing the admiration of all on that side of the 105th, all stood at rest eating hard bread and drinking water. Meanwhile, new supplies and sustenance were being passed out, ammunition from the Stores Orderlies, bread and cold meat from Sedgwicke and the followers, which he had organised throughout the morning. Shakeshaft looked at Fearnley, now chewing bread and cold boiled pork. The question was now in the mind of both, with their Company now having been pulled pack and now relatively at peace, 'Wonder how the Captain's doing?'

Were he there to see for himself, Shakeshaft would be in a state much more uneasy. With the windows boarded up, bar a slit, Drake's Section were fighting in a nightmare of near dark, made more so by the smoke from their flintlocks. Drake had pulled lightly wounded men away from the firing slits to clean rifles passed back to them, these being too fouled to be fired until their barrels were cleaned and their flint changed. Losing the fire of three men was little compared to many muskets failing to fire. There was little talking, beyond curses, practically the only sound there was came from their musketry, but then Ellis pulled back from his firing slit.

"They're tryin' again, Sir!"

Carr ran to the slit and looked out. More French were running to the porch at the front, carrying anything considered combustible for a second attempt at setting the building alight. All around lay dead French soldiers. At less than 100 yards range, the Light Company's rifles were exacting a fearful toll on the French Voltiguers, while their main column was pushing on past and up to capture the series of stonewalls, but with noticeably less than usual Gallic vigour. They were well within range of Carr's farmhouse and the rifle assault from Carr's men onto their flank, killing Officers and NCO's, was such that the French Commanding Officer had all but abandoned the effort to take every wall up to the British line. Seeing the damage inflicted by Carr's Lights, he had detached men onto the farmhouse which stood like a blockhouse fully in the path of the right hand files of his column. They had tried to fire the farmhouse once, but it was all stone with a tiled roof, even the front porch was of stone. The first attempt had been a total failure, not enough material that would burn could be set in place, but now they had gathered all that could be found from the nearby houses of the village. Carr had to quickly decide as he looked out, whether or not to draw men away from the back and the sides to strengthen the defence of the front of the house? A volley of shots from the floor above made the decision for him and many carrying armfuls of fuel for the fire fell dead. Some material was thrown into the porch, but anyone carrying a burning torch had given themselves a death sentence. Carr was satisfied. His men were holding well and providing grave problems for the French before they could advance further up the ridge, but then a cannonball, fired by a weak charge, penetrated between two houses, ricocheted off a wall and passed through the roof of Carr's farmhouse. Red tiles showered down.

***

Although perhaps they should have been, thoughts on the fate of Carr and his men were far from the minds of Lacey and O'Hare as they stood together. Lacey was looking along his line, estimating his losses so far, whilst O'Hare was watching the backs of the Guards as they advanced forward to the 105th's old position above Elvina. Suddenly, there was a roar of musketry as the 1st Guards Battalion delivered their first volley, then O'Hare's face changed from one of relieved pleasure to one of horror.

"Oh God! No!"

He had been watching Moore leading the Guards forward, when he saw his Commander suddenly sent spinning from his horse, the epaulette from his left shoulder flying through the air and, sickeningly, his left arm swinging back as though it were the arm of a corn flail. O'Hare touched Lacey's arm.

"Moore! They've got Moore!"

Shock also came into Lacey's face and he spun around to follow the direction that O'Hare was indicating. The Guards had advanced on, leaving Moore and a group of anxious attendants behind. He was on the ground, barely moving, but a blanket was being spread beside him. After being carefully lifted onto it, with Guardsmen and men of the 42nd on the four corners, he was carried back and soon he reached the 105th, but little could be seen within the blanket. Lacey looked at O'Hare.

"What do you think?"

O'Hare shook his head.

"No. He was hit by a round shot. I'd say it took away the whole of his left shoulder."

He continued shaking his head.

"He'll not survive that."

Lacey had recovered slightly.

"So, now it's Hope. He leads us now and where he is, God only knows, and I doubt that Hope himself knows that he is now Commander in Chief. He'll be running his own bit of this affair, just like us."

O'Hare managed a slight wry grin, but Lacey was continuing.

"Right, we've lost Baird and now Moore. I suspect we're going to be on our own, Bentinck has more general concerns."

He paused in thought.

"Get over to the 4th and see what's happening there. If they're holding their slope, then I'll feel much happier. Don't come back yourself, take charge over there and send back runners."

O'Hare saluted and ran off, across the front of the 105th's line. Many questions were thrown at him as he progressed along, but all were of the same nature.

"Are we done, Sir? Are we to fight any more?"

His answer was always the same.

"Could be, boys, could be not! 'Tis more up to the Johnnies than anyone else. But, whatever, we'll give it the same go as we did before!"

Nelly Nicholls noticed him and pushed through the ranks.

"Major O'Hare darlin'. Will you not take a bit a meat, for yourself, here now?"

The 'darlin' part was rank insubordination from a follower, but O'Hare ignored it. He fully appreciated what they were doing and was quite moved, so he reached out to take the dough cake with a slice of pork inside it.

"Who organised this. Mrs. Nicholls?"

"Well, Parson thought we should and so we took it up, we bein' me an' Bridie an' the rest. Are not the men all now a needin' of somethin', would you not think, your Honour?"

O'Hare's mouth stretched with emotion as his jaw pressed closed.

"Indeed they do, Mrs. Nicholls, and may God bless you for it."

He took a bite to show his appreciation and then ran on, this time to be accosted by Captain Lord Carravoy and Lieutenant D'Villiers, stood with their Grenadiers at the far right of the line.

"Sir. Will we be called forward again?"

O'Hare's answer to Carravoy was more stern than his answers before.

"Assume so!"

He stood still.

"The 4th, on our right. What can you say?"

"Very little, Sir, other than the sounds of a conflict much as our own."

D'Villiers felt he had a contribution.

"Such as to suggest that we may have been turned from there, Sir."

O'Hare became slightly annoyed.

"But it hasn't happened yet has it, and you've seen no French from that direction, have you? Have you taken a look?"

D'Villiers recoiled somewhat.

"Yes, Sir, I mean no, Sir."

O'Hare frowned.

"Give me two runners."

Two men were indicated and they followed O'Hare to the top of the slope that overlooked the Monelos valley. O'Hare looked all around and felt much more reassured. Their old comrades of Paget's Reserve Division were pushing the French back, across the whole stretch of the valley, even the French cavalry, who could not manoeuvre owing to the stone walls also creating their own maze across the valley floor. More immediately for O'Hare, the 4th were plainly seen down the slope, intact, but now too far down to play any part in the defence of Elvina. However, most importantly and reassuringly, there would now be no threat to them from out of the valley. If the 105th were to be called forward again, their right flank was safe; the French attack up the valley of the Monelos had been wholly repulsed and thrown back. He then turned left to look over to the French slope opposite Elvina. From there he could see beyond the village and what he saw there was a complete counterbalance. Two more French columns were descending down the French slope opposite, very much in the direction of Elvina. O'Hare turned to one of the Grenadiers.

"Get over to the Colonel, he's over on the left. Tell him that there are now no French on our right. Is that clear? No French on our right."

"Yes. Sir."

"But, when you've told him that, tell him that the French are reinforcing their attack on the village with two columns, from across the valley. Finally, I've asked the 4th for support."

He looked at the man, just a ranker but the face looked more than halfway intelligent.

"Three things. Repeat!"

The Grenadier did so, perfectly, then saluted and ran off. O'Hare reached into his pocket and scribbled a note, which he handed to the second Grenadier.

"Go down to the 4th, there. Find their Colonel and give him that. His name is Wynch. If you lose it, simply say that Elvina is being attacked again and that Colonel Lacey would appreciate any help he can spare above on the ridge for its defence. Is that clear with you?"

The Grenadier saluted.

"Sir."

O'Hare took one last look into the valley at the lines of red pushing back those of blue but he soon turned away from this luxury and ran back to his own men.

Lacey received the message quickly, then, with equal urgency, he set off on his own journey, to find the Colonel of the Guards and tell him what he had heard from O'Hare. Finding the Colonel was not difficult; he was stood before a knot of very well tailored Officers gathered behind their Colours. There was a line of casualties behind the ranks of Guards and the roar of musketry was incessant as Lacey ran up to stand just in front of what was evidently their Colonel.

"Lacey. 105th."

The Colonel looked at him as though Lacey had just disturbed him whilst engaged in fly-fishing.

"Anson! First Foot."

"Have you heard from Bentinck, or Hope or anyone?"

Anson's brows came together.

"No. Should I?"

The last "I" was long drawn out, which irked him somewhat, but Lacey ignored the question.

"There is another French attack coming up from below. Two columns, as I've been told. They will come against you! My men are stood ready in support, but we've taken heavy casualties. Also, the 4th, over on the right have been asked for support, but they are now somewhat detached from us and are involved in the valley."

He paused, but Anson was still regarding him quizzically, so Lacey continued.

"If you want my opinion, this is Johnny's last throw. The day's wearing on, and he's got nowhere. We hold this; we've won!"

At last Anson answered.

"And I should do what?"

Lacey took a deep breath.

"If nothing else, make preparations. The 42nd on your left have been in action since the first shot. They can have very little left. Expect them to withdraw."

At that point help came to Lacey, for arriving on horseback surrounded by Staff came Bentinck and another high ranking Officer whom Lacey took to be Hope, which was immediately confirmed by Bentinck.

"General Hope. This is Colonel Lacey of the 105th. Colonel Anson I'm sure you already know."

Hope was looking directly at Lacey, but it could have been Bentinck looking down, were it not for a slightly fuller face.

"Ah, Lacey! I've been hearing. Both to you and your men, well done!"

"Thank you, Sir."

"Are those yours in that farmhouse?"

"They are Sir."

Hope looked at Bentinck.

"Hear that, Bentinck? With such as that we've no business losing, what!"

"No, Sir."

However, Hope was now looking at Anson and, whilst he spoke Lacey previous words, Lacey had to suppress a wry grin.

"You are about to be attacked. Two columns are coming up. One, the largest will hit over on your left, beyond the 42nd, but they are Manningham's problem. A smaller is coming to reinforce them in the village. They know they've got it, so they hope to spring on from there. The 42nd are almost used up, so I've placed your 2nd Battalion behind them in reserve. But, in the first instance, there's a column coming out of that village and it's all yours, Anson. If we hold here, Johnny will give up the game, of that I'm sure."

A pause. Then he looked at Lacey.

"What is the state of your men?"

Lacey took a deep breath and gave the only reply he could.

"If required, we will support."

Hope nodded.

"Good luck to you both!"

As the horses wheeled away, Lacey looked at Anson.

"My men can come up on your right, if you need us."

Anson nodded. His mood had changed, but Lacey had another enquiry.

"The farmhouse. Is it still holding out, can you say?"

Anson's face changed again, to one of respect. He had seen for himself the incessant conflict around the farm and knew full well how its occupation was wholly discomfiting the French trying to close up from the village to his own line.

"It is, and whoever's commanding in there, give him my compliments, if he lives!"

Lacey nodded.

"Thank you. His name is Carr and he has a habit of coming out of scrapes, albeit with a few changes to his appearance."

Anson grinned.

"An occupational hazard!"

Lacey nodded.

"Right. I'll get back. My men stand ready for your call, should you choose. Good luck!"

Anson offered his hand and Lacey took it, then he ran back, avoiding four soldiers carrying back a wounded Officer, him with no left leg below the knee. More cannon shot added to the furrows already plentiful all around.

Lacey arrived back with his men, determined to support Anson if asked, but he was convinced that they would be and he went first to Shakeshaft.

"Give me four runners!"

Four were picked and soon paraded.

"Go along the line and make sure that every Captain knows that we will be called forward again. Every man to have a full cartridge box, water in his canteen, a good flint and a clean musket carefully loaded! Go now."

The four saluted, which Lacey returned, then he stood before Shakeshaft's Section, listening to the conflict, hoping that he was conspicuous, should a Guards' runner need to find him. One of the runners despatched was Byford and it was he that came to the Colour Company and found Captain Heaviside.

"Sir. Message from the Colonel, Sir."

Heaviside looked at him, saying nothing, but Byford evidently had his full attention.

"We will be rejoining the conflict, Sir, perhaps quite soon. The Colonel is anxious that each of your men has a full cartridge box, Sir, and some water. Also that their weapon is fully serviceable, Sir, clean, and with a good flint. Then loaded, Sir."

"Imminent?"

"Possibly, yes Sir. Any minute."

Heaviside was more than a little surprised that the word 'imminent' should be included in the vocabulary of a mere "ranker"; he wasn't really sure why he used that word, perhaps instinct, but, nevertheless, Byford received a stock Heaviside response.

"The Lord bless thee, and keep thee. Numbers six, verse 24."

Byford laughed, but he remained within the bounds of respect.

"Yes Sir. Thank you, Sir. The lot is cast into the lap; but the whole disposing thereof is of the Lord."

Heaviside nodded gravely and finished the quote.

"Proberbs 16. Verse 33."

With that reply, Byford saluted and ran on down the line, to check that the remaining Captains knew. Meanwhile, Heaviside was calling on his two Colour Sergeants.

"Deakin! Bennet!"

Both were already within earshot.

"Did you hear that?

"Sir!"

"Go now and ensure that all is as required."

Both saluted and left their places to check down the line. All had by now, from the efforts of Sedwicke and others, a full cartridge box, and most had automatically cleaned their musket and checked the flint, therefore but a few now needed to set about the short task. Meanwhile, Lacey was dividing his attention between his men obeying his order and the conflict that was being played out above Elvina, evidenced by smoke, noise, wounded men falling back and the unending cannon shot droning above. Thankfully his men were formed below the height of the ridge.

Suddenly the sounds of battle doubled and doubled again. Lacey listened and decided that all was far from over. Beyond Anson's Guards, to the left of him, the sounds were probably from the largest column contesting with Manningham's Brigade. He could do no more than look as far to the left as was possible and there he saw the 42nd falling back, but then stop. He could see, even from that distance, that the Highlanders were fixing their bayonets. Their Colours were prominent and then he heard the wail of pipes, the tune much broken by the sounds of battle. Plainly, the 42nd were not quitting the field, if needs be they would confront the French at the point of the bayonet. However, Lacey had concerns for his own men, they had been engaged for almost four hours, the same as the 42nd, prior to that they had endured a cannonade for two hours and faced two French attacks, repulsing the first alone. Also, unlike the 42nd, they had cleared Elvina, as ordered, then had been forced back out. His casualties had been heavy. He asked himself the anxious question; did his men have anything left in them to yet again face up to what would be fresh French troops? Would they stand? Would they give way when they came under sustained fire once again? It would not be the first time, in his own experience, that troops so tried, had melted back when asked for so much, once too often.

He brought himself back to the immediate situation when he saw the four runners coming back. He motioned one over, Byford again.

"Name?"

"Private Byford, Sir."

"Get over to Major O'Hare on our right. Ask him if anything has come from the 4th. If no, he is to request again. Clear?"

Byford saluted and ran off, to yet again cross the front of the whole battalion. Whilst the minutes passed, if anything, the sounds of the battle on Manningham's front intensified, then also did that before Anson's Guards. Lacey walked to the summit of the ridge to make himself more prominent and soon discerned a figure leaving the ranks of the Guards, carrying a sword. Anson was sending an Officer and Lacey thought it right to walk forward to meet him and the distance closed. The Officer, a Subaltern, halted and saluted, this returned by Lacey.

"Colonel Lacey, Sir?"

Lacey nodded.

"Colonel Anson asks that you support us on our right, Sir. The French are emerging in force from the village, on that side, beyond our line. Unopposed, Sir. They may well turn our right."

He paused.

"What shall I say, Sir? In reply?"

Lacey took a deep breath.

"Tell Colonel Anson that the 105th are on their way and he can be assured of our best support. Our very best. Tell him that."

The subaltern grinned, saluted and hurried back. Lacey took another deep breath and hurried to the centre of his line, this being Heaviside's Colour Company. He took a good look at his men. They all looked utterly weary, fifthy, black faced, many bandaged and all with their uniforms practically falling off them. Behind them was a line of prone figures in red uniforms, some still, some writhing in agony attended by the followers and the bandsmen. Perhaps they did not need inspiration, but they were going to get whatever he could give. More cannon shot droned overhead. He walked to Heaviside.

"Captain Heaviside."

"Sir!"

"Advance The Colours."

Heaviside turned to Deakin, Rushby, Neape and Bennett.

404

"Twelve paces forward."

As the four marched forward, Lacey looked for Gibney and found him, in the gap left by the Colour Party.

"Sar' Major. Call the men to order. We're going back!"

Gibney left the line, turned to face the ranks and sucked in a huge amount of air.

"Paraaaaade!"

Out from the centre, beginning with those who heard, all came to "order arms."

Gibney took another huge breath.

"Shoulder! Arms!"

As a wave from the centre, all muskets came up to left shoulders. Lacey looked both ways, left and right. His Officers had taken the signal that, if The Colours were advanced, then so should they be, stood out before their men, to lead them forward, and with swords drawn. Soon all stood before the line, at the attention. Feeling his mouth quivering, Lacey clamped his jaw together, then he drew his own sword, sloped it back, and walked on to stand ready to begin the advance. He waved his sword forward and marched on and then the tears really did start to flow. His men were singing 'Brighton Camp'.

In the farmhouse, his situation was as desperate as Carr imagined it could be. Their ammunition was almost used up and so the French had at last managed to light a fire in the front porch, which added to the choking smoke and Stygian gloom that was the ground floor. With the rifle fire from the windows now slackening, the French were attempting to pull away the boards that covered them. This was resisted mostly by rifle butts being slammed onto any fingers that came through the slit, but the French were using their own musket butts to smash the boards backwards. At one window, the lowest, the boards were gone and, if any defender showed himself, a volley of musket shot came through the opening, such that one dead and two wounded had been added to the list of casualties. Drake had taken charge at this vulnerable point and, with careful timing, when French uniforms appeared, himself and two others either side, used their bayonets to force them back, the French bodies preventing musket fire from being used against them. In that way they were holding, but it was desperate work.

Carr took stock. Ammunition was low, but they could fire from the upstairs windows unhindered. Riflefire was their best defence, he concluded. He shouted through the gloom and above the noise.

"Fire until you've three cartridges left, then butts and bayonets only. Keep them out!"

He picked up the final box of cartridges, already a quarter empty and carried it upstairs to distribute the cartridges himself amongst the 30 men at the windows. That done he took off his own cartridge box, to drop it besides Davey, then shout above the noise.

"You've all the ammunition we have, men. Make them count! When you're down to three, then come downstairs."

He walked back to take a look out of a side window, over towards what he knew was Manningham's Brigade. There his hopes for survival fell a mark or two. On that side was a sea of blue uniforms and shakoes, with several Regimental Eagles raised high. A huge column had pushed up, past that side of Elvina and was engaging the British there, whose thin, two deep lines could barely be seen through the smoke and beyond the foremost French heads; they were so close to each other. He concluded that it must be a conflict of appalling ferocity, over less than 50 yards. He thought he could hear bagpipes, but they were faint and drowned by the sounds of the conflict, the column's drums and shouts of "Vive l'Empereur." His one comfort was that none from the column had detached themselves to add to those assailing his own men.

"Pike! Here, keep watch. If any come our way, I want to know."

Joe Pike took Carr's place and looked through the firing slit to see a target he could not miss. He looked into his cartridge box and counted seven.

"John. How many did Mr. Carr say we were to keep?"

Davey fired his rifle, then answered as he reloaded.

"Three!"

Pike reloaded, and pushed his rifle between the boards. The French column was so wide and thick he had but to pull the trigger. He sighted, then stopped. He had lost count of how many he had killed that day and these were no threat to him, nor his comrades, at least not so far. To use the four cartridges would just be a mindless killing, murder of men whose backs were to him. He released the trigger and set himself to merely keep watch. His rage was gone. He thought of Mary and the children they would have, then he pulled out his spare four cartridges and tossed them to four men defending the threatened side of the farmhouse.

Meanwhile, Carr was nothing like as at peace. He had reached the back wall of the ground floor, fixed a bayonet and added himself to the defence of one of the windows.

"Hold them out there, men, it can't be too long now. The lads above will drive these back! Then I'll get you all a brandy! Best French, off some dandy Officer!"

A volley of musket balls smashed through two boards, drowning out the faint laughter, then came the musket butts. Carr brought up his bayonet and waited for the first Frenchman to come through.

The 105th were advancing into cannon fire, taking casualties with every yard, but the first ball sent their way by the French gunners damaged no one, bar giving Rushby a huge problem. It severed his Colour shaft one foot below the cloth. Rushby did not hear the passage of the ball, but he felt the impact on the shaft and then the weight leave his arm. The Colour itself toppled back onto the heads of the rank behind, but they soon passed it forward. Rushby looked at Deakin in utter consternation.

"Sergeant?"

Without a word, Deakin drew his bayonet and fixed it to his musket. He reached for the top of the shaft, thrust the point up through the top corner of the cloth, next to the shaft, and then handed the whole to Rushby.

"Just do your best with that, Sir. I'll find another musket."

With that he was gone, but he was back within a minute. There were muskets aplenty left on the ground, which were once the weapons of men now dead, or with the Surgeon. They marched on; cannon shot ploughing the ground before them, whistling overhead, or killing men. Their ranks already thinned, the 105th could now leave an interval between files, a gap of less than two feet but the space saved lives.

A file went down on Deakin's right, but his attention was intent on what had just happened before him. Heaviside was down; the ball that had hit him had passed on to take out the same file. Almost immediately they came to his prone body on the turf and Deakin was partly relieved to see his Officer still alive, but gasping as though his last. Nevertheless, Heaviside managed some words.

"On! Go on, boys. Keep the line."

Then they were over him and marching on, with the battered Elvina churchtower coming plainly into sight. Deakin looked at Rushby, awkwardly holding up The Colour from the butt of the musket.

"Sir!"

Rushby looked at him.

"You command the Company now, Sir. Now that Mr. Heaviside is down."

Rushby looked at him in horror.

"Yes Sir. You're our Senior Officer now!"

Without waiting for a reply, Deakin continued.

"I'll call up Mr. Farquharson to take over The Colour. He's just over there."

Farquharson was called over and seemed delighted with the role he was being asked to play, the exact opposite of the emotion now torturing Rushby, but Deakin had some understanding, as Rushby walked forward to take his place out before the line.

"Just stand with the lads, Sir. In the front rank; that'll do! After that, they'll take it on themselves."

Rushby nodded and drew his sword, but, seeing how much it was shaking when he held it up, he sloped it back over his shoulder.

Lacey was looking all around and deciding. Anson's Guards were above the uppermost wall, in fact some way up from it. This he did not like, because the French were there and firing at the Guards from cover. Also, and worse, beyond the Guards right flank the French were over that same wall and were advancing up the slope unopposed, in sufficient numbers that would soon force that right wing of the Guards to curve back and "refuse a flank". This would take men away from firing at the French directly opposite them. These were coming out of Elvina, which must be full of French, such were the numbers advancing up from it and coming on confidently, evidenced by French Officers capering about, encouraging their men to advance into this obvious gap in the British line.

Lacey took his bearings again. The centre of the 105th, that being himself, would come onto the far right of the village, its top corner, so his right wing, commanded by O'Hare would have no opponents. He looked over to see O'Hare in the distance, leading his men and carrying a musket, but he felt able to trust him to wheel his men left and inwards to assault that side of the village. An assault from that direction would do as much to make the French pause in their advance as any amount of fire from the front. On his left, his files would just clear the Guards' firing line. That was much to the good; however, once past, they must first push the French back to the topmost wall and use it for themselves. He gave his first order.

"Fix bayonets!"

He and his men marched on, bayonets flashing in the dull light as they continued forward, to then increase the height of their line as muskets were returned to "shoulder arms". His line came to that of the Guards, but progressed comfortably beyond it. Saunders looked over and saw the Corporal who had identified his Regiment earlier, him being obvious from his prominent whiskers.

"Hello, mate! We've come up to give you a hand!"

The reply was a black scowl, but Saunders marched on, then he heard, just, Lacey's first order.

"Full volley!"

Lacey heard the order being passed down the line and he then gave it time to travel. The French before them were disordered but numerous, many discharging their muskets at them, but not from effective range. Lacey took a deep breath. His tiredness allowed his emotion full rein.

"Right up to them, boys! Right up! Then a volley, let's see if they can take it muzzle to muzzle. Then the bayonet!"

From the far right O'Hare looked over and had divined himself what Lacey intended. O'Hare had seen that he did not have the density of numbers before him as those that were outflanking the Guardsmen but the French were fanning out before him and so that made them entitled to his finest offering. He turned to face his men, causing him to walk backwards as he encouraged them on.

"We're going right up to shake hands, boys! To see how well they've shaved!"

He looked at Carravoy and D'Villiers, just behind him, but plainly neither appeared too reassured. O'Hare marched on, his men following.

Before Lacey, the French had halted their advance when they finally realized the threat of the long firing line advancing towards them and now their Officers were screaming at them to form one of their own. Lacey watched carefully, he wanted close range but priority was to hit them when they were reloading. French orders rang out. Lacey saw them come to the present, but at nearly 100 yards. He decided to himself, 'we'll take this, then get right up into their faces.' A French Officer raised his sword then brought it down. The volley was thin and ragged, but he heard grunts and screams from behind him and a sudden pain in his own left arm. He checked that it still worked and it did. He looked both ways and saw no change in the firm rhythm of his line's advance, they were still moving forward. 'That won't help their peace of mind.' He took his men on until they were even in the French smoke. His orders came quickly and were quickly obeyed.

"Halt."

"Present."

"Fire."

The shortest of pauses after the fearsome crash.

"Charge!"

He ran forward, sword aloft.

"Charge, boys, charge! Right into them, boys! Right in. Show 'em they're not welcome!"

His line surged up to him and carried him forward. Immediately they were through their own smoke and amongst the French, stepping and jumping over the dead and wounded. The French, having suffered a crippling blow from the volley at such short range and with so many down, both Officers and men, turned at the first sight of the bayonets emerging from the smoke and ran, taking back with them those that were immediately following. Lacey led on his men in pursuit. The French must be given no chance to reform.

Rushby checked that The Colours were following, then ran forward, following Lacey, his men either side. It seemed that the French had halted, yet his men pitched straight in with butt and bayonet, straight into what seemed a dense mass, but this was because the French were trapped against the wall. It was now an obstacle that many could not surmount in time to escape. Any who surrendered were hauled out, disarmed and thrust to the rear. Rushby had used his sword but once and that blow had been parried, but now he had the stonework at his chest. He looked over to see beyond a mass of French, disordered but still standing to put up a fight.

Saunders was five yards from the wall, but could not see it. The French were massed before it, not in an ordered line, but here the French were thickest and preparing to defend themselves, even before the wall. He heard Shakeshaft give an order.

"Halt! Reload!"

The line halted and all reached down for a cartridge, but men were still dropping from French fire whilst they hurriedly reloaded.

"Make Ready!"

The muskets came up and cocking hammers were pulled fully back.

"Fire!"

An explosion of sound, then again all was smoke. Saunders needed no order, what came next was obvious, to get to the wall and hold it. Through the thinning smoke he fought his way to the wall, using his height and great strength to smash the butt of his musket into any French face standing before him. There were but few remaining after their volley and soon he was stood at the bloodied stonework, then the order came along the line from the centre, 'Half Company volleys', so he reloaded quickly, then stood at the 'Make Ready'. Byford and Bailey were 'locked on' behind him. Almost immediately he heard the rolling volleys come down the line towards him. Shakeshaft shouted, 'Present', then 'Fire', then he pulled the trigger to have his face singed yet again by the exploding firing pan and the musket butt to yet again kick heavily back into his bruised shoulder. He dropped the butt to the ground and began another reload, biting into a cartridge to have the grains of powder add to an already raging thirst.

Over on the far right, O'Hare was leading his men in a turn, wheeling left to close with his side of the village. Against him the French had not even stood to receive a volley, but fell back at the sight of the line of bayonets, parade ground straight, steadily advancing with irresistible intent. Here there were fewer walls for cover and O'Hare bade his men advance towards the houses, but from within the village the French before him had been quickly reinforced, so, O'Hare held his men within the jumble of structures on the outskirts of the village; huts, pig sties, animal pens and garden walls. He allowed his Officers to direct their own fire and, as the small volleys crashed out, he took himself down to his far right to see if any threat was coming up from alongside the village, but instead came a very welcome sight. Running towards him was a Captain of the 4th, with a good number of that Regiment behind him. The Captain halted and saluted formally.

"Captain Bentridge, Sir. 4thFoot. I have three Companies. Sorry it took so long, Sir, it's just that our own Colonel has been wounded. Where would you like us, Sir?"

O'Hare grinned openly in reply.

411

"Form on my right, please, Captain, and hold the French within the village. Do not try to enter, not yet, and watch your right, some more may arrive."

Bentridge saluted as O'Hare continued.

"And you're very welcome!"

Meanwhile, at his wall above the village, Lacey stepped back to see where the Guards were and was pleased to see that they had advanced forward to align themselves with his own men. Satisfied, he stepped into a gap in his own ranks to look over the wall, to see that there were now fewer French; the volleys being sent their way were doing their work. He looked over to see O'Hare's men closing from the right, then he noticed Rushby shouting his head off and now on the far side of the wall, waving his sword towards any group of French that were good targets. 'Time to join him' Lacey thought. 'One more push should do it.' He found a foothold in the stonework, and then pulled himself up to stand on top.

"Come on, boys. The walls are ours to take, one at a time!"

He jumped down followed by his men and ran forward, sword raised. Butts and bayonets cleared away those before them, until they reached the next wall, then, using it for cover, his men began firing again. Soon he saw that the French were falling back and soon there were none to be seen, even between the houses and the last wall. He looked both left and right, to see the same. Everywhere above Elvina, the French had fallen back. Lacey knew that they must now get into the village, but to take it all was asking too much and too much of a risk.

"Reload. Hold you fire!"

The order ran down the line. Lacey waited for his men to reload, then he climbed the next wall and advanced on, climbing each in succession with his men. They entered the village and advanced cautiously down through the alleyways, now choked with both French dead and their own from earlier in the day, until they again saw threatening blue uniforms. Lacey held his sword aloft, although it was seen by but few of his men.

"Halt! Hold here."

O'Hare and his Captains had thought the same. They had penetrated until contact was renewed and then halted. Some bickering fire began, then died away, both sides had had enough. With the fading of the light, then the cannon fire that had caused so many British casualties, itself finally died away.

Carr was sat on the stairs. He had a raging thirst and his head ached, as did his upper right arm, from a part of the house hitting it. Also, for some reason, his left eye was closing and there was a buzzing in his left ear. He picked up his rifle, examined the firelock absentmindedly, then stood up, just in time to see his men open the back door, having cleared away the stone that had held it shut. As it opened, in came a figure, incongruously clean and immaculate, as though he were stepping into a ballroom rather than into a building that had been the centre of intense fighting throughout the full course of a vicious battle. However, a welcome waft of fresh air pushed back the screen of smoke and then the figure spoke.

"And you are?"

Carr hefted his rifle back onto his right shoulder, to look disdainfully upon this less than gracious newcomer.

"Carr. Light Company. 105th Foot."

He paused.

"You?"

"Von Witberg. First Foot Guards."

Carr exhaled, emptying his lungs. He was more than a little annoyed.

"Is that the Von Witbergs of Baden Baden?"

He was talking nonsense and he knew it, but he was not, in any way, going to allow either his men or himself be treated haughtily by some overblown Guards "aristo". However, things then changed, for Von Witberg exited through the door, but remained within earshot, as Carr could hear.

"I think I've found him, Sir."

Then through the door came a full Colonel, followed by much more, a Brigadier.

"'Shun!"

All his men came to the attention, as did he, but the Colonel was advancing forward, hand extended. Carr saluted, then took it.

"My name's Anson. First Foot. Felt the need to convey my thanks to you personally. Your fight here saved many of my men. Made our job easier, I can tell you."

He was still pumping Carr's hand and grinning openly.

"Well done, Carr! To you and to your men!"

With that Anson turned and was gone, before Carr could say anything, only to be replaced by a Brigadier, who did not introduce himself, but merely began speaking.

"My thoughts also, Carr. I endorse that fully. Your stand here broke up their attack. Attacks, even!"

He nodded, then released Carr's hand.

"Well done! Now, try to get some rest. We'll be embarking soon."

Carr was still bone weary and what came next was from a dulled mind.

"So we won then, Sir?"

The Brigadier grinned, then his face fell.

"Indeed we did, but at a cost. Moore's been killed. He was hit up on the ridge, just above you."

With that he also turned and left, leaving just Von Witberg, who looked around.

"A hard business!"

Carr nodded.

"Can't remember much worse."

He took a deep breath, then sighed.

"Who was that? The Brigadier, I mean?"

"Bentinck."

Carr nodded acknowledgement, then Von Witberg grinned.

"Well!"

He paused.

"Hoorah for the Hundred and Fifth!"

Then he actually saluted whilst fully at attention. Carr returned both.

***

Within minutes, the 105th in the village were relieved by Anson's Guards, amongst whom casualties had been surprisingly light. Thus, with their relief, the 105th trudged back, but with every step came realisation of what the day had cost them, every body in a red uniform that they came to had their own green facings. So, they set about their mournful duty, to carry them up from the buildings to then be arranged in a long line between the walls so bitterly fought over. Meanwhile a burial party began the next task; to dig the first grave. The village was

414

now completely in British hands and so the bodies recovered included Simmonds, found in the field on the French side. The four soldiers carrying him approached Carravoy, busy trying to divest his boots from contamination within a pigsty.

"Sir, this is Mr. Simmonds. Which grave, please Sir, the one just up?"

Carravoy looked up.

"No, there will be an Officer's grave on the ridgetop. Take him there."

They walked on, to pass the open mass grave now dug in the space between the nearest two walls, this for the lower ranks still being brought up from Elvina. Similar was happening just behind the ridge, where as many again of the 105th's casualties had been brought. All assumed that tomorrow they would embark and so, within the growing dark, this was their only chance to give their comrades a decent burial, for all knew what came with the night after a major battle, and the scavengers would be no respecters of uniform. Deakin himself took the time to find the bodies of Alf Stiles and Sam Peters. Their arms and legs were entwined from where they had been hit back by the cannonball and Deakin had to cut through the straps of their knapsacks and backpacks, then he stood and waited until the burial party came for them

"These two is to be buried together, in one blanket. You don't try to pull them apart. There's no point and they was best mates. Clear?"

The four men nodded, but Deakin still watched to ensure that his orders were obeyed, then he followed to watch where they were placed. They were lowered down into the grave and the blanket gently eased from beneath them. Satisfied, Deakin turned to leave, but his Colonel was approaching him, so Deakin came to the attention.

"Deakin! Have you seen the Chaplain?"

"No Sir! In fact, not for some days, Sir."

Deakin looked at the worried face of his Colonel.

"In fact Sir, it was Parson, I mean Private Sedgwicke, who said a few words for the lads last night. Sir."

Lacey looked puzzled and not a little concerned.

"Really! Sedgwicke?"

"Yes Sir. We put a few drums together, Sir, and Mr. Heaviside helped out, like."

"Where is Sedgwicke now?"

"I think he's with the wounded. Sir."

Chaplain's Assistant Sedgwicke was indeed with the wounded, with one in particular; Captain Joshua Heaviside. Sedgwicke was looking concernedly at Heaviside who was sat up, bare chested, apart from a thick swathe of bandages that were encasing his lower ribs, then both of them gave their full attention to Heaviside's Bible, particularly the oval indentation disfiguring the front cover.

"It must have hit you a glancing blow, Sir, and hitting your Bible saved you a much more serious injury."

Heaviside held The Bible before him to better study the damage, but more to see the gilt cross on the front.

"I will spare them, as a man spareth his own son that serveth him. Malachi 3, verse 17."

"Amen, Sir. The Lord will protect his true servants."

Heaviside remained looking at his Bible as Sedgwicke had further thoughts.

"Now Sir, should we not get you into Corunna and onto one of the transports, to make you one of the first?"

Heaviside lowered his Bible.

"No! I stay with my men. I can walk, so we leave this place of sorrow together. Them and myself.

"As I was with Moses, so I will be with thee. Joshua 1, verse 5."

Sedgwicke sighed, brought some water to him, arranged his tunic around his shoulders against the coming cold and walked away.

Davey, Miles and Pike trudged wearily up the slope, their chosen destination being the Colours, still prominent, where they knew that Jed Deakin would be, but he was not there, only Harry Bennet. Davey arrived first.

"Where's Jed?"

Bennet pointed behind him.

"Back over. At the grave."

The three trudged on to first see the long line of arranged dead and then the grave, then they saw Deakin. This time Miles spoke first.

"Jesus, Jed! All these!"

They all stared over the open grave as bodies were carefully lifted in. Miles was still stunned.

"There's hundreds. How many?"

"Don't know, as yet, but a lot. It was a rough go, worst I've been in. They got Alf and Sam. Same cannonball took both out together."

The three looked at him, their faces showing both shock and sorrow, but Deakin continued.

"They've gone in. I saw to it myself."

He paused.

"But they're still bringin' lads over."

They all looked to see bodies still being delivered to the edge of the cavity in the hillside, brought up from the village and between the stone walls.

"Some may still be out there wounded, some still down below the village, an' scavengers'll be arrivin' soon. 'Tis not long 'till full dark, an' there's still some not brought in. Maybe wounded, so there's plenty of the lads as is stayin' out, on guard, like."

Davey hefted his rifle.

"That'll include me!"

He walked off to be immediately followed by Pike and Miles.

Lacey had at last found O'Hare. They shook hands almost absentmindedly, for each could see what had just pre-occupied Deakin and the others. O'Hare began, he was seeing for the first time the extent of the cost paid by the 105th, the cost of their place in the centre of Bentinck's Brigade.

"How many? Do you know?"

Lacey shook his head?

"No, but I'd guess at a hundred plus. We won't know until the Rolls are called, and that's for tomorrow. Priority must be given to burying the dead, now! The French will come forward again, perhaps at dawn, and we must prepare for another defence, perhaps even before the town walls."

Lacey changed the subject, partly.

"Have you seen, Prudoe? I haven't."

O'Hare shook his head and then spoke.

"Not even with the wounded?"

"No, not even there."

Then Lacey thoroughly changed the subject.

"They got Simmonds, out below the village. Who do we make Brevet?"

O'Hare replied instantly.

"Carr!"

Lacey was surprised at the immediate reply.

"Not Heaviside?"

Then he answered his own question.

"No, he's wounded. Lucky to be alive."

He paused.

"Not Carravoy?"

O'Hare released a breath, halfway between laughter and sarcasm, then he walked off to make a final check of his side of the village for casualties. Lacey walked to the mass grave, now filling up, wondering what to do about the burial service. Something must be done, before they withdrew back to the city. He went to the lines of wounded and found there, thankfully, fewer than he had feared, but the sounds coming from the Surgeon's tent chilled him as badly as anything that had happened throughout the day. Close by he found Sedgwicke, there with the orderlies and bandsmen, administering to the wounded. Seeing the Colonel, Sedgwicke sprang to the attention.

"Sedgwicke!"

"Sir."

"I hear you gave a service last night?"

"Yes Sir. Chaplain Prudoe was heavily indisposed, Sir, and so ….."

"Yes, yes, Sedgwicke, I hear you, but I want you to do another."

Sedgwicke's forehead knitted in puzzlement.

"Sir?"

"That is, if Chaplain Prudoe cannot be found."

"No Sir. He cannot. At least, not by me. His tent is empty, Sir."

"Right. There's a grave about to be closed just above the village. I'd like you to say a few words there. I'll conduct something at the one up here on the ridge. And there'll be other another after that."

"Yes Sir. I'll get what I need. Should I go now, Sir?"

"Yes Sedgwicke. And well done! I've been told about last night."

"Sir."

Sedgwicke turned to leave, but Lacey had more to say.

"The men call you "Parson", do they not?"

"Yes Sir. I was one once, and the name has stuck."

Lacey nodded.

"Seems appropriate."

Within an hour both were stood at their allocated graves, but both stood waiting, bodies were still arriving, emerging from the almost full dark, but a fitful, yet merciful moon had emerged to give some light. A lantern was found for Lacey, but not for Sedgwicke, but it made little difference. Both spoke of commending their comrades into the hands of God, both spoke the Lord's Prayer, but their voices were lost in the midst of those around. There was no singing, both Lacey and Sedgwicke judged the men to be too tired. Finally both graves were closed and both men wondered how long they would remain as hallowed ground. Not long, both silently concluded, then both walked to another grave, Lacey for the Officers', Sedwicke to another of the men, still being filled.

At the lower end of the village, Davey, Pike and Miles had returned to their wall, now eating and drinking what they had found in French knapsacks, but Davey and Pike were more than half asleep as they automatically fed themselves whilst sitting down leaning against the wall. They had checked that all that remained beyond their wall were French casualties, both dead and wounded. Pike had spent some time giving water to those he found still alive, until ordered back by Davey. Now all three guarded the village, still being searched for men of the 105th. At the last grave, to receive these last being found, stood Sedgwicke.

Miles, with his rifle poised and balanced on the wall, angrily chewed some kind of sausage, washed down with rough Spanish wine. He was looking over and, sometimes revealed by the scudding clouds across the moon, he could see the shapes of those who had emerged to rob the dead and the sounds of the murder of those still living were reaching his ears. One shape, definitely large, even though the dark hindered judgment, came to his attention. He heard the cry as the knife went in and at that point he decided that he had witnessed enough. He took up his rifle, cocked the hammer, sighted and fired. Smoke and also cloud covering the moon hid all. Davey looked up, now fully awake.

"What? French?"

"No. A scavenger bastard murdering some poor sod."

"Did you hit him?"

Miles took another bite of his sausage.

"Dunno! But I heard a shout of some sort."

***

419

The old couple had remained hidden in their cellar, hearing well enough the sounds of battle, but grateful that, so far, the conflict had not spread to their poor building. With the return of peace and the dark, they had sat waiting, but hearing nothing. What they did know all about, was what happened in the dark after a battle and they feared that such thievery would spread to them. Once in the mood, any pillager could as likely rob from the helpless living as the helpless dead. The old man had loaded their ancient family blunderbuss and they sat long into the watches of the night, hoping for no sounds, which would mean all was at peace and had remained so, but suddenly came the sound of violent squealing from the pig pen. One of their animals was either being slaughtered or stolen. The old man; old, yet not unused to violence as part of the family feuds in the hills around, crept out of the cellar door, his mighty weapon cocked and ready.

At the sty he saw a huge shape just straightening itself from some activity in the pen. The old man raised the blunderbuss and fired. The figure, hit between the shoulders by a dreadful mixture of nuts, bolts and pieces of chain, snapped forward to drape itself over the side of the pen, then remain still. The old man ran back into the house to reload, then came out again to examine the scene. There was no change, so the two, he and his wife, resumed their vigil from within their house until morning. With the daylight they both took themselves over to the pen to find the body of a huge man, legs outside, but head in, at least what was left of it and what was left of the top half of the body. Both the sows and the boar had eaten their fill, from even inside the chest cavity, having used their combined strength to drag the body further over the pen wall. One thing they noticed, was that the figure wore British grey trousers and what remained of a British tunic. Also there was a neat bullet hole in the outside of his right thigh. His hands were massive, as were his boots, but these, too big for either of them, went straight onto the fire, now kindled up for breakfast. The handy knife, however, went into their drawer, whilst the British bayonet was stored for a family heirloom. Finally, with full daylight, the remains were dragged out of the pen by a carthorse, when they could then see a large 'D' branded onto what remained of the chest, then all was dragged away down a track to be dumped in the Monelos. It rolled on downstream, then for scavengers of the sea to finish what the pigs had started. Thus passed Seth Tiley from beyond this world to his judgment in the next.

Ten miles away a weary but delicate hand was knocking at a huge door in a towering wall. He had wandered all night, unguided and confused, until he had blundered into the long high wall and then, by pure chance, turned in the direction that brought him to the door, this with a Cross prominent on both sides. It took some minutes before the knock was answered and then Septimus Prudoe entered into the Monastery. He had never felt more peace or relief in his life as the door swung heavily closed behind him and the sound of the Monks' prayers at Vigils washed healing through his disheveled mind.

***

Chapter Ten

Home Shores

John Davey was lolling on the stonework of his own construction, that built to his own design merely two day before, but him barely more awake than the stone he rested on. However, so far he had resisted the waves of sleep by quietly reciting Psalms, as taught at Sunday Dame school. His head was swimming with fatigue, but he still managed to hear the heavy footsteps from the alleyway behind him, which brought him back to a better, but nothing like total, state of consciousness. Before he was fully aware of what was happening a hand came onto his left shoulder.

"'S all right, mate. We'll take over now."

Davey turned to look into a face barely discernible in the dark, it being so dirty, bearded and sunken.

"Who're you?"

The reply came in a broad West Midlands accent.

"Sixth Foot! First Warwicks. Best get back on up the hill, yours'll be marchin' off soon. We've to set fires, make the Frogs think we're all still holding up here."

Davey nodded, for all the good it would do in the dark, then he reached down to rouse Pike and Miles. The response was a sigh from Joe, but a "What's up?" from Miles.

"These lads is takin' over. We're to get back up."

Davey hefted his rifle sling over his right shoulder.

"Thanks mate. What was it like for you?"

There was no need to name the subject.

"Didn't fire a shot! We was back guarding the town, which is why we're now up here on rearguard. That's my guess."

Davey nodded, he was too tired for more, but Tom Miles managed a typical distinguished comment.

"Well then, you was well out of it, well out of it, 'cos that was as rough a do as I never wants to see the likes of again!"

It was Davey who responded.

"Take no notice! He treats everyone the like. At least he spreads it about even. He's no favourites."

The soldier nodded, equally uselessly in the dark

"I'm not arguing, and that's why it's now for you to get on back up!"

Davey pulled Miles up.

"Come on! Perhaps there's tea!"

The soldier also cheered up.

"Send us down a drop, if any's spare!"

Davey patted him once on the arm.

"Count on it, if there's spare, like you say."

The three filed back up through the village, walking on one side of the alleyways they used, for the other side was filled with French dead set back against the walls, waiting for their own comrades to reclaim the streets with the coming of daylight, when they expected the British to give up the village. Half way up, Miles stopped.

"What's that? Sounds like a wench!"

With no further word he disappeared into the inky black of an alleyway. The two heard words of belligerence, common as from Tom Miles,

"What're you doin' 'ere, you daft bugger? You can't stay here."

Out from the alleyway came Miles dragging a French drummerboy, complete with drum.

"What'll we do with this?"

Davey examined the small shape, located in the dark by whites of a pair of terrified eyes staring upwards. It seemed that Miles had relented somewhat.

"May as well leave 'im here. The French will sweep him up come mornin'"

Davey had used the time for thought.

"If he don't get swept up by the bloody scavengers! They'd kill him, just for what he's wearin'."

"What then?"

Davey paused for more thought.

"May as well chuck him in with ours. We've lost a few, saw them in the grave! He can beat a bloody drum for us same as he can for the French!"

He resumed their climb upwards.

"Bring 'im up. Let some Officer decide, when they gets round to it. Meanwhile, put him in with our drummermites."

"And the drum?"

"Leave it. Belongs to the Frogs!"

But the boy would not relinquish the drum, putting up a fierce struggle with Miles. At the sound, Davey grew impatient.

"Oh, let 'im keep it! What's so different to a French drum as one of ours?"

Once through the series of gates in the walls so viciously fought over, with their new charge, who didn't seem at all inclined to attempt an escape, they sought The Colours, for this would show the whereabouts of Jed Deakin and, most likely their own followers, particularly Joe's Mary.

That they were successful was announced with a shriek and through the crowd of exhausted soldiers and attendant followers came a hurtling Mary to clasp herself to Joe and press her face to his chest. Tom Miles stood and looked at the spectacle, mildly astonished, before using his thumb to point in their direction, Miles' face now pulled sideways in puzzlement as he looked at John Davey.

"He ought to marry that girl!"

Nelly Nicholls both saw and overheard all and, whilst torturing a piece of cloth for all the water contained therein, she looked askance at her sworn enemy.

"Now, just what would you know about a thing like that, Tom Miles?"

Miles returned her gaze equally belligerently.

"Enough! I had a mother and father, didn't I? An', afore you asks, they was married! Church married!"

Jed Deakin wanted an end to the exchange before it started. To him, the arrival of the Warwicks told its own story.

"That'll do! We has to move back, now. These Warwick lads is takin' over, and'll need our fires."

Then he saw the addition.

"What's that?"

Davey answered.

"French drummerboy. We winkled'n out from a back alley. Too likely to be killed with what's goin' on down below."

Deakin must have gone through the same train of thought as Davey.

"Give to our Drum Sergeant. Let him decide."

Bridie had heard all.

"Ye'll not! Not yet, not till he's fed."

She looked at the boy, now easily seen in the firelight and looking utterly forlorn. She held out her arm in welcome.

"You come this way now, honey, and we'll find ye something to eat."

The lifted arm and the welcoming face told the boy all he needed to know and he walked forward for the arm to go around his shoulders and the remains of the stew to be placed in his lap, after he had been sat down on a log.

Deakin paid no further attention, but returned to his original subject. He indicated the squads of ordered soldiers moving around in the light of the more numerous campfires, all having been set to make the French think that, throughout the night, the ridge would be occupied in force.

"Look, things is happening. Get packed up! When the order comes we has to be off, an' sharpish."

All immediately set about obeying such an instruction from Jed Deakin and the three Lights departed, with their drummerboy, now devouring a doughcake.

Deakin was right. It was but five minutes before the bugles blew 'assembly' and the men found their Company Captains, to then join the column whilst the followers took their place to bring up the rear. They set out, blessedly downhill, on the road that led down from Elvina to the main road North to Corunna. They needed no guide, which was fortuitous, for many slept as they marched, often needing to be pulled back from the road edge by their comrades, waking only from their torpor when they reached the highway at Eiris and there they made the turn for Corunna. For two hours there was no sound save a monotonous, rhythmless tramp. Past the column in the dark, as if its constituents were marching on a treadmill, came the shapes of houses and farms, all now dark, silent and shut down, as though withdrawing into themselves and holding to the hope that the forthcoming onslaught would pass them by, or even perhaps that they would awake from no more than a bad dream.

At the rear of the column, Carr, Drake and Shakeshaft were sustaining themselves with French brandy, but conversation was non-existent, bar the one question from Carr to both his Lieutenants.

"Casualties?"

Drake answered first.

"I've not counted yet. This is the first assembly since, but I think, for us, surprisingly light, seeing as we've just been through a battle."

Shakeshaft gave his answer.

"Same here. Spending most of our time running around in the village saved us several, not like the Companies who stood on the ridge. They've suffered."

The same topic was occupying the mind of Colonel Lacey, marching at the head of the column with O'Hare. Between these two, similarly few words had been exchanged, except one simple observation from Lacey.

"If we're below 600, we could be back as Detachments!"

O'Hare tried to be reassuring.

"I don't think we're that bad. We'll call the Roll when we stop, and then we'll see."

***

At midnight, Lacey was faced with an Aide stood alongside and illuminated by a large fire, waiting between some buildings, which turned out to be the village of Santa Lucia and there, they could not only smell the sea, they could hear it. However, Lacey's thoughts were pulled away by the Aide.

"Sir, my orders are to allow no more into the town. It is already too crowded, for the embarkation is going more slowly than was hoped, Sir. So please would you hold your men here?"

Lacey did not argue.

"Very good! Are these buildings occupied?"

"No Sir. They can be used by your men."

"Thank you!"

He turned to O'Hare.

"Call the Rolls, then rest. We won't be out of here before daylight, but make sure you get some sleep yourself. Give Carr the job of getting rations back to us. He's our new Brevet; he can earn it!"

"Have you told him?"

"No!"

"Then this will be his confirmation!"

Both chuckled exhaustedly as Lacey walked off to find his Sergeant Orderly to require him to find the Rolls and O'Hare walked back to tell each Captain to obtain the document which applied to his Company. Eventually, O'Hare found his way to Carr.

"Henry!"

Carr turned and saluted, which gave O'Hare the opportunity for a small joke.

"No need for that anymore, at least not to me. You've heard about Simmonds?"

"Yes Sir. I'm very sorry!"

"Indeed, but we need a new Major and we're making you Brevet."

Carr took a step back in astonishment.

"Me Sir?"

He paused to indulge in some serious eye blinking.

"Yes Sir. Thank you, Sir."

"Forget the thanks, it is well merited and may only last until we get home, as you well know, but meanwhile, get Drake here to call your Roll. I, we, want you to get into Corunna and get us some rations for the morning. You're a Major now, so throw your weight about a bit!"

As O'Hare walked off, Drake and Shakeshaft, who had overheard, ran up; Drake to arrive first, hand extended.

"Most sincere congratulations, Henry!"

This was quickly followed by Shakeshaft.

"Same from me, Sir. Well done!"

Carr managed a rictus grin, made more sinister by the firelight, but he was genuinely shocked and could only mumble thanks in reply, then he brought himself back to the immediate task and left. He had no idea how far Corunna was from Santa Lucia so, having quit the village, he counted his paces to keep himself awake and had just reached the significant figure of 1000 when he arrived at the main town gate. Inside he saw what could well be described as one of the quieter portions of Hell. At all street corners were fires or braziers and both into and out of their light came animated and hurrying figures, the fires throwing their shadows onto the pale walls as manic shapes, small, then large, but always frenzied and always accompanied by urgent shouts and cries of near panic. The light also cast its yellow and inadequate light on columns of men, some shuffling forward towards the quaysides, some slumped down on the hard cobbles, whilst the civilians of the town ran hither and thither, carrying timber, sacks, bales of anything and buckets of sand. The people were preparing for a French siege. Carr walked on and, by pure chance, found the square. There he calculated which building was both the largest and the busiest and he walked into it.

Meanwhile had come the sombre calling of the Rolls, each Company Commander, by lamplight, calling out the names and making their mark to show dead, alive or wounded. Now, being stood still and in their ranks, fatigue fully overcame many and these had to be shaken awake to give answer when their name was called, for, even though stood with ordered arms, sleep finally claimed them. With the last name they were dismissed to find any wall or corner to there slump into disordered subconsciousness. The Captains gave themselves the final task of reporting to O'Hare who, in the light of a poor candle lantern, added up the ten reports and then entered a low door, to pass the final numbers across a bare, rough table to Lacey. O'Hare looked at him as he read the four figures; two Officers dead, three wounded, 210 rank and file, dead or seriously wounded. Current muster 570.

"We came ashore at Montego with 880 men! We lost more in the battle than we did on the retreat!"

O'Hare nodded.

"Yes Sir. I'd hazard that yesterday we lost more than any other Regiment."

Lacey released a long, sad breath, then placed the paper in a satchel and blew out the candle.

Soon, both were too far into deep sleep to hear the rumble of supply wagons arriving in the darkest reaches of the night. Carr had found the supply warehouse and kicked everyone awake, then he had told them of the numerous battalions coming off the ridge that could not enter the city and that they all now needed supplying, having come off the battlefield with nothing. The Senior Sergeant of Commissariat complained of their being neither orders nor requisition but Carr had looked him fully in the face, but three inches from his nose, accompanied by his right arm waving expansively.

"And what do you think will have happened to all this, come this time tomorrow?"

The Sergeant had shrunk back.

"In the harbour, Sir."

With that every available wagon and mule had been loaded and pulled and driven to Santa Lucia from there to be quickly unloaded in the midst of the sleeping battalions along the road. Thus, at the first light of day, the 105th at least awoke to plentiful supplies that provided dough cakes, pork, peas, bacon, tea and army biscuits. For the Officers there was coffee, which did much to enliven Carr and other Officers, although he needed it most, having been the last to awaken.

As full daylight arrived, Deacon again made his calculations and formed his conclusions, which he passed onto Bridie and Nelly.

"We'll be aboard ship come Noon. Time to shed all as we don't need. Too much an' they'll make us heave it into the harbour anyway, without the chance to make much choice."

Both women looked at him, both puzzled and concerned, as he continued.

"What you can't carry won't be going'. Including him!"

He pointed at Pablo.

"Time for good-byes!"

Eirin and the other six children had been stood nearby, hearing all and holding the mule's tether. Nelly's Sally, Trudy and Violet immediately began to cry, quickly followed by Sinead Mulcahey. Without a word, Bridie took the halter and led all, children, mule and Nelly, up the most likely road inland. It was not long before they came to something like a farm and so they bodly entered the gate. By great luck there was a woman stood in the doorway of a decent stone built house, who quickly shouted through the door to instigate the arrival into the yard of what was probably the owner of the whole. Bridie tugged the mule up to him, accompanied by Nelly, to then offer the halter to the farmer, who stood confused and so did not take the offered halter. Bridie, knowing of the hurry, spoke up.

"This is our mule, Pablo, and we'd like you to have him and take care of him. He's a good animal, so he is, did he not get us over the mountains to here?"

Gradually, with the halter still held out in offering, the farmer began to understand.

"Usted me da este mulo?"

Neither woman understood a word, but the man had at last taken the halter. Nelly was encouraged.

"He's a good mule, so he is. He'll give you fine service!"

The farmer nodded.

"Un mulo bueno, dice?"

That was simple enough for Bridie to understand.

"Bueno! Yes, much bueno mulo."

They turned to leave, but the woman of the house had been watching and she called out.

"La espera, espera por favor."

The farmer joined in.

"La espera, por favor"

His vigorous pointing at the spot where the women were standing carried enough signal that they were to wait. They did and within a minute the woman emerged with a large jug, which she offered to Nelly.

"Para los niños!"

Nelly looked perplexed, but the woman was pointing at the children.

"Para los niños! Los niños!"

Nelly took the jug and identified its contents. It was full of milk.

"Gracias! Thank you. Saints be with you."

The reply from the woman of "Los santos por con usted" seemed to make sense to them. The children were all weeping and hugging the mule's neck, but two good swallows of the milk each did much to soothe them. So, with understanding smiles and friendly waves they parted company, but it was then that Nelly noticed that Bridie was limping.

"What ails your foot?"

"Ah, 'tis nothing'. Nothin' a few days off me feet won't cure!"

They made a hurried return to see the last of the followers disappearing down the road and so, with all they now possessed carried on their backs, they hurried on and after the disappearing column. Once inside Corunna, they saw for themselves the ongoing intense preparations for the coming of the French, which Carr had seen the previous night. Both population and Spanish soldiers laboured up with materials to strengthen the weak points of walls already strong and surprisingly modern. The soldiers carrying supplies of war, the civilians that with which to build yet higher, but the women and followers took little notice of either. Not so the Officers of the 105th, particularly one who expected a Brevet promotion at any time, this being one Captain Lord Charles Carravoy. He was in ebullient mood about himself, because, now being fully aware of the death of Major Simmonds, he fully expected, as the Captain of the Senior Company for his elevation to be confirmed at any moment. Thus, he felt well fortified within his own self-esteem to pass comment on the furious activity around him, to his inevitable companion, Lieutenant D'Villiers.

"Pity they couldn't have stirred their stumps as well as this when it would have made some difference, such as back before Christmas when they still had Madrid and we were at Salamanca."

D'Villiers looked around, barely nodding, too tired to think of a reply diplomatic enough to appease his Captain and to encompass his own thoughts that the Spanish people themselves were practically blameless. To his mind, responsibility for their plight lay utterly with their ruling classes and their petty rivalries. He was content to simply shuffle forward when, suddenly, their pace quickened as the battalion turned left into a side street, then turned right along another to be greeted by the myriad rigging of a large transport, its countless ropes and "all angles" spars clear against the grey January sky. Soon they were on the quay and halted beside not one transport but two.

D'Villiers looked around at what was not there, rather than what was. All was little more than wet, grey cobbles, bare of anything harbourlike, for anything weighty and removable had been hurried inland for the defence of the walls. The warehouses that terminated the surface of bare stones stood plainly empty, both doors and windows gaping, because those of the citizenry that had been broken in to divest the buildings of their bales and barrels had neglected to close what would hide their emptiness. But such thoughts were soon ended when he heard the noise accompanying the approach of the lowly followers, whose existence who could barely bring himself to acknowledge. They had seen the second transport and were mightily anxious that they should be on the same ship as their men and so they hurried, both up and down, to find their husbands and mingle themselves with the Company they belonged to. Thus, Mary plunged in to stand beside Joe, whilst Nelly and Bridie, their broods in close attendance, did the same for Jed Deakin and Henry Nicholls.

At The Colour Company, Number Three, Jed Deakin became anxious at the mingling of women and children within the ranks, such was a minor breach at least, but he need not have worried. Captain Heaviside supported by a walking staff, a common crutch with crosspiece hurt his side too much, was slowly progressing along his lines and had not failed to notice those of his men and their wives, now joined together.

"In thee and in thy seed shall all the families of the earth be blessed. Genesis 28. Verse 14."

Deakin knew what an indulgence was being granted and looked down at Patrick and Kevin.

"You two! Get in them buildings and find the Captain a chair, or at least something for his comfort!"

The two boys scuttled off to enter one warehouse, then another, then a third, from which they appeared with a somewhat ornate wooden armchair, such as would be not out of place in the office of the overall proprietor of the business that had been conducted therein. The two boys brought it up, placed it just so and dusted it off. Heaviside sank gratefully down upon it.

"My presence will go with you, and I will give you rest. Exodus 33. Verse 14."

All grinned, but Henry Nicholls had something extra and he offered it to his Senior Officer.

"Will you take a nip with us, Sir, against the cold, like?"

Heaviside took the flask.

"I will, and I thank you, but for the pain, you understand, rather than the cold."

"Yes Sir! Medicinal purposes only. Of course, Sir."

Heaviside raised the flask in the direction of Bridie and Nelly, then drank of the brandy. He returned the flask, but then they were moving. Deakin pointed again to the chair whilst looking at the two boys.

"That comes with us, for the Captain!"

Their transport was called the Dauncy and her consort the Teignway and within an hour all were aboard one ship or the other. First aboard were the wounded, either stretchered, helped or supported and, as the rest of the Regiment came on board, all on the decks were soon wracked by the frequent harrowing shrieks and screams as the Surgeons resumed their interrupted work. All that were able, closed their ears and thanked their God for their fortune, which had favoured their thin and exhausted, but essentially whole, bodies not to be carried down to join the growing number in the improvised sickbay.

With her family now placed on the lowest deck, Bridie was returning to the quayside with a pot and a saucepan, because an Officer of the Royal Wagon Train had decided that she had too much baggage and these two items had to be left behind. When back on the cobbles, she noticed Beatrice Prudoe, alone and bereft, and in somewhat of a daze, such that Bridie had to pull her to the edge of the quay to clear the way for the next Battalion arriving for embarkation.

"Mrs. Prudoe! Are things all right with you? Is your good husband near at hand, now?"

Mrs. Prudoe looked down at the concerned face looking up. She took a deep breath, its passage broken twice by sobs of emotion.

"My husband I have not seen for two days. I have no idea of his whereabouts, which makes me reluctant to board any ship without him."

She took another deep breath.

"I really am wholly unsure as to what to do!"

Bridie took her arm and did her best to laugh it off.

"Ah sure, now isn't that the army all over! On such as this, I've been split from me husband for days, weeks! You get poked into any old hole, without so much as a word or even a quick good-bye."

She nodded up at her, smiling for extra emphasis.

"The thing to do is to follow the drum, like, and not get left behind. Once at home all can be sorted out, whilst here, right now, there's more order in a herd of cats than the job they're makin' of getting' us out of here! He's most likely been pushed onto some other ship, havin' been told that we all was on it."

She saw the hope come into Mrs. Prudoe's face.

"There, I'm sure that's the truth of it, now!"

She took Mrs. Prudoe by the elbow.

"You come on board and you're more than welcome to mess down with us. Then, a few days, and we'll be home. We can have a better sort out there."

With her hand still on Mrs. Prudoe's elbow, she lay down the pot and the saucepan, each grating on the hard stone.

"I've to leave these behind, as ordered by some gombeen Officer!"

Beatrice Prudoe smiled.

"No need! I'll take them as my own!"

With that, she reached down herself and picked up both and then, grinning, both women supported each other up the narrow gangplank.

Elsewhere, on the same ship, Carr was doing anything other than stood with spirits uplifted. He was looking over the ship's side, across the whole of Corunna Bay and urgently pulling out his telescope to study its furthest point. A quick adjustment confirmed his fears; the fort there was flying the French flag and between there and the opposite arm of the bay, across the full entrance, was but 1000 yards. He looked around and saw what he took to be some rank of Officer of the Dauncey. He pointed to the, now French, fort.

"That fort, there. What's it called?"

The Officer did not need to take too careful a look.

"Fort San Diego. Titchy little place, but it covers the bay."

Carr paused for thought.

"How far from the main channel out of the harbour?"

Again the man did not need to look.

"Two cables! That's 400 yards to you, give or take!"

"What've they got there?"

The man looked puzzled, so Carr expanded.

"Calibre of gun?"

"That Sir, I could not say."

Carr hurried away to the cabins below. He was looking for Lacey or O'Hare but found only Sergeant Bryce, Lacey's Clerk.

"Where's the Colonel?"

"Still ashore Sir."

"Major Lacey?"

"With him, Sir. They're handing over our report, Sir, of the battle and our casualties, Sir. They was woken early to write it, Sir."

Carr's lips narrowed with impatience.

"Where?"

"General Hope's Headquarters. Sir."

Carr hurried back on deck and down the gangplank. He did not know if anyone else had seen the Tricolour and whether General Hope may already know, but he was not prepared to take the chance. Therefore, using his knowledge from when he first entered Corunna to obtain food and supplies, he went straight to the same building. Entering the building, the first thing that hit him was the heat, whether from numerous bodies running numerous vital errands or from the huge fires of paper burning in the grates, he did not know, but all was a hive of activity, perhaps not frantic, but certainly on that side of urgent. The one area of calm was a Major sat at a desk at the foot of a magnificent staircase, all being of white marble; steps, risers and balustrade, at least what could be seen of it, such was the traffic both ascending and

434

descending. Carr went up to him.

"Excuse me, but I'm now aware of something that the General should know."

The Major looked up and folded his arms, his whole demeanour a study of effortless superiority. Carr appeared, after all, to be a mere Captain.

"And what might that be?"

Carr stared insolently back at him.

"That the fort covering the passage out of here, the exit from the bay, is now in the hands of the French. Their flag is flying over it. Its name is Fort San Diego. If it has naval guns, we are in trouble. If all that the Frogs can put in there are their own field guns then we will probably get away with it. So, it would be a very good idea to find out the calibre of whatever's in there. "

He paused as the Major's face changed to one of deep concern, but Carr continued.

"I'm willing to pursue the matter myself. I would have thought that the Spanish Officers now preparing a defence for this place would have some idea, but to find that out I would need an interpreter."

The Major was now organising a pen and paper.

"I will compose a note now and see that it is passed in. Right now, as you can imagine, the General's a bit up to his eyes, as it were. If you would find out what you can, and then return, by then I may have a reply. Or perhaps I should wait for what you discover. You are?"

"Carr, Major Carr. Brevet from yesterday."

The Major nodded, already writing. Carr stood waiting, for clamouring seconds, then his patience ran out. He leant on the desk and over the Major.

"An interpreter?"

The Major stopped writing and looked up, then looked around, to fix his gaze on a young figure in a Spanish uniform, whom he motioned over. The young man came to the attention and saluted, then the Major made the introduction.

"This is Teniente Luis Da Costa. He has been giving good service with any translating we require."

Carr wasted no time in explaining what he needed, but he was disappointed that the Lieutenant did not already know.

"I am not of this town, Senor Carr. I am not knowing. We need Officer of this soldiers, how you say, 'garrison'?

Carr nodded

"Garrison, yes. Very good. Let's go find one."

Finding one meant joining the columns of men and material going to and from the walls, but the journey was short. However, once on the ramparts, even Carr's single mindedness was distracted to see several red uniforms working to fill in what looked like a deep hole in the walkway. Da Costa noticed Carr's puzzlement.

"Your General, General Moore. He has been buried up here! There!"

As Da Costa pointed, Carr nodded and, curiousity satisfied, allowed himself to be guided to an Officer supervising the placing of a gun. 'At least he should know something of the subject,' thought Carr. Da Costa plainly thought likewise, this evidenced by his next words to Carr, now that they were stood close to that very Officer.

"This Capitano is the Artillery Officer for Corunna. He should know the size of the guns."

Da Costa immediately began his questioning.

"El fuerte San Diego es tenido por el francés. ¿Qué es el calibre de los fusiles allí?"

The Officer looked at Da Costa, then at Carr. It was plain he had realised that it was Carr who wanted to know, so he spoke directly to him.

"Veinte cuatro de sus libras."

Da Costa translated.

"El Capitano says twenty four of your pounds."

Carr looked at Da Costa. 24 was not good.

"Ask him if the guns have been spiked. A nail through the touchhole!"

Da Costa looked wholly flummoxed. He had no idea what Carr had said, yet alone have to translate it. Carr saw the helplessness on his face, so he went over to the gun the Capitano had been supervising, pointed to the touchhole and went through the motions of hammering a nail down through it.

"A nail! Hammered into the hole for fire!"

Da Costa did his best.

"Un clavo martilló en el hoyo del fuego."

The Capitano's face changed to "how should I know?", with his arms outspread, palms upward. The message was obvious, so Carr turned to Da Costa.

"Are the guns still there?"

The translation brought a repeat performance of the outstretched arms, but he did add some words.

"Yo no pienso, pero yo no estoy seguro."

Da Costa now looked equally helpless as Carr listened.

"He is thinking not, but he is not sure."

Carr nodded.

"Thank you. I must now return."

He bowed in the direction of the Captain then descended from the ramparts, but not before executing a very respectable salute at 'eyes left' towards Moore's grave, then to hurry back to the Headquarters and the bottom of the marble staircase. The Major was still there, with the note to one side, evidently waiting for Carr's addition.

"24 lbs. They may be spiked, they may not. They may still be there, they may not."

The Major added the information, called over a Corporal and proffered the note to him, but Carr intervened.

"I'll take that up myself!"

He seized the note and ascended the stairs, two at a time, his boots making a solid connection with each step. He went to the tallest, busiest door that led from the landing and approached another Major. This time he thought proper introductions would help.

"Major Carr. One Hundred and Fifth! I have something here that the General should see. It's important."

The appearance of Carr, black eye, its eyebrow with a livid scar, another scar emerging from beneath his shako and his uniform in the same state as it was at the final minute of the battle, upper right sleeve ripped open, had some effect on the Major, but also did the Regimental Number on Carr's shako. The conduct of the 105th 'Rag and Bone Boys', had circulated. He opened the door.

"Go in. There's a bit of a queue, but that'll get where it needs to be"

Carr entered what was plainly a library. Tens of large sets of books filled the shelves from floor to ceiling, creating a oddly calming patchwork of several different colours all around each of the four walls. The ceiling displayed paintings of neck wrenching breadth, of galleons being tossed in a gale, sails and rigging in such disorder as to do more harm than good, but the British Officers in the room did full justice to its academic ambiance, all studying maps and documents like so many distinguished antiquarians. All were dressed in red, save two in Naval blue. However, despite the lack of movement, the tension in the air was palpable.

There was a room beyond the library, its entrance guarded by another desk and a full Colonel sat commanding its extensive surface. At the desk was a queue, but the Colonel was dealing with each person waiting with speedy efficiency. Soon, Carr was stood before him and, eschewing the need for words, the Colonel stuck out his hand to take the note and read it. Then he looked up at Carr.

"I'll see that this is sent in."

Carr's concerns, which had formed themselves whilst standing in the queue, were immediately aroused. If the fleet was sunk by heavy cannonfire as they left the bay, they, these being the powers that be, would look for a scapegoat and that would be him, unless he had proof that the information was delivered.

"Thank you, Sir, but I would very much appreciate being able to leave here with an acknowledgement."

The Colonel seemed to have read Carr's mind. What Carr had discovered was, indeed, of the highest importance. The Colonel stood.

"Wait here!"

Carr stood aside, allowing the next in the queue to take his place and within five minutes the Colonel had returned, still with the note. He handed it back to Carr.

"Thank you, Captain. The General is grateful for the trouble you have taken."

Carr took the note and saluted.

"Thank you, Sir."

Carr stepped away and opened the note. At its foot it said, simply,

"Information acknowledged. Gen. Hope. 17th Jan."

Carr thrust the paper into his side pocket and hurried out of the building and back to the ship. Once down amongst the cabins he found Bryce again, still alone. Bryce looked up at Carr's sudden entrance and questions.

"The Colonel? Major O'Hare?"

Bryce dropped his quill at the ferocity of Carr's questions

"Still not returned, Sir. Sorry Sir."

The last words Carr did not hear, he was out and hurrying back on deck. There he found someone who, at last, could be of help, this

438

being Sergeant Major Gibney, who sprang to attention and saluted.

"Major Carr, Sir."

Plainly Gibney had heard of Carr's Brevet.

"Sar' Major. I want 40 men, four from each Company, to go back ashore and return with anything they can find that will float; float well enough to keep somebody up, in the water. Get them back up here to me."

"Yessir, but we could be leaving soon, Sir."

Carr's temper resurfaced.

"Then it needs to be done as soon as possible, doesn't it!"

Gibney saluted and hurried off, to disappear down a companionway to where the men were placed, all now asleep or soon to be. The first Officer he came to was Carravoy and, as was his nature, he was blunt and to the point, wasting no words to explain an order dispensed from a superior.

"Sir. Four of your men are needed to go back ashore, Sir."

Carravoy looked at him, more than a little annoyed.

"By whose orders?"

"Major Carr's, Sir."

The words impacted upon Carravoy as though he had just been punched. He leaned forward towards Gibney. The first word heavily accented.

"Major Carr?"

The pointed emphasis was lost on Gibney.

"Yes Sir. I believe it to be urgent Sir. He needs a squad of 40."

Both Carravoy's brows and jaws came together, his mind churning in all directions. He had expected the Brevetcy to have arrived by now and he had put the delay down to the turmoil of their evacuation, but now his hopes were totally dashed and, on top, he was in receipt of an order from Carr. Carr, of all people! He exhaled a deep breath.

"Four, you say! You pick them out!"

Ameshurst had been sat close by and he had heard all and fully understood all, including all implications.

"I'll be one, Sar'Major and I'll find three more. You get on to the other Companies. We meet on deck?"

Gibney saluted, grateful to be no longer dealing with a very irritated Carravoy.

"Yes Sir. On deck Sir. Thank you Sir."

He saluted and hurried off. Ameshurst saw no need to converse with Carravoy and so he did not, instead he chose three men still awake and led them up the companionway to the weather deck where Carr was waiting. Within 10 minutes the 40 were back on the quayside and within an hour they were back carrying all manner of possible flotsam. The best were small barrels, the worst, bales of cork, whilst in between were small bouys, fishing net floats and a type of bottle wrapped in straw. Carr looked at the collection with more dismay than hope. It would be enough for the followers and the children, plus the wounded. The rest of the battalion would have to take their chances and, after all, a sinking ship releases all kinds of bits and pieces that float! Carr looked at his men looking at him, each wondering at the reaction they could expect from such a paltry assemblage, but they should not have worried.

"Well done, men! You've done your best. Now, get these down to the wounded and the followers, and you are to spread no alarm. Tell them that it is just a precaution, just in case they find themselves in the water. Well done again! Go now."

The men each gathered into their arms and hands whichever of the buoyant items they could and began the first of several journeys below decks. Carr walked to the stern, climbed the companionway up to the quarterdeck and viewed the French flag, thinking to himself, 'I wonder, just what are you cooking up over there?' He pulled out his telescope and focused it. On the slope above the fort, significantly above, the French had ranged a long battery of field guns. 'Why there?' he questioned. 'Was it because the main embrasures were filled with potent 24's or were they higher up to add to their range? Were the embrasures blocked by spiked guns, or was it the extra height what mattered?' He lowered the glass and shook his head. This was mere conjecture, only time would tell. He walked back to the first companionway down; he would spend the waiting time by checking on his own wounded, did they have their own piece of 'flotsam'? It did little for his mood when an orderly came up past him carrying something in a bloody blanket. The orderly went to an equally sanguine blanket, folded it back and added a severed arm to the gathering of shattered limbs already there. These would be jettisoned overboard once they were fully out at sea. The orderly replaced the fold of blanket and chased away a seagull perched knowingly nearby.

Meanwhile more waiting was taking place but with a greater degree of impatience. Captain Lord Carravoy was waiting for the return of either of his Superior Officers, either Lacey or O'Hare would suffice and in the event he got the latter, him hurrying back to his cabin. Carravoy straightened himself at O'Hare's approach.

"Sir, might I have a word?"

O'Hare looked at him testily, showing no small measure of his own impatience.

"I know the subject, Charles, but I can spare you only a little time. Nevertheless, you'd best come in."

They both entered and Carravoy closed the door, but O'Hare was already speaking.

"It was a joint decision! By both the Colonel and myself. We needed a replacement Major and no-one can say that Captain Carr is not worthy of full consideration. Since we landed at Mondego he has barely put a foot wrong and it is for that reason that he got the nod over you."

He paused.

"We know that Carr will do the job! It's a question of trust, and that's built by performance, and his has been good enough. Well good enough!"

O'Hare paused again.

"I've just heard that Captain Carr has noticed that the French have occupied a fort that commands the bay, found out all he could about the guns there, made sure that Headquarters knew, then organised floats and such for the wounded, should we be sunk. Why didn't you think of that?"

Carravoy was utterly dumbfounded and had no answer. He knew that he had spent the first of his time aboard defending his space in the tweendecks. Instead he cleared his throat.

"I am the Captain of the Senior Company, and by the common practice of seniority, the Brevetcy should come to me as the next in line."

O'Hare leaned slightly towards him. At that moment he did not much like Carravoy.

"We decided that Carr would not let us down. With him, what needs doing will get done, from his own initiative, and that is what we need, now, as we speak!"

He paused again, but, despite the next words being encouraging, he did not smile.

441

"However, it is a Brevetcy, it's temporary. Everything may change when we reach England. It does not lie within the compass of either the Colonel or myself to confirm it. You know that, as well as I do."

He turned to pick up a document, which was his real reason for coming down to his cabin.

"Now, leave it there. We have a battalion, our wounded and our women and children to keep alive on the journey home. It would be the biggest tragedy if they were to starve having come through the retreat and a battle! Would you not agree?"

O'Hare left the room leaving Carravoy standing there and not caring if he remained in the sparse, even desolate, cabin or not. Carravoy, for his part, felt the blood rising from his collar to his hairline. He was seething with anger, firstly at being compared so unfavourably to Carr, with evidence, and secondly, to be spoken to thus by an "Irish come-by-chance" as he viewed him. He did remain for some minutes in O'Hare's cabin, before recovering himself to leave and then shout at Binns, that their clothes and equipment were still in disordered piles on the deck of their allocated place.

At that moment, in nothing like a cabin of their own, but in the hold directly above the stinking bilges, Nelly was examining Bridie's foot in the hopeless light of a candle lantern, but even in so worthless a light she could see that the first section of the centre toe on Bridie's left foot was black, with an open sore at the very end. Nellie did not look at her friend, she was too worried.

"Bridie, that's mortifying! You should see the Surgeon, so you should."

"And how much time do you think he can spare me? Sure is he not the busiest man on the ship right now? Perhaps tomorrow, when we're out at sea. I'll ask for a moment of his time then."

Nelly knew that it was useless to argue. Bridie had made up her mind, both about what would happen to the toe and about the timing of her visit to the Surgeon. Nelly responded in the best way that she could.

"Well, at least we can put a clean sock on that foot, now!"

With that she pulled a clean pair of stockings from her own bag and carefully put them on, noticing, but pretending to ignore, the pain on Bridie's face. With the stocking in place, she looked concernedly at her good friend's face, but it was Bridie who spoke.

"Sure, 'twas just a bit of frostbite. I've had worse. We'll just

442

have to see."

At that moment the thoughts of both were distracted by shouts and yells, seemingly from above and from the side, then both could feel movement in the hull. This time Nelly spoke.

"Looks like we're on our way. Now, would it not be nice to go up above and take a last look? I'm sure the children would find that a desirable thing."

Bridie nodded, then scrambled to her feet, wincing slightly as her weight went onto her left foot, then she turned to the brood sitting patiently in their portion of the hold, sat beside Beatrice Prudoe.

"Come on up, you lot! We'll go and take a last look at this place. Perhaps not of Spain, but surely of this Corunna boghole!"

Beatrice grinned at the blunt language as the children eagerly gained their own feet and it was Nelly who led them up a companionway from the hold, then to another ladderway to take them from the tweendeck to the weather deck, but only Nelly emerged into the daylight. She was confronted by an Officer she had little experience of, but he was giving her a strict order.

"Back! Go back down."

Then he asked a question, which caused her to pause on her return journey.

"Have you been given anything that floats?"

He paused.

"Just in case."

Nelly nodded.

"Yes, your Honour."

"Then keep it handy, or better still, get it fastened to you."

He saw the crowd just behind.

"That applies to you all!"

He saw her anxiety and smiled and nodded.

"Just in case."

Carr watched Nelly turn again on the companionway; he heard the protests of the children, then he lost interest. A strong and very convenient Southwesterly breeze was easing them perfectly away from the quayside, as it had done for several ships before them, all now almost out into the anchorage of the harbour. Carr took himself to the bows, drew out his telescope and focused on the object of his fears. The angle was too shallow for him to see through the main embrasures of the fort, which was what he particularly wanted to do, but he was distracted by the roar of gunfire from beyond the houses around the harbour. From the walls, he concluded, meaning that the locals had begun the defence of their town, or, more likely, the French had begun their attack. His attention was further drawn away by the sight of many of the local women, stood on the rocks where the town wall met the sea, stood there to wave farewell. Many men of the 105th were along the ship's side and giving an answer, waving their battered shakoes in slow circles. Carr himself was moved to go to the side and wave his own bicorne. Was it really a genuine goodbye and thanks for their efforts? He liked to think so, then came the answer to his uppermost question; an immediate volley from the fort. He immediately looked forward beyond the bows and, even from what he could see with the naked eye, he felt justified in giving a long sigh of relief. There was no smoke from the embrasures, only from the field guns positioned above. However, the French guns were well served by their crews and soon sent another volley quickly across the bay. Carr was joined by O'Hare. Force of habit added the first word, which Carravoy often omitted.

"Sir, where's the Colonel?"

"On the Teignway! What do you make of this?"

"Not much, Sir. They're only firing field guns. We should get through and away."

O'Hare was looking forward.

"Don't speak too soon! What's that all about?"

Carr looked himself and saw the leading four vessels going through a drastic change of course. Instead of standing on Northeast for the harbour entrance, they turned Northwest to sail directly away from the fort. Both men watched, but O'Hare voiced the thoughts of both.

"I hope they know what they're doing!"

The anxiety in both grew, it not being eased by Carr's thinking aloud.

"Perhaps this is their first time under fire. If not, then surely they would know that their ship could take a few knocks from a six pounder, even a French nine!"

O'Hare was still looking. The four were still holding to a course West of North, then they finally made their alteration to resume Northeast. This lasted but minutes, before, one by one, each came to an abrupt stop.

"It would seem not."

They had run aground. Carr was incensed.

"The French must be laughing themselves silly! For a bit of popping off with their field-guns they've sent four ships into the shallows!"

Their own ship, the Dauncy, was altering course towards the stricken vessels, with their consort, the Teignway, following astern. The crews of both were in frantic activity, altering the set of the sails for the new course. The French were, unsurprisingly, serving their guns for all they were worth and Carr heard a loud thump from the hull of their own vessel. He ran to the side, but could see nothing, but then another hit just down to the left of him and there was a splash in the sea. The ball had bounced off! Of whatever calibre, at that range it did not have the power to penetrate the thick timbers, but, impervious to the cannonballs as she may be, the Dauncy was sailing to where four ships had already run aground. However, their Captain was taking no chances, for one hundred yards from the closest stranded vessel, the sails were cast loose and the anchor dropped. The Dauncy immediately swung on her anchor into the wind, to face back to Corunna, however, unused as they may be to being under fire, these sailors did know their business. Longboats were being launched from all ships involved and, sheltered from the French cannonfire by their main vessels, they were taking out towing cables. Some cables led back to vessels still afloat; others remained with the longboats, which would attempt to tow off the stricken vessels from the rocks using oar power alone. After an admirably short time, there was a towing cable from the stern of the nearest transport aground to the sterns of both the Dauncy and the Teignway, and the transports' own longboats were out, poised and positioned.

Orders rang out, utterly unintelligible to the soldiers, but the sails fell again from their yardarms and were sheeted home to draw the strong wind. The anchor was rapidly regained, mostly using the muscle power of some men of the 105th and the Dauncy turned in the wind, then moved ahead to take up the slack. Carr looked over at the vessel they were attempting to pull off to see the stern crowded with red coats; the soldiers were using their weight to lower the stern and lift the bows. Two shots sailed between the rigging, parting one rope, to then land in the water beyond, bounce across the surface and sink. No other harm was done and the parted shroud was quickly re-joined. The slack of the towrope was used up and rose out of the water with a hiss, for the water contained within its strands to then be squeezed out in fountains as the thick cable tightened. The Dauncy stopped with a judder and did not move, and the wind, now trapped in the sails which did not yield into a forward motion, began to sound alarmingly louder in the rigging, but worse, the sails began to shiver and strain. On the quarterdeck telescopes were ranging all around, not so much to look at the ship they were attempting to tow, as to look at reference points on the shore to indicate if they were moving, or not. The Dauncy was held for an age, whilst more cannonshot hit her sides, then, at first imperceptibly, she began to move and they all saw the angles change between themselves and the ship in their charge. Ingloriously and ignominiously stern first, they towed their rescuee to open water. The tow rope was cast off and gathered in, but by then it could be seen that they alone had been successful. The other three were stuck fast, on a dropping tide. There was now but one recourse and that was to take off the crews and soldiers and transfer them to other ships. All to be done under fire from the shore.

Carr drew O'Hare's attention to a whole cloud of signal flags that had broken out from the yardarms of the largest vessel in the fleet back in the harbour, she being yet to leave the quayside to reach the anchorage. Then they heard words that they both understood from voyages past.

"Start all sheets and furl! Drop anchor."

The Dauncy was once again being brought to a halt, obeying the order sent by her Commodore. Another ball hit her side and another hummed through the rigging, its sound the only sign of its passing until it skipped across the surface, just missing a longboat. Within minutes there was a flotilla of longboats either heading for, or arriving at, the stranded three vessels and, but minutes later, boats full of soldiers, their followers and sailors were being sculled over the intervening water to pull up beside any convenient transport. D'Villiers and Carravoy were looking over the side, the latter still in a foul temper, made worse by the continued sound of French gunnery and cannonshot either hitting the side or splashing across to sink amongst the hurrying longboats; gunnery that was beyond their power to answer. Whilst Carravoy could scarcely speak on any subject other than himself being passed over for Carr, the good D'Villiers could, at least, discern some merit in the scene before him as yet another six pound ball sang past, just off the stern where they were.

"I'd say the way they're evacuating those ships is about the only admirable thing within this whole affair."

Inane as it was, it was well meant and plainly held some truth, for rescue boats were arriving and departing at the three vessels, with admirable speed, but Carravoy was in no mood to dispense praise upon anyone, well merited or otherwise. He pushed himself away from the rail.

"I'm going below. Why should I watch any more of this farce, and risk losing my head to a half-spent cannonball?"

He left and, having absorbed Carravoy's last words, D'Villiers took himself below decks also.

Not so the 105th Light Company, now with Drake in command as Senior Lieutenant. They were required to be on hand to help onboard their share of the evacuees and these could now be identified as three longboats that had remained heading directly for them whilst others had steered elsewhere. As their tillers finally came hard over to bring them alongside, they could see that one contained mostly soldiers, but contained therein were also some followers, whilst the remaining two carried only men in uniform. Cargo nets were thrown over and the occupants of the longboats began their climb. Miles positioned himself above the one containing followers, in that place being able to justifiably lay his hands on female flesh was too great a temptation, but first came the soldiers, most with nothing but a haversack, all else had to be left behind. The first soldiers ascended and Miles took himself over the side to position himself at the top of the net, the better to 'help out' as he termed it to himself, but as he and others climbed out over the rail, their bright green facings, albeit dirty and faded, were plainly to be seen by those below and the first comment came up.

"It's the "rag and bone" boys! We're back with the "rag and bones"!"

Miles was immediately incensed.

"That's right! So you'd better come up on the cart, but scrap iron first, 'cos it's worth more than all you lot of pauper waifs and strays put together!"

He drew breath for a good shout.

"'An if you don't fancy a voyage in our company, then you can bloody well paddle yer ownselves home!"

Drake had heard all.

"Miles! That's enough. Get these aboard. There'll be a second helping and no-one has time to waste on such as that."

By now the first were coming over the rail, the very first helped

447

by Davey and Pike. They recognised the facings of these through the dirt, a light buff, almost yellow, but their number was plain on the badge of their crossbelts. Davey, as was his nature, managed a far more warm and considerate greeting.

"20th! All right boys. Come aboard. Come on up!"

The affinity between the 105th and the 20th was sound and genuine and so, with obligatory and routine insults finally dispensed with, once on deck, canteens of brandy were passed around and men already exhausted from marching and battle were helped up the steep climb of the cargo net. Miles went down into the longboat to help the followers, but even he was moved by the fear and exhaustion he saw on their faces.

"Come on, girls! Yer safe now! Let's get you all up into the warm and dry."

He helped many over the longboat gunwale and onto the net, but several had to be hauled upwards by Lights positioned at various levels, being pulled up by the army crossbelts that most of the women wore. Immediately that the boats were empty, they left to obtain more from the three grounded vessels, but they soon returned with a complete mixture, 52nd, 56th, and 14th. Also some wounded, but none that were amputees, at least so far. There were fewer followers in these last and so all were soon embarked and sent down to find a space in the already overcrowded decks. The Dauncy's two longboats were recovered, by this time in peace and quiet, for the French had finally realised that their cannonfire was doing no damage, merely wasting shot and powder. They were content with the damage that had already been done, three British transports gone aground and lost. This was confirmed when they were fired, their burning hulls adding a second sunset to the horizon, as the British fleet, three vessels fewer, finally made a heading Northwards and the coast of Spain slipped away.

***

Another waterfall came over the edge of the companionway, cascading onto the steps, which further enabled it to spread further the soaking it administered, the remainder finally joining the noisome mixture of stale water and vomit that swilled around the sodden planking of their deck. The deck was a dark, foul smelling dungeon of moaning, slumping figures, most stood erect, for lying on the flooded deck meant an immediate soaking in the chilling, noisome water. The one aspect to the good was that the deck was warm; stuffy and noisome, but warm. The Light Company, customarily close to the Third were in the bows of the hold, one level above the bilges and one below the tweendeck, which was one below the weather deck, but still the sea found its way down. Surprisingly, there were few complaints, but at frequent intervals came curses as people found soaked what they had hoped to be dry, both clothing and food. However, the sounds most discomfiting came down through the deck above, from the surgeon's sickbay, him still at work at his table, if not now so frequently.

They had not been a day out of Corunna before the wind veered and strengthened into a vicious Southwester that they could only run before. The one saving grace was that the motion of the ship affected only the most vulnerable of those prone to sickness, for the hull was moved by a long succession of waves that lifted the stern, then passed on, dropping the stern and then lifting the bows. For three days now there had been nothing but the noise of the wind, the crash of the waves and the clank of the pumps. No-one went on deck, other than the brave few to answer a call of nature and that was perilous enough as each had to inch their way to the bows using the hand ropes and several lost their footing on the pitching, slippery deck to be almost swept out through the scuppers by the retreating water, many owing their lives to alert sailors. So, instead many used the buckets at the bow end of their deck, which added to the reek, but it was safer and no-one concerned themselves with lack of privacy; this was barrack life, albeit at sea and in a storm. However, the deck was kept as clean as possible at the insistence of Beatrice Prudoe, this being well supported by the likes of Jed Deakin and Cyrus Gibney.

There were some hammocks, but by common consent, these had been given over to the children, one now holding the combined brood of Bridie and Nellie, bar Eirin, now counted as grown up and required to shift for herself. As did all other adults, sleeping on their feet, slumped against each other for support and leaning against the hull, which never seemed to be at the same angle for more than two seconds at a time. Others had resigned themselves to being wet, so they simply lay down on their packs and haversacks and went to sleep, simply too exhausted. Many had lain there for all of the three days so far, emerging up into half consciousness to take some food, which was merely water and biscuits, plus whatever they had themselves in their haversacks.

On the fourth day the wind relented to a mere howl and many felt able to venture onto the deck, whose pitching had much lessened, and lesser again was the roll from side to side. The waves still surged up through the scuppers, but at least they no longer came over the rail as solid water. What did come over was a stinging, soaking spray, but mostly up at the bows. The four women, Eirin now being the fourth, and the children, were stood out of this, at the break of the quarterdeck and on the starboard gangway, the higher quarterdeck giving shelter above them and they were on the starboard side, which was in the lee of the wind. The children were with them, as were Saunders, Byford and Bailey, all there to give their strength to any emergency that may imperil the children. Behind them, a second serving of hot food was descending the companionway, but, with the children having been given priority with the first serving, all were now sucking in draughts of fresh air. The very first task, insisted on by Beatrice Prudoe, Nelly and Bridie, and backed up by Jed Deakin, had again been the communal effort to thoroughly clean and freshen their deck and carry up the foul buckets of slops, to be emptied over the side and washed. Done at the very earliest opportunity, all was now completed; therefore all was not of the best, but at least a great deal better. Better to the extent that Eirin's attention was now much more mindful of where she was and who she was with. So much so, that Nelly and Bridie had noticed her giving several sideways looks to a young sailor working nearby, so frequent that, in her role as irate and protective Mother, Bridie put a hand each side of Eirin's head and turned it to look seaward.

"Keep your eyes for your own business, you shameless hussy! Setting your cap at any young passing bhoyo as happens to be round and about! Shameless! Just wait 'till your Uncle Jed hears."

However, the sailor, not wishing to be the cause of any offence, felt the need to mollify his nearby companions if he possibly could and, in any case, he well recognised the potential of the situation, so he pointed over the side to a faint line of green, just visible through the variating murk.

"See that, ladies! That's Quessant, the last bit of France. You've crossed the Bay and so, from now on, you'll be in the waters of The Channel."

Eirin, concentrating more on the sailor than what he was pointing at, was the first to react.

"How long till home?"

The sailor delayed his answer, as though thinking, but more as an excuse to look carefully at Eirin, something not lost on Bridie, who looked fiercely from one to the other, but the sailor did finally answer, after looking at the pennant at the top of the mainmast, then grinning openly at Eirin again, which further aroused her good Mother.

"Well, the word is that we're for Weymouth."

He paused for more gazing at Eirin, then brought himself back to the question.

"Well, with this wind, we're about as set fair for that place as you could be."

Eirin was now openly grinning back, which was rousing Bridie to a state of apoplexy, with Nelly not far behind, but it was Eirin who again spoke.

"Yes! But how long?"

A pause for more fond gazing.

"A day! Or so. We'll be there tomorrow."

With that answer, Bridie placed her hands squarely on Eirin's shoulders and thrust her and her now sullen look, off towards the companionway.

"Thank you, Sir. I'm sure that we're all very much obliged for your kind attention and information."

The full stop to this was provided by a push forward on Eirin when she tried to turn her head for a final glance at the sailor, such that Eirin may even have fallen down the companionway had she not finally given full attention to the stairway before her. The children followed, shepherded down by the badly limping Bridie.

Above them, on the quarterdeck, O'Hare and Carr were keeping company with their Captain, a Captain Gavell, a stocky, weatherbeaten man, of a shape that included the desirable low centre of gravity desirable to remain upright on a heaving deck. Both Officers held him in some high regard, for he had brought his ship through the storm with the minimum of damage, certainly none to the precious spars, nor the even more precious masts. All three had telescopes over the stern and slightly off to starboard, where three more ships were keeping them company and all Officers were hoping for good visibility to identify who they were. It was Carr who asked the question, asking it of Captain Gavell, for he had a more powerful, naval telescope.

"Is that the Teignway, Captain?"

Gavell lowered his instrument, but only to wipe the glass of the eyepiece, before sighting again.

"I believe her to be so, yes."

He paused for another view.

"And undamaged, if I'm not mistaken. That's Wetherby for you. I'd trust him to get across the Bay without too much damage to his top hamper, and then set about The Channel in like manner."

He lowered his telescope and looked at both Officers.

"We're a pair you see, the Dauncy and the Teignway. Sisterships! And on top, he's my brother in law!"

Both Army Officers laughed then raised their own telescopes for further examination, but it was Gavell who spoke next.

"But there is something that I view with some concern!"

Both looked at him, but nothing was forthcoming, whilst he held to his studying, then he took his telescope away from his eye and pointed.

"Off beyond the Teignway. I can't identify her, but she's flying pestilence flags, a yellow and another, in this case a red and white quartered."

It was Carr who spoke first.

"That means she's got plague on board?"

Gavell nodded.

"Disease of some sort. Of what sort, well, take your pick. Typhus, dysentery, any one you like, caused by hundreds confined below decks for days on end, all weak and worn out as you are."

He turned and regarded both.

"Take your pick!"

O'Hare voiced the concern of both.

"Are we clear?"

Gavell nodded.

"As far as I know! I'd say so. Thank your people for keeping their decks clean and the water we shipped that swilled all down through the ship."

He paused to allow both to listen to the clank that both had come to recognise, then spoke further explanation.

"That's the pumps sweetening the bilges. At the insistence of your Chaplain's wife, Mrs. Prudoe, I'm pumping in clean water at the bows and pumping out the foul at the stern. That woman is one piece of work, and she's right! Disease makes no discrimination between crew nor passengers. A clean ship is everything, that's how I view it."

Both nodded mechanically, but Carr was now curious.

"What will happen to that ship?"

"She'll be the last in. Then they'll march 'em all off to some barracks or similar and separate the sick from the well. Those that lasts a week without catching it gets to go, and those that recovers gets to go. For those that dies, well, that's obvious."

He resumed his studying through his glass.

"If I've not got it wrong, t'other's clear. So, as we are, we'll be first in! Landfall tomorrow."

***

For a crowd, a large crowd, they were strangely silent, as if some sixth sense told them that all was not well. Shrouded in a thin, but cold drizzle, both ships had slowly emerged to take shape, each being carefully towed into the harbour, but all the while, the absence of cheerful faces lining the rails of each told its own story. The crowd, gathered at the first news of troopships back from fighting Napoleon, had stood at first in eager anticipation. The news had been of a victory, but the sight of the two ships with blank, far from joyful faces, gazing in haggard stupor across the brown and green water, told its own story, and so anxiety grew. There were many of the town population on the hill over the water from the main unloading quay and also in the upper stories of the harbour warehouses which marked the back edge of the quay's surface and from their higher vantage point they could see down onto the decks of the transports crowded with humans dressed in all colours, but mostly pink, faded from red, with a black shako. These were obviously the soldiers, but all had the look and bearing of people in an advanced state of exhaustion and starvation. No seemingly victorious army this, at least not one celebrating, or even raising a cheer to at last be home.

The impression formed within those waiting and watching of the arrival of people at the edge of endurance grew, as the ships were slowly warped closer to the old, worn stonework of the ancient quay and, as the distance lessened, so increased the pity and concern of those stood watching. This was even heightened when, after the gangplanks had been hoisted in place, those on board finally filed off, some limping, the worst to collapse on the quayside, to be helped up by their comrades and then carried forward, to be allowed to lie down at their assembly points. But, equally as alarming for the watchers, was the appearance of all, their clothes and boots worn out to the point of disintegration, all with holes, at least, at one of the four points of knees and elbows and all with rents caused by the simple fact of the cloth coming apart from sheer wear.

All seemed to be wearing a uniform that was too big, even allowing for the fact that, as a garment, it had lost its original shape, but it was the expression on the faces of all that gained the deepest sympathy of the people of Weymouth there assembled. Sunken eyes carried a hunted look that a homecoming had not yet dispelled and all faces were hollow and haggard, thin lips closed tight by jaws too tense to loosen into any kind of warmer expression. However, notwithstanding, the people knew that this was a victorious army, that despite their appearance and evident exhaustion, they had fought the French to a standstill and were deserving of whatever could now be done. There were a few hearty cheers, the odd "Well done, lads!" and sporadic applause, but most concern was attached to the sorrowful state of the men; thus pity vied with admiration.

Both were uppermost in the thoughts of the good Mayor of Weymouth, Councillor James Bower. He stood stock-still, as aghast as any, but he was waiting for the first Senior Officer to disembark. This proved to be Colonel Lacey, coming off the Teignway and Bower immediately moved forward.

"Colonel. Welcome home. Permit me to introduce myself. Councillor James Bower; I am the Mayor here."

Lacey took the proffered hand.

"Your servant, Sir."

Bower looked into the tired eyes. This Colonel looked as worn down as his men.

"Colonel. If I may make so bold, you and your men look in poor shape. Is there anything that we can do, the people of Weymouth, that is?"

Lacey looked into the face of the good Councillor and saw only genuine sympathy and concern.

"I thank you for your most kind offer."

Lacey paused for thought.

"We must return to our barracks in Taunton, but, as you yourself have concluded my men are in none too good a condition to undertake such a march, at least not immediately. If we could remain here, not necessarily under cover, for a day at least, then they could recover somewhat and repair their uniforms and kit. In addition, although the possibility is far from certain, it is possible that new boots and uniforms may arrive. To help with that, I would care to send two letters. You have a mail service from here?"

Bower cheered up at Lacey's positive tone, all the more so for it being so unexpected.

"We most certainly do, but we can do better than that! At our expense we will provide a courier, well mounted, to deliver whatever you wish."

Lacey brightened at the kind offer, which would certainly speed up what he most wished for.

"Thank you, that is most kind. I need to send one to my barracks and another to our Regional General, that being General Perry. I could write both immediately, if that does not inconvenience you."

The good Mayor straightened and leaned back as if the pose would give extra weight to his words.

"Indeed it would not! I will not leave this quayside until you have given me both letters and I will see to their dispatch as my priority on my return to the Town Hall."

He drew in a deep breath.

"Now. Your men! What are their needs? Food, would be my immediate guess. Hot food."

Lacey was genuinely moved by Bower's earnest desire to provide significant help.

"Mr. Bower. I cannot thank you enough for ...."

Bower had held up his hand, his face closed in serious thought.

"This is what we will do. I will martial the Town Council and we will all traverse the town with a piece of chalk."

Lacey was perplexed, but Bower continued.

"We will knock on each door and enquire how many of your men can be billeted there and we will pay a bounty to each house for every man, as an incentive. And within each house, each man must first be given a hot meal."

He thought further, as conveyed by his mouth working from side to side.

"Your Officers take your men around the town, both sides of the river, and send in men according to the number chalked on the house."

He paused.

"It is not yet Noon. With reasonable speed, we can soon have all of your men in some kind of shelter. I don't expect this weather to improve."

Bower nodded, evidently pleased with his scheme.

"Now. If I am to get away and gee up my Council, that means I need your letters!"

Lacey's eyes widened in both agreement and surprise. He looked behind him, somewhat anxiously, but he was there, as duty dictated, he being his Sergeant Clerk Herbert Bryce, carrying all that was required in his satchel.

"Bryce! I need to write two letters."

With no more words, Lacey walked into the shelter of the wide entrance of a ship's chandler. He begged the use of the counter and began writing, one letter to their barracks ordering that all available uniforms and clothing be brought to Crewkerne to meet them on their march, if not there, then Ilminster, closer still to Taunton. That done, he penned another, equally brief, informing General Perry that they were marching to their barracks, that their muster was 570 and requesting clothing and supplies. Both were sealed with an unfussy blob of wax into which Lacey impressed the Regimental Seal, this being the simple number "105". Bower had been patiently waiting and he eagerly took both letters and was forthwith on his way.

Lacey then returned to the business of his men, now mostly off the transports and assembling down a side street, where the tall buildings gave at least a modicum of shelter from the thickening rain. Lacey found O'Hare and both thought it best for them both to remain where they were, on the quayside, and there supervise the final disembarkation, which would soon involve the wounded. Thus, neither noticed a young, well dressed man, energetically progressing up and down the lines of men, these not in any ranks, but sitting in the wet road or propped against a back wall. However, he was noticed by Regimental Sergeant Major Gibney, as the young man frequently stopped to ask questions and then scribble on a paper that he had sheltered beneath a square of tarpaulin. Gibney did not like what he saw, this eager chap, bothering his charges. He strode up, which massive motion immediately gained the attention of the object of Gibney's short walk.

"Now then, Sir, is there anything, now, that we can help tha' with?"

The young man looked up, then further up, to notice first two very piercing eyes above a hooked nose above the parapet of an imposing moustache, but he was about his trade and was not intimidated.

"Why yes, Sergeant. I'm trying to gather a few facts for my newssheet, the Dorsetshire Echo. I'm asking your men a few questions, now that they have triumphantly returned to the safety of old Albion."

Gibney found this very irksome; firstly because this youngster had reduced him in rank, secondly, because he was trying to "butter him up" but thirdly, and the most important, he was bothering the men, who plainly just wanted to sit quiet.

"Questions? Questions, tha' say? About what?"

The stern tone and impatient gaze began to have an effect on the young man, who began by nodding his head.

"Yes. On what happened, how it was, what they think. That sort of thing."

Gibney nodded.

"Right. Well tha' can ask that of me."

"You, Sergeant?"

"Aye. I were there. What thee can get from them, thee can get just as well from me."

The reporter's face considerably brightened as he turned back the tarpaulin.

"Right!"

However, before he could either ask or write, Gibney's huge hand was covering the page, already written on.

"I'll tell thee this, an' no need to write it down, for thee'll remember it."

He leaned down to stare directly into the young man's eyes.

"For six month, we fought the French. Two battles, won, then a march, a retreat, then another at t'end of that, a battle, that is, again won! It were bloody awful, beyond what thee can imagine. Come final, the French could put out more than us could handle, an' so we had to get out an' go, leaving' hundreds of good lads and their followers behind. Women and children."

He leaned back.

"Now, thee can write that down an' use it, but the rest th'gets from what thee can see. Just take a good look, that'll gain thee all th'needs to know."

457

He paused, for effect.

"This is what fightin' the French amounts to! An' if I sees thee botherin' these lads some more, I'll see thee out of here, with my boot up tha' backside!"

The young reporter remained transfixed under the stoney stare for some seconds, but, eventually, the unrelenting condemnation of that baleful gaze had its effect. He tucked the tarpaulin shield down over the paper, nodded and sidled off, taking the first side street that took him out of Gibney's hostile attention. With that, Gibney took himself off to supervise the placing and comfort of the wounded, now arriving in the street.

Lacey and O'Hare were watching the last of the wounded come ashore. The condition of his men, hidden from him by the crowding and gloom of the lower decks was now in view apparent and he was both angry and saddened that the men of his battalion, for whom the last few months had engendered within him the deepest respect and no small affection, were now brought to so low a state. This was not in any way ameliorated by what came next, this being a Sergeant of Transport approaching and standing at a respectful distance at the attention, but plainly wishing for Lacey's attention, who did finally notice him and so turned to receive a full salute.

"Sir! I'm hoping that you can help me, Sir!"

Lacey clasped his hands behind his back and leaned forward.

"With what, Sergeant?"

Remaining at rigid attention, the Sergeant continued.

"Sir, I have four wagons with new kit, Sir. Sent down by General Perry, to be issued to whatever troops landed here, Sir, this place here being within the General's command."

Lacey turned towards O'Hare, for both to share a look of pleasant surprise, but it was O'Hare who asked the question.

"Kit! Does that include new tunics, trousers and boots? Plainly, those are our greatest need."

Fear passed behind the eyes of the Sergeant.

"No Sir. 'Fraid not, Sir."

The Sergeant's jaw clamped shut. O'Hare leaned forward.

"Then what do you have on your four wagons, compliments of the General?"

The Sergeant's mouth opened, then shut, then he began speaking.

"Crossbelts and shakoes, Sir. Crossbelts with new bayonet scabbard and cartridge box."

The look the Sergeant received from both very superior Officers did nothing to ease his discomfort, but he was veteran enough to know that brevity was the safest course. He said no more. Lacey and O'Hare looked at each other again. Lacey look appalled, which emotion and appearance was mirrored in O'Hare, but it was the former who now spoke.

"How many men do you have, Sergeant?"

"Seven, Sir, besides myself."

Lacey nodded. He did have some sympathy for the plight of this NCO.

"Very well. Draw up your wagons onto the quayside. Issue what you can to whomever you see, then take the old, which you are replacing and load it onto your wagons for your return journey. Take good care of it."

He smiled, which did much to ease the anxiety of the Sergeant.

"Can't have you losing your stripes for destroying the King's property, now can we?"

The reply was a rapid salute and the Sergeant executed an about turn and gratefully marched off. O'Hare watched him go, then turned to Lacey.

"Whatever was he thinking?"

The subject was plainly General Perry rather than the Sergeant, now departing. Lacey shook his head.

"Who knows? Parade ground appearance, as we march back through the towns and villages. Perhaps he thinks that it's only our brightwork that's suffered and that's all that needs to be brought up to standard."

O'Hare laughed cynically.

"What was it I read somewhere: "Our gayness and our gilt are all besmirched by much rainy marching in the painful field"?"

Lacey laughed in reply.

"Yes! Perhaps our esteemed General should re-read Henry V. If he's read it all!"

Both men folded their arms together, savouring the thought, but it was Lacey who first stepped away.

"Right! Let's see if the good Mayor's chalk marks are making any kind of appearance. I'll do the far side."

However, they did not yet leave the quay, for anxiously coursing up and down the wide expanse, looking at anyone in a uniform, could be seen Beatrice Prudoe, the level of her anxiety plainly increasing as the worry on her face spread its contagion down to her hands that were now wringing themselves together into fretful shapes. Lacey went over to her, but despite his words, she continued her peering into any door or window within sight.

"My dear Mrs. Prudoe. Is there something amiss?"

She nodded, then continued her desperate searching. Lacey immediately became almost as anxious as she.

"What is it? I would like to know. Please calm yourself and tell me."

At last she turned to face him.

"My husband, the Chaplain. He is not here and I have not seen him since before the battle."

Her voice rose to a tremulous sob.

"I cannot see him!"

Lacey called up O'Hare.

"The Chaplain. Have you seen him?"

"No Sir. Not since we got onto Corunna ridge."

Lacey released a long sigh.

"Was he in the casualty lists?"

"No Sir. I compiled them myself."

Lacey looked from one to the other, twice. The Reverend Prudoe had disappeared. He may show up, he may not, but for now his wife must be cared for.

"All is not lost Mrs. Prudoe. He may be on another ship, all was very confused. Meanwhile, please accompany us to Taunton, where you will be well cared for. You have my assurance."

The evident concern of the Colonel, no less, did much to raise her spirits and Lacey continued.

"I'm under the impression that you have strong acquaintances amongst the followers?'

She nodded.

"Then may I suggest that you re-attach yourself to them. They are all good and hardy souls whom I'm sure will get you up to Taunton with us."

He pointed to the 105th 's side street.

"I believe they are down there, with our men. I'm sure that's the best place, if only for now."

They watched her leave and were re-assured themselves when, by pure coincidence Bridie and Nelly emerged from around the corner with their children, to immediately welcome her back into their company. Lacey turned to O'Hare.

"Right. That's done, now let's see where the men are to go."

For the next hour and more, both toured the houses on both sides of the river Wey until they decided that sufficient numbers had now appeared on the doors of the surrounding houses, then both returned to the quayside. A quick gathering of Officers soon explained all and soon after that groups of soldiers were marching around the town, the Officers leading not only their own men but a mixture which included many of the 20th, 52nd, 56th, and 14th.

What was also a mixture was the reception that they received. In the terraces and cottages around the harbour, the men who came to count themselves as lucky, were made as welcome as the poor means within these dwellings would allow. However, at the more affluent houses, these now extending fashionable Weymouth both South to Portland and North to Dorchester, the beggarly sight of the soldiers, accompanied by the sight of crawling lice and a strong human odour, resulted in the 'guests' being immediately sent through the house to shift for themselves as best they could in the cellars below and the sheds beyond. The lucky one's were soon stripped of their clothing, notwithstanding the livestock contained therein, for both to be washed, while the men stood in one portion of the yard, using buckets of water to clean themselves. In another portion of the yard their uniforms were now in the tubs, being pounded by the women of the house, pounded to the point which threatened their final disintegration. Yet, come the evening all were before the fire, the uniforms drying and the soldiers sat wrapped in blankets, talking with the men of the house, whilst the women, including the followers, attempted some repairs to the first clothing to dry. However, whatever the warmth of their welcome, all had been well fed.

The following day, all formed up on the esplanade to the East of the river, which terrace carried the Dorchester road, going North, the column with the 105th in the lead, then an assemblage of the various Regiments behind. All paraded there were now much improved, from a good night's sleep, a wash, of sorts, and a clean uniform, albeit one barely hanging together. Over this, in much contrast, which emphasised the poor state of their uniforms, were the brand new, bright white crossbelts, with a shiny new shako above that, whilst between the two, could be seen a still very haggard face. However, what was not new on either was the regimental badge showing on the belts, nor the faceplate on the shako. These had been retained by all, replacing the "other" that had been found there.

As much in appreciation as in sympathy, for the Mayor's bounty had been very generous, many of the population had turned out, 'to give a cheer' and bid farewell, which was done with gusto as the soldiers marched off, but there was no singing; only the rhythmic beat of weary feet. Soon they were amongst the hills inland and onto the straight Roman road that led directly into Dorchester, but here far fewer of the townspeople were making an appearance to view their passing or to pass on supplies or even sympathy. The "affair of Moore" was turning into a scandal and few wanted any part of it.

**

Chapter Eleven

Homecoming

"Henry! Wake up!"

Carr opened his eyes and, after a third effort, managed to keep them open. Merely an hour previous, he had again submitted to the overwhelming lethargy that had fallen upon him since they left Corunna harbour and this time sleep had been deep and irresistible, all the more so for the fact that this time he was lying on a proper bed. However, now he was being violently woken by Nat Drake. Carr managed to focus on the deeply concerned face, to which he eventually cudgelled his brain to pay attention. It was speaking further.

"They're here!"

Carr blinked twice, imagined himself to be lying on the ground and consequently rolled forward to almost fall off the bed, only to be saved by Drake pushing him back.

"They're here! Cecily and Jane!"

The last name cleared his brain like an April wind clears fog.

"Jane? Here?"

Drake showed his impatience.

"Is that not exactly what I have been saying?"

He found Carr's shako and placed it on the bed before his face.

"They heard that we've landed, and have come down in a carriage, to meet us. Which includes you!"

He stared intensely at his superior Officer; superior in every way except coping with such an occasion as was now upon them both.

"You've got to come downstairs, they're waiting"

He paused whilst the bleary eyes before him blinked in continued amazement.

"Jane can't come up here, much too unseemly! So, smarten yourself up! This is your prospective intended whom you are about to be re-united with."

Carr swung his legs off the bed, which only caused him to see his horrible boots. He stood and did his best to obey Drake's order, using the corner of the carpet on the boots. He picked up his shako and made for the door, but Drake stopped him, took the shako and placed it on his head.

"You need something to remove, from your head, to show good manners and respect. Very important!"

Carr nodded dumbly and followed Drake out onto the landing and down the stairs. At the bottom was a door and Drake, albeit in the gloom, turned to make a final examination of Carr. There was much that was wrong, but it could not be helped. He opened the door into a room well lit by wide windows either side of the front door. Whilst Drake walked forward to close with his beloved Cecily, Carr stood stock still, inevitably failing to remove his headgear. He was brought to a halt, for, framed in one of the windows, was a pure vision, of two figures, both radiating sweet grace and healing tranquility, both achingly lovely and both smiling at him. His view of Cecily was immediately blocked off by Drake running forward, so that only Jane remained in view, she in a full length slate grey coat, edged with maroon, with a matching bonnet and hand muff, but he saw her face quickly change from her own joy at seeing him, to one of deep concern, almost dismay, as she examined him further. She had looked first at his face, with the now extraordinary, and alien, colours of his black eye. The lack of flesh on his face accentuated both his scars, one bisecting his left eyebrow, the other in the centre of his forehead, exactly on his hairline. His uniform was a poor fit, with many sags, folds, patches, and sewn up rents. She did not fail to notice that the worn notch on his belt, the point where it was most commonly fastened, was now two places outside the belt buckle.

"Henry, you look terrible!"

He nodded without thinking, his mind still taking in what was before him, waves of delight and surprise almost overcoming his ability for conscious thoughts, which came, when they did, in staccato phrases; 'It's Jane! She's taken the trouble to come down to meet him. He still mattered to her, a lot. Is that what it showed?' Then, particularly for this occasion, the clumsy societal requirements broke into his thoughts, to produce the necessary greeting, which brought the automatic action of him finally baring his head.

"Miss Perry. I trust you are well and also your parents, that they are well."

Jane could not help but giggle at the absurdity of such a speech in such circumstances, but she curtsied and replied, in proper form.

"Captain Carr. I am quite well, thank you, as is my Father."

Gloom and confusion overtook Carr. Jane had not mentioned her Mother, but she spoke no more. She simply stood regarding him with her familiar half smile, but there was full happiness in her eyes;

464

deep brown, both staring at him with an unbroken gaze. By now Carr had at last thought of something better to say.

"Did you get my letters?"

Her head went to one side to now regard him from the corners of her eyes and she lowered her shoulders; it was almost coquettish.

"Yes. Three! Was that how many you wrote? Cecily got four!"

At last the spell was broken and Carr managed an embarrassed smile and rubbed the side of his face.

"Well, there was a fourth, but I had to post it at Sahagun, where the retreat started. My only excuse is that sacks of mail soon lost any importance. Nat got his last one off at Mayorga, some days before."

At the mention of the retreat, her face fell.

"Was it that bad? The papers are full of it. And nothing complimentary."

Carr nodded.

"Yes, it was bad. The papers won't convey the half of it."

Then he angered.

"And I've seen no report of Corunna! Was there nothing of that? Of the men standing for hours under fire, and yet having enough in them to fight the French to a standstill? If not, then it's a disgrace!"

She smiled. This was her old belligerent Henry, but always in a good cause. She started across the room and he moved also, his hands hesitatingly reaching out in the hope of meeting hers and they were not disappointed. He felt her hands, soft and warm; she had deliberately removed her gloves. She was now but feet away and his affection for her welled up, to arrive intensified from the wonderful surprise, but this was as much because of what her presence signified. Not only was she someone whom he could believe cared about him, but she also confirmed his return to that other world, the one that he could barely dream about over the past months. He wanted to take her in his arms, but he dared not chance offending her by moving any closer, unlike Drake and Cecily who were in the corner breaking all the rules, faces well within kissing distance.

She tugged his hands, but did not release them.

"We can't stay long. Father thinks we're out for a ride. For a bit of fresh winter air! We have to get back. Crewkerne's a bit further than Father has in mind after we'd told him what we were doing."

Carr wasn't listening. His senses were only tuned to what his eyes could see, his cherished Jane, but he had a concern of his own, a much deeper topic and he voiced it quietly.

"What are we going to do?"

She tugged his arms again.

"We, are going to write to each other, and, we, are going to meet, whenever we can!"

Her next words sent his hopes soaring further, each phrase spoken with a pause in between, to give added emphasis.

"I am over 21. I have obtained my majority. I am of independent means. In fact….."

She held up her head at a haughty angle.

"………… I am now described as a spinster of this parish!"

The significance of that was left hanging, but then Drake broke in, still holding Cecily's hand.

"Did you know that he's now a Major? Only Brevet, that's true, but that's the first step."

He turned to Cecily.

"And, if it's confirmed, then I have a very good chance of being made up to….."

He struck an absurd, ham acted, heroic pose.

"Captain! Already am I acting as such!"

Cecily beamed and Jane finally released Carr's hands to clap delightedly, whilst looking frequently from one to the other.

"Oh, that's grand. We're so proud of you both. Aren't we Cecily?"

Cecily was in Seventh Heaven. If confirmed, then she and Drake would marry. She made no reply, other than to clasp her hands under her chin and beam up at Drake. Jane now decided that she must break the spell.

"We must go!"

At that, Cecily and Drake came together to kiss each other very carefully and chastely. Jane had turned to Carr and he, very slowly, reached out his rightt hand to offer to take Jane's leftt and she raised the hand to marry with his. The pressure on his own fingers told Carr everything and he raised her hand to kiss it, which gesture caused Jane to take one small step towards him. It was the closest they had ever been, but the moment was broken by Drake and Cecily opening the door. However, Jane did not release Carr's hand until they came to the waiting closed carriage, this with the crest of Cecily's family on the door; the Fynings. They were using a carriage owned by her aunt, this being all part of the subterfuge, but Jane did not enter the carriage; instead she turned to Carr, closer still than before and raised her face to his. He looked deep into her eyes for the briefest second, before lowering his face to hers, to meet his lips with hers for the merest, slightest touch. Then she was into the carriage and the closing door broke the shared look between each couple and then they were gone, and then all too soon trundled around the corner of the street and out of sight. It needed an elbow in the ribs from Drake for the stunned Carr to wake up to his duty to wave at Cecily as the coach turned the corner.

***

At the same time, the 'other ranks' of the Regiment were marching towards their new camping grounds. The previous day, Lacey had sent a messenger ahead to the Mayor of Crewkerne, gathering places and numbers of billets for all his Officers. Luckily, the billet allocated to Carr and Drake was in the town, whilst the most of the rest of the Regiment marched on and off, to a variety of places, mostly somewhat Eastwards. Joe Pike was marching just behind Miles and Davey and the lanes and the houses began to look very familiar, this confirmed by their finally passing between the ornate stone pillars of a very impressive gateway. The gates themselves, painted an ostentatious gold and shiny black, each thrust right back to the edge of the driveway, this well defined by its junction with the edge of manicured lawns.

"Tom, John. I know this place. I used to work here, I was a fencer. This place is called Farslake Hall, owned by the Coatsleys. At least it was back then. It was my last job."

Both his audience knew that this 'job' had nothing to do with swords, it was merely concerned with posts and rails, but Miles was unimpressed. His temper was not improved by the fact that the soles of his boots were finally giving out, to the extent that he could feel every prominent stone he trod on up from the pale shingle drive they were now marching over.

"And what good's that goin' to do? You'll be able to march up to his Lordship or Earlship an' he'll give you a joint of beef an' a bottle of brandy, for old time's sake?"

Davey felt irked himself by Miles' harsh rejection.

467

"The boy's just sayin'! Knowin' the place can't do no harm. May even do some good."

In the event it did, but first came new uniforms. Modesty required that the men march deep into the woods, followed by a train of supply wagons and there, using a convenient stream, they washed and put on new, committing all their old, bar their already new cross-belts, cartridge boxes and shakoes, to a smouldering pile in a clearing. General Perry had finally performed his duty and provided what was needed most. Whilst this was taking place, Lacey and O'Hare approached the main door of the grand house, this being double leaved, black, with crystal glass in the upper half, all beneath a barbican sized portico. Lacey rang the bell. Nothing happened and so he rang it again. The door was opened by a stern faced woman, she being identified as female solely from the severe iron grey dress that she wore. Had she been in men's clothing, they would have taken "her" to be of the male gender. She spoke no greeting and so Lacey filled the silence.

"Good morning. We are the 105th Foot. I am under the impression that Mr. Coatsley is expecting us."

There was a pause and both men felt the chill.

"You are correct."

Nothing more was forthcoming and so Lacey continued.

"Well, perhaps you could tell him that we are here, and that we would like to offer our thanks for his kind hospitality."

Again, a frosty pause.

"No need. I am his Housekeeper and you may deal with me. My instructions are to have you conducted to the West wing, which has been prepared for you, your Officers only, that is."

The omission of "your men" told its own story. Plainly, they were to camp out in the open, but Lacey was still engaged with the formalities.

"Then permit us to introduce ourselves. I am Lieutenant Colonel Lacey and this Officer is Major O'Hare.

O'Hare half bowed, oozing Irish charm.

"A pleasure to make your acquaintance, Ma'am."

The stony visage remained.

"I am Mrs. Gimlet."

Both men laughed inwardly at so appropriate a name, but Lacey did not let it show. He took a deep breath to help.

468

"We have wounded. Also women and children. Where may they stay?"

There was a long pause and the movement of her eyes told that she had no answer as pre-delivered by Coatsley, instead she had to think for herself, but an answer came, spoken testily.

"I have no instructions, but behind your wing are stables, for both horses and carriages. Any space you find there, you may use."

With that she turned away and beckoned a black liveried servant.

"Thornby here will show you the way."

She stepped back to allow Thornby to exit and, with that, she closed the door, leaving Lacey and O'Hare staring at their reflections in the bright crystal. O'Hare turned to the obediently waiting Thornby.

"Lead on. Please."

The servant took them on a 100-yard walk across the left half of the imposing frontage of Farslake Hall, to eventually turn a corner, which showed a side entrance. Thornby went to the doors and opened them to reveal what was a ballroom. Remaining obediently silent, he allowed the two to enter. Lacey looked at him.

"Are there rooms above?"

Thornby nodded.

"Yes Sir. You have the next floor up. This is the guest wing."

Lacey nodded.

"And the stables?

Thornby went out through the still open doors.

"This way, please, Sirs."

They followed him again, for merely a 50 yard walk this time, to an imposing building with six high, wide doors, evidently to allow the storing of carriages. While Lacey and O'Hare examined the impressive frontage, Thornby bowed and took his leave. Lacey looked at O'Hare.

"So, let's take a look."

'A look' revealed just three unoccupied spaces, these in the carriage section, but the rear of these parking spaces was merely a wooden stable partition behind which was a space fully occupied by horses. For a stable, it was clean, but the place stank of the animals it contained in such close proximity. Lacey looked at O'Hare.

"We can't keep wounded in here, so close to these horses."

He looked around further, as if to confirm his decision.

469

"They can have the ballroom. Any followers with ailments go in with them."

He looked at O'Hare.

"Use this as a store. Put a guard on it."

They then left to return to the ballroom and soon messengers were running around the grounds giving orders and information. When Lacey returned to his now erected command tent he was greeted by Bryce, his Clerk Sergeant.

"Sir. A letter's arrived for you, Sir. It's on the table inside."

Lacey went straight in to see the letter on the table. It was on mean, cheap paper, barely above newsprint quality. Lacey carefully opened it to find it was from General Perry and sparse to the point of discourtesy, summoning Lacey and O'Hare to a meeting. Lacey placed the letter to one side and then sent for O'Hare. He had no illusions that the meeting would not be a pleasant one

By the afternoon, the men were setting up camp in the woods, using the trees for both shelter and fuel. The Officers, bar the lucky few such as Drake and Carr in the town, were settling into the rooms above and the wounded were being carried into the ballroom, including several followers each with their own "wound" or malady. The midday meal was soon prepared over good fires, there being plenty of fallen branches and, with both now fed, Joe took Mary's hand.

"I used to work here. Let's go see if any I knew are still here."

They set out across the lawn and went first to the kitchens, where Joe was immediately recognised amid squeals of delight and all there, bar the Cook herself, incongruously, but by force of habit, curtsied to Mary when she was introduced. The Cook seeing Mary and the obvious affection that Joe had for her, soon made up a sizeable parcel of 'left overs', which Mary gratefully slipped into her haversack. With that they left, but rounding the corner, they bumped into another figure familiar to Joe.

"Mr. Tilsley!"

The stocky figure halted and looked up at Joe, somewhat puzzled.

"It's me, Mr. Tilsley. Joe Pike! Perhaps you remember?"

Tilsley's face changed completely to one of genuine pleasure.

"Joe! Yes of course."

He looked Joe up and down.

"So you joined up! And look at you now! Ha!"

He seized Joe's hand, which he pumped up and down. Between jerks of his right arm, using his left hand, Joe pointed at Mary.

"This is my Mary. We've been together now over a year."

Tilsley let go of Joe's hand to take one of Mary's in both of his.

"You are welcome my dear. Most welcome. Joe was a fencer here, before the army."

He left out the circumstances of Joe's leaving and turned his attention back to Joe. Then his face changed.

"Are you back from Portugal? Part of Moore's army"

Joe nodded.

"I am, that's right."

"Bad business Joe, the country's in uproar about it."

He leaned forward, consolingly.

"But don't let that worry you. You beat the French before you left, and that's to be proud of, and more's the pity that no one seems to want to dwell on that. Very much more's the pity, in my view!"

He beamed at both, then slapped Joe on the arm.

"Well, I must be about my business. Come and see us before you leave."

Both nodded and left to cross the lawn again, but this time under the baleful gaze of RSM Gibney, but they were wandering back, not wandering off and so he held his peace. As did Tom Miles when Mary delivered the parcel from the Cook. It contained offal that the 'Family' would never touch, kidneys, sweetbreads, liver and a sheep's heart. All added a significant amount of meat to their stew that night and much was left over for the frying pan for breakfast. However, Nelly did not pass up the opportunity to discomfit Tom Miles over the extra.

"Now then, Tom, would you like a nice fried kidney as was brought back by Joe, like, him bein' in the good grace of the cook an' such, havin' worked here, you understand."

A black scowl was the only reply, but his bowl was lifted up to receive, nevertheless.

That very good breakfast brought a smile to the faces of all, bar Bridie, whose pained expression was soon noticed by the arch worrier, Jed Deakin.

"What's up?"

"Oh 'tis nothin'. Nothin' at all."

Nelly had also noticed and was having none of such an answer, she knowing the true cause.

"'Tis and all! 'Tis her foot, Jed. A toe 'as been painin' of her since we come off that hill at Corunna. If y'ask me 'tis mortifying."

At the dreaded word Jed sprang forward and knelt before Bridie, who was sat on an empty barrel.

"Let's have a look."

The boot and sock were removed and Jed did not fail to notice the wince of pain that came across his Bridie's face as both came off. He held the foot gently in his hand and but a short glance told him all he needed to know.

"Nelly's right. It's mortifying and crossing the knuckle."

He looked up into her face, hers matching the deep concern in his.

"You'll have to see the Surgeon. If he wants coin to see you, well that's no problem."

They shared another look, before Deakin gave his verdict.

"We'll go now. And you'll not walk another step on that foot!"

He turned away, a Colour Sergeant again, to address the onlookers.

"John, Joe, Toby. Get a chair or some such, or some poles. We've to carry Bridie here across to the house."

A chair, unsurprisingly, was not to be found, but five poles were and John Davey, the woodsman and ex-poacher, expertly lashed the five together to form a seat, with four side extensions for the carrying of. That done, Bridie was scooped up onto the platform and all set off.

Meanwhile, Mrs Gimlet was stood at the ballroom door, dispensing a look that would have frozen a waterfall. Instead of the dancing floor being occupied by the polite society of relaxing Army Officers, it was a hospital ward, including women, scandalously in the same place as the men, with children running around. There was even an operation taking place, thankfully not an amputation, but plainly, from the cries of discomfort, it was very painful. She looked around for a target and found one. She would have much preferred to vent her spleen on 'that Colonel or his Major' but her anger was too full to accommodate any delay. Lieutenant Ameshurst was stood nearby overseeing the wounded from his section of Grenadiers and Mrs Gimlet did not recognise any need to approach him, she merely raised her voice to 'servant ordering' mode.

"Are you an Officer here?"

Ameshurst had no idea who was being spoken to but it could be him and so he turned around to see for himself and then experience the harpoon gaze of Mrs. Gimlet, but, nevertheless, he smiled and set about being as pleasant as he could.

"For the want of a better word, Ma'am, yes."

Mrs. Gimlet paused momentarily to dismiss the cheerful response, so that it made no impact on the composition of her own reply.

"This is a disgrace. I specifically stated that this was for Officers and that means Officers only. You have turned it into some kind of infirmary for whomsoever you choose, including, these, these …………"

She sought for the right word.

"………… attached people!"

She drew further breath.

"You have broken the terms of the agreement we made with the Mayor! Be certain that I will be informing Mr. Coatsley!"

Ameshurst took a deep breath himself, now somewhat saddened at such a verdict upon what he saw as an obviously good use of the space."

"Sorry you feel that way, Ma'am. Our Colonel intended no sleight upon you or your Mr. Coatsley, I feel sure. It's just that, well, wounded, sick and injured are a special case, Ma'am. At least that's how we see it. To be given the best treatment we can come up with."

"I spoke specifically of the stable for such as this!"

Ameshurst screwed his mouth sideways, as he always did when being required, against his good nature, to contradict someone.

"Well a stable, Ma'am, is hardly the best place for those with open wounds and the like. Flies and dirt and all. To be frank, I'm not surprised that our Colonel moved them here."

Joshua Heaviside, having his bandage changed, had heard all and was not at all impressed. On top he was possessed of a character far less diplomatic or ameliorative than that of Ameshurst. He began in typical fashion.

"The merciful man doeth good to his own soul: but he that is cruel troubleth his own flesh. Proverbs 11. Verse 17."

Ameshurst smiled and backed away one step, whilst Mrs. Gimlet's brows knitted together perplexed, but Heaviside was in his pulpit, fixing Mrs. Gimlet with his own dark eyes under black and overhanging brows, his eyes seeming to come from some smouldering pit.

"These are our wounded and of our own. We, together, have journeyed through trials, dangers and hardships beyond your understanding. These have succoured me as I have them. Are we now, at your mean behest, to cast them out, into a cold and bare stable, to live and heal or die amongst the beasts of the field? I say thy mean spirit does thee no credit. Go then, to thy Master and inform him that sick and wounded lie within his house and ask that they be cast out, but first examine your mission. How Christian? The Lord deal kindly with you, as ye have dealt with the dead, and with me. Ruth 1, verse 8."

Into this heavy discussion suddenly came the four carrying Bridie. Mrs. Gimlet took advantage of the timely interruption and gratefully crossed the floor to leave at the far door, but, whilst Heaviside watched her go, Ameshurst was paying attention to the new arrivals. Deakin chose him for his request.

"Sir. Bridie Mulcahey, Sir. She has mortification in her foot, her toe, to be exact. Can she see a Surgeon, please Sir?"

Ameshurst looked at Heaviside and then back at Deakin, then back at Heaviside.

"Do you know the whereabouts of our Surgeon, Sir?"

Heaviside raised his walking stick and pointed.

"There! But in the middle of what I take to be surgery."

He then rose to his feet.

"You are to take this seat, good lady. I need to be about my business. Because of laziness the roof caves in, and because of idle hands the house leaks. Ecclesiasticus 10, verse 18."

With that he lurched off, within the good gaze of all, but Ameshurst was thinking.

"Our own Surgeon is busy, but there is a local doctor who has arrived, and is visiting."

He peered around.

"There he is. I'll fetch him over."

Deakin nodded.

"Thank you, Sir. Very kind, Sir."

The Doctor duly came over and immediately his face changed when he found that he was not dealing with an Officer but with one of the lower social orders attached to the Regiment. However, the face changed again when he saw a silver sovereign held out in Deakin's palm. This was duly taken and pocketed before he examined Bridie's foot. He took but a second to earn the coin.

"Gangrene! My advice would be to lose the foot. The toe alone may suffice, but it's a risk."

He stood to look sternly at Deakin, as though it was all of his doing, both the ailment and the fact that he was forced to deal with followers and their kind. Deakin divined the look and returned one equally as stony.

"We'll settle for the toe!"

Bridie looked horrified, but Deakin continued.

"How much?"

The Doctor remembered the shiny sovereign.

"Two sovereigns."

"And what will you do?"

The Doctor immediately took umbrage at being so questioned.

"Why, to amputate a toe: a knife and a small saw. Being a middle toe, there could be some difficulty."

"Making it slow? And very painful?"

Now the Doctor was angry.

"Yes. Of course! There must be some skin left on either side to sew back over the bone!"

Deacon stared fully back.

"Thank you, Sir, but no."

Bridie remained horrified.

"But Jed ……..."

Deakin had decided. He took hold of one handle of the "chair", now leant against the wall.

"We're taking her back."

The others copied Deakin, the seat was proffered and Bridie sat on it. On their way back across the lawn, Bridie was still protesting with the same words, but Deakin was outlining their own course of action.

"We needs pitch, a mallet and a chisel."

Bridie wailed, but Deakin now talked to her as tenderly as he could.

"It needs to be done quick. That way you'll feel hardly any pain, like you would with that sawbones goin' through the full palaver to earn his two sovs! I've seen such as this done before, an' done quick and it works, too, in every case I've known. 'Tis the best way, love. It is. The least pain and the best chance of savin' your foot."

He looked up, evidently now speaking to his companions.

"Where do we get what we need?"

Within ten minutes Joe Pike was knocking on the door of Jacob Tilsley and, to his relief, it was quickly answered.

"Joe! Hello to you."

Joe Pike waited for more words but none came, so he spoke his own.

"Mr. Tilsley, we have to do an amputation on one of our followers, a woman. We have to take off a toe, and we need a chisel, a mallet and some pitch. Can you help?"

If Tilsley was shocked at so crude a request, for so evidently brutal an event, he did not show it.

"You know where the tools are, Joe, and I'll see to the pitch. They are in the same place, anyway, that you'd know from back along."

He reached behind the door for his coat and, whilst putting it on, closed his door and began to cross the yard.

"I assume you'll be getting her drunk?"

Joe had no idea, but it seemed likely and so he nodded, then Tilsley continued.

"Right. Whilst I'm boiling the pitch, you take the tools and, by the time she's rolling, the pitch'll be ready. You're where?"

"By the tent, by the great oak."

That was all Tilsley needed. They ascended the stairs into Joe's old workshop and all was as Joe remembered it, thus it was easy for him to find a small chisel and a heavy half sledge. As Tilsley stoked up the coals in the fireplace, Joe touched up the chisel on a whet stone, then he left. Back with Bridie, he was greeted by Deakin.

"Perfect, Joe. Just the job!"

He examined the narrow chisel, half an inch across.

"Right. Let's get this clean."

The chisel was given to Mary for cleaning, whilst the process of anesthetising Bridie had begun, Nelly practically force feeding her rum and brandy in equal measure. Bridie was sat in a chair, her foot on a piece of planed timber. Jed Deakin watched events, studying Bridie for all the signs of being "in another place." Tilsley arrived, having been waved over by Joe, and he and Deakin shook hands.

"My thanks to you, Sir."

"No thanks needed. Pleased to help."

Tilsley leaned forward to examine the foot and nodded. No

words were required. He placed the pot of pitch onto the fire to keep it hot and liquid.

When Bridie started singing in Gaelic, Deakin knew that this was the time. Mary brought over the chisel from the boiling water that it had been standing in. She knew not why she did that, but it seemed as good a way to clean the chisel as any. Deakin took the chisel in his left hand and hefted the mallet.

"John, Toby, hold her foot."

The two did as they were asked.

"Mr. Tilsley, Sir, if you could be ready with the pitch?"

As Tilsley fetched the pot, Deakin knelt and placed the chisel. He looked up at Bridie, who was wailing some lament about what, only God and herself knew. As Tilsley returned, Deakin struck. The chisel went cleanly through into the wood. Deakin first saw that it was bleeding clean with no pus, then Tilsley dapped on some hot pitch. As chance would have it, both Bridie and Deakin stood up together, Bridie howling from the hot pitch, but, whilst Jed concerned himself with being concerned for her pain, Bridie unleashed a ferocious right hook to his jaw that sent him spinning to the ground. Then she passed out. Deakin managed to bring himself to his knees whilst hoots of laughter came from all around. He straightened himself and checked that his jaw was not broken. It was not, and so he shook his head and stood, to then look tenderly at his "wife", as Nelly wound on the bandage. He picked up the toe and threw it on the fire.

That night, Joe Pike knelt in front of Mary and proposed. She uttered something between a gasp and a laugh almost hysterical. Then she said yes. Both hurried off to see Jed and Bridie. Bridie wept and Jed embraced both, before setting out a small dampener. Joe would need the Colonel's permission.

***

The following morning, the Lord of the Manor, the yet-to-be-ermined Mr. Coatsley, after a good breakfast, hurried to the ballroom, with Mrs. Gimlet's complaint buzzing in his head. She followed on, not so much to provide moral support, more to witness what would surely be a relished comeuppance. However, although ready for a ferocious tirade against the senior Officer therein, Coatsley's ire was blunted by his being prevented an entry. Out through the doors came a seemingly endless queue of wounded, both civilian and soldier and, towards the end, came Major O'Hare. Coatsley pounced on him by forcing himself into the stream of sick and wounded all making their exit as best they could.

"I want a word!"

O'Hare smiled.

"At your service, Sir."

"My agreement, the conditions on which I allowed you here, have been broken. You have brought wounded and others into my house! Under my roof, into my ballroom!"

O'Hare allowed his face to go blank.

"Sounds terrible! What would you have us now do?"

Coatsley's voice rose to a shout.

"Why, get them out! All of them!"

O'Hare did no more than smile ingratiatingly and walk on, to then ease a child back into a group of followers. Coatsley looked this way and that, into the ballroom, then out across the lawn. Finally, out came the last; two Orderlies and then the Surgeon, who paid his respects to the now apoplectic Coatsley.

"Obliged for your help and the use of your dance floor. I'm sure we're very grateful. Best treatment room I've ever used!"

Coatsley, although it was wholly subconscious, for he was so angry, beat his riding crop against the leather of his high boot, before stalking off, too fast for Mrs. Gimlet to follow, off to find some workers to shout at.

The wounded were on their way to wagons organised and sent by the Mayor of Crewkerne and, whilst they were being tenderly helped up onto the bare boards, but at least under a canopy. The rest of the battalion assembled and within minutes the whole cavalcade was on the road, leaving a heated Coatsley at the end of his drive, with no more outlet for his anger than to watch the closing of his gates upon the final wagon.

Once their men were on the road, Lacey and O'Hare hurried on ahead, mounted on hired horses. The summons had arrived the previous day and they had no choice but to hasten on and repond. As soon as they had achieved their own company, O'Hare asked the question the possible answer to which had long been preying on his mind.

"What do you think he will say?"

Lacey already knew the answer. General Perry's thoughts on the future and proper role of their men had been made abundantly clear since they first assembled as a battalion of detachments back in 1806 and it was this that formed Lacey's reply.

"The usual, that we're of no worthy count as a Regiment. That, as trained men and veterans, we be spilt up to bring other Regiments up to strength. I'd put my house and personal fortune on it, such as it is, but then I never bet for high stakes!"

O'Hare laughed. That was indeed the very predictable outcome of their forthcoming meeting, but he knew the implication.

"So, this could be the end?"

Lacey nodded.

"If Perry has his way, the answer's yes. Very much so, especially with our Roll now what it is."

O'Hare nodded and changed the subject and for the next two hours they talked of the campaign they had just returned from, a topic which lasted until their arrival in Ilchester, which was greeted by the Noon chimes of the clock in the tower of St. Mary Major. Lacey looked around, looking for what he assumed to be the premier inn of the village, this being the Denys Arms, to which place General Perry had ordered their attendance. A sign showing the three Danish battleaxes gave him all the lead he needed and soon the pair were inside, to look for the landlord. The said individual was the first they found of the inn's staff, him being most notable by a series of bulges at the front, of ascending size for any observer working his way down from his nose, to his chin and to his stomach. However, the former two were immediately separated by a wide, cheerful grin.

"Morning Sirs! How may I help?"

Both Officers grinned widely themselves at the avuncular response, and Lacey quickly replied.

"Good Morning. We understand that General Perry is waiting for us, somewhere here?"

The reply was an expansive gesture with his left hand towards the stairs, which hospitable hand remained prominent and inviting as he walked towards that very staircase.

"Indeed so, gentlemen, indeed so. The first room on the left. He has been here some few minutes now, and I have sent up coffee and rolls. Please go up."

Between rough plaster walls, both men climbed the stairs, which became narrower as they ascended and each step more of a challenge to the builder's true perception of horizontal. Eventually, a corridor stretched before them and they approached the stated room, for Lacey to then knock on the door. "Enter!" sounded through the thick panelling and Lacey twisted the handle for the door to open. It was plain that Perry had been there for some time, for he had arranged the room to his advantage. The largest table was in the corner, with him behind it, and two chairs placed in lonely isolation some yards away in the middle of the carpet. The said rolls and coffee were behind Perry, within only his reach, on a small chest of drawers. He looked his usual irritated and liverish self, but said nothing other than to indicate the two chairs, inviting them to sit. At this, Lacey picked up one of the chairs and moved it to a more social distance and within reach of the table, which brought a look of annoyance to Perry's face, at Lacey's effrontery to make changes to his arrangements already made. Having sat, with O'Hare beside him, Lacey opened the discussion; coldly, for there was no love lost between these two.

"Good afternoon Sir. You have some business with us?"

Perry nodded, ignoring O'Hare, and came brutally to the point.

"Moore's army is now disbanded, so, you are now under my command! Your muster is now even less than 570, I assume, now that your surgeon has been at work?"

Lacey's face remained deadpan.

"Yes Sir, some more wounded have now been discharged unfit. We are now on 553."

Perry nodded, evidently well pleased with this confirming information.

"Right! I'm classing you again as Detachments. I have two Regiments of the Line to bring up to strength, both the 14th First Somerset and the 40th Second. Two hundred and more men each will bring both to almost full battalion strength. You two can return to pasture."

He paused to allow the relished sentences to sink in. When he discerned no particular reaction he felt the need to add more, perhaps that would have the desired effect.

"Horse Guards are assembling another army, this time under Wellesley to return to the Peninsula. I expect one or both of the 14th and 40th to be called for."

Again no reaction; two blank faces remained, regarding him from the short distance across the table. He continued.

"I expect you both to divide your force into two equal parts and await orders at Taunton."

Lacey draped his right arm across his belt, sufficiently to support his left elbow as his left hand rubbed his chin. That done, he exhaled. This had been a wasted journey and he disliked Perry all the more for it. This could have been done via a letter, but Perry, for his own personal satisfaction, too much delighted in the idea of saying it face to face. Lacey reached back to the insolent tone he has used as an Ensign to superiors he had little liking for and even less respect.

"Will that be all, then, Sir?"

The sight of Lacey's deadpan, 'see-if-I-care' face, turned Perry's irritation to anger, but Lacey's acceding attitude gave him no excuse to make it manifest. However, it did help for him to repeat again, forcefully, the reason why he had summoned the two.

"In two wings, Lacey, then ready for further orders."

He fixed him with a malevolent stare.

"Report back here in four days time, same hour. Back here, Lacey!"

Lacey nodded and rose from his chair, as did O'Hare. As Perry's wishes would mean both leaving the army, neither bothered to salute, for Perry's evident satisfaction with the moment had not been lost on O'Hare. Then they turned and left the room, if this irritated Perry further, he did not say so. They would have ignored him anyway.

***

It was a hullabaloo that grew with every step, as the two parts of the whole came closer together, the 105th climbing the last hill, the families and the friends of the Regiment's men coming down it. They had marched out themselves with the dawn to hasten the meeting. When a collision seemed inevitable the wave of joyful and cheering humanity parted around the Colour Party to flow on, down either side. Sergeants on the flanks, out of the marching files, looked more than a little concerned at the threat to their precise parade, especially Gibney, but these were all families of the military and knew well enough the form. They stood to the side to wave and shout, allowing the march to continue and, thus, even Gibney mellowed into softer feelings as the women and children cheered their men home, forming a noisy escort on the flanks. However, there was the inevitable anxiety present. The people had not the smallest inkling who was alive or dead and two sentences were most often heard "There he is!", or "I can't see him, can you?"

Once halted inside their old barracks, at "fall out" the men flew into the arms of family, then friends, but amongst the cries of joy were heard cries of heartbreak as many learnt that their men remained in a faraway grave, often in a place too bleak and obscure to have ever been given a name. Soon, these could be seen sat or slumped against the parade ground wall, often with but a small bundle of possessions brought home by the comrades of their men now dead. It was small compensation, but of some measure, when Jed Deakin handed the bundles of Alfred Stiles and Sam Peters to their relations, for each contained their small hoard of the rescued sovereigns. They thanked him and shook his hand, before joining the mournful procession now exiting at the gate.

For some there was a genuinely sentimental and tearful parting. Nelly, Bridie and Beatrice Prudoe embraced amid heartfelt sobs and kisses, but Beatrice had remained steadfastly hopeful.

"My husband may yet arrive, who's to say, but whatever, I'm going to start a school. Nearby. So, whilst you are here, you must send your children. And you must come around yourselves. I insist! We must keep in touch."

At that Nelly and Bridie looked at each other, both immediately knowing the problem, but Bridie spoke it.

"Well now, if you're sure, Mrs. Prudoe. I mean, the likes of us, y'see. Well, we wouldn't want to be causin' you any embarrassment from the likes of us turnin' up, sort of thing, amongst your own fine acquaintances."

Beatrice Prudoe put her arms around Bridie.

"It won't be in that part of town. It will be where the children are, and you will be very welcome. You will! So, do come. I will let you know where."

Nelly felt the need to counter such maudlin thoughts at their time of parting and flicked Bridie's arm with her fingers.

"Now sure, it'll be alright, will it not, if you puts on your best frock?"

All three laughed and Beatrice Prudoe left, giving a last wave at the gate for Nelly and Bridie to enter the barracks proper and re-acquaint themselves with their old barrack room. As they entered the outside door, Nelly stood back to watch Bridie enter and she observed a walk with barely a limp.

"So, how's your foot?"

Bridie turned and then spoke softly.

"Fine. 'Tis healing well. Very little pain and the pitch is about to fall off, but don't tell Jed. I fancy the excuse to hit him once again!"

To gales of laughter the two women pushed open the once familiar door of the dark, dingy and stable like room that was to be their latest home, but also a repeat from the past. It looked as it always had, the small "crib' cubicles around the sides, the tables and chairs in the centre, the cooking fires and the high windows, fixed closed. Thus, it smelt as it always had, stale from much used air and the left over smells of dust, damp, and too overcrowded humanity.

Once settled in, John Davey took himself over to the crib occupied by Joe Pike and Mary.

"Joe, you can write a bit, can't you?"

Joe Pike's face screwed up.

"What for?"

"A letter."

Pike shook his head.

"Not well enough for it to get there."

He paused.

"Parson's the one."

Davey nodded.

"Parson! Who else!"

A run and trot took Davey to the quarters of the Regimental Chaplain, to find Sedgwicke carrying vestments and the few other trappings of office that he did have, into the spacious rooms reserved for the Chaplain.

"Parson! I needs you to help me write a letter. To Molly and the family."

Sedgwicke's face split into a wide grin at the thought of Molly, the beauty who had been so kind to him, when others would not, back in the days when this barrack had been their long term home. Sedgwicke and Davey had shared a crib together, before Molly moved in, but even then she had treated him kindly despite the overcrowding.

"Of course! Do you have their address?"

Davey pulled out a much worn envelope containing a letter. Whilst he could not read each word in its place, he knew them all, in the correct order, off by heart. This had also been written by a Cleric, the incumbent of the church of Far Devening where Molly had moved to join Davey's Mother whilst he was away. She had taken her own daughter, 'Tilly, with her and had there given birth to John, Davey's own child. Sedgwicke fetched paper, quill and ink and led Davey into his own cubbyhole of a room.

"Now, what do want to say?"

"Dear Molly, I am writing to say that I am now home. Safe and well."

"No! It should begin: Dearest Molly, I have returned and cannot wait for us to be back together again, now that I am home, safe and well."

Davey nodded, grinning broadly.

"Yes! Yes, that's it, Parson. That's the start."

Sedgwicke began writing.

Elsewhere, another leftover piece of business was being tidied up. Sergeant Major Gibney was at Lacey's office door, announcing his presence with a rapid knock of three.

"Sir. Private Joseph Pike's here, Sir, waitin' outside, and would like to see you. Sir."

Lacey did not look up, although he was stood up, examining the appearance of two imposing letters, one of several pages with a Horse Guards cover, the other with the crest of the Prince of Wales, no less. He felt certain that both would reveal good news and was immediately ebullient.

"What's he want?"

"Permission to get married, Sir."

Now Lacey did look up.

"Married?"

He threw back his head, his brows coming together, as if to help his memory.

"He's took up with Mary Mulcahey, has he not. Pat Mulcahey's daughter."

"Sister in law, Sir, and she's an O'Keefe. Bridie Mulcahey's youngest sister."

Lacey's face showed his puzzlement at the mistake, but nodded anyway.

"Tell him yes."

He paused but not for so long that Gibney had the time to leave.

"She lost her baby."

It was not a question, more a statement of fact.

"Yes Sir. On the retreat."

"We make a show. And help out, too. The Regiment! Yes?"

Gibney took that as an order, which required a salute, which was duly delivered. Within a minute Joe Pike was running across the parade ground, but his mind registered not one step. Within one minute more, all on the parade ground could hear a loud cheer emerging down the corridor from one particular barrack room. None there in the square knew the reason, save RSM Gibney, who permitted himself a rare smile, but that only to himself.

Meanwhile, Lacey had opened both letters. He read both and his mood improved immeasurably still further. Almost laughing, with one in each hand, he turned to his office door.

"Bryce!"

Herbert Bryce appeared in seconds.

"Fetch Major O'Hare. Tell him immediate, that nothing could be more important!"

Bryce saluted and left. He found O'Hare in his own office and so it was but two minutes lapsing before O'Hare was entering the door of Lacey's office. Lacey thrust both letters at O'Hare simultaneously, so he took both and read the one in his right hand first, then the left. As he did so a huge grin spread across his face.

"What are you going to tell Perry?"

Lacey's eyebrows reached his hairline.

"Nothing, as yet. Damnable man! He can find out when we go back to report. 'On progress', as he put it."

O'Hare grinned a reply, then both settled down to re-read the letters several times more.

***

The forthcoming marriage may have been the cause of the rare smile from RSM Cyrus Gibney, or it may not, but whichever, come the next day several yards of white muslin arrived at one particular barrack room. It was immediately pounced on by every women there who had any level of seamstress skills and, for the next days, and for lengthy and unscheduled periods, men were banned from the room during daylight hours, Nelly Nicholls being the first to intercept any interlopers, by working close to the door.

The due day arrived. The ceremony was planned for the afternoon, which gave time for attention to the last details of dress and uniform and all were determined on the full observation of every propriety and tradition. Mary put a blue trimming onto the hem of her dress, but not quite completed, giving her something to sew up immediately prior to leaving the room. There were no mirrors anywhere to worry about, so that concern did not arise. She left the barrack room on the arm of Major O'Hare to leave the building itself over a swathe of smashed crockery on the doorstep to the parade ground. A fiddler bandsman led her to the point where stood a local chimney-sweep, who upon kissing the Bride released a black cat from his sack, who immediately hared in through an open window. Leaving behind both the fiddler and the sweep, both Bride and her Escort proceeded to walk across the front of the Light Company. These were all drawn up in full dress uniform, providing a guide into the gap of an open square, these Lights being drawn up at right angles to this opening. Carr was at the front with Drake and Shakeshaft, both stood before their respective Sections, all now immaculate in their new uniforms. As the pair came level with the first files, Carr bellowed "Present Arms!" The clash of hands on muskets, accompanied with the flash of polished swords and bayonets had quite an effect on Mary, something between shock and embarrassment, that such was being done for 'Little me', but everyone knew of the tragedy that had befallen her and Joe. Nothing was too much trouble, which sentiment was shared by the rest of the Battalion waiting drawn up in a hollow square.

Within, stood the local Vicar, beaming, chubby and avuncular, "apple cheeks" thoroughly port stained, waiting joyfully, in conjunction with all others present, and unarmed; the Vicar would not countenance any weapons at any religious occasion officiated over by him. The followers formed a small block off to the left, whilst Lacey and the Colour Party were off to the right, lined up in front of that side of the square. Behind the Vicar, was a stack of drums, but these were not to be used, instead before him he had a small lectern, portable, it being constructed for just such an occasion as this. Stood facing this was Joe Pike, quaking but not from the cold. Beneath his tunic that had been carefully prepared by Tom Miles, he was wearing a thick shirt made by Mary over the days before. Three paces from him, O'Hare halted and Mary walked on to take her place at Joe's side. The Light Company, having piled arms, marched smartly in to form Joe's part of the

486

congregation, which, until then had merely been the lonely figures of his Parents, brothers and sisters, summoned by horsed messenger and brought by coach, paid for from battalion funds. Mary's section was from amongst the followers who were all close friends or relations, with their husbands, which included Jed Deakin and Henry Nicholls; 'excused parade" for this occasion.

The Vicar opened his Book of Common Services and placed his hands on the leaves to prevent them blowing shut. Thus came the first words.

"Dearly Beloved. We are gathered here together, today, to witness this day the joining of ...."

He got no further before his words were almost drowned by the wailing of Bridie and Nelly, both wholly overcome, both clinging to each other for some form of support, even the simultaneous shout of "Shut up, woman," from Jed and Henry had no effect. Eventually their suffering subsided enough to hear the Vicar ask the first vital question.

"Who brings this woman here to give her away to this man?"

The reply from O'Hare contained more than a small hint of challenge.

"I do!"

The exchanging of vows that followed set off more wailing but not as disruptive as before and the Service continued until the triumphant pronouncement from the Vicar, spoken as though he was being witnessed from above by the full complement of the Heavenly Host.

"I now pronounce you man and wife!"

They kissed as Gibney organised the rousing three cheers, then Lacey and O'Hare both shook Joe's hand and kissed Mary, she still in a state of shock at all that was being done in her honour. The pair, pelted with rice, then made their way back to their barrack room, where a fine meal was being prepared of beef, pork, potatoes, peas, and, a thorough exception, greens and parsnips, all followed by a fruit suet pudding, but first came the breaking of the bride's pie, and all broke off a small portion, as it was passed around. Although not enjoying the same meal, the rest of the Regiment gave thanks to the pair as a double ration of rum came around. With the Wedding Breakfast dispensed with, so began the celebrations. Mary danced the wreath dance, surrounded by some 'married' women, which was sedate enough, but finished somewhat before due time when the men grew impatient at the prolongued delay of their drinking time and so Zeke Saunders reached into the ring and seized the wreath. Tradition then returned when he broke it up and scattered the remains. Finally, Mary put on a matron's cap to mark the end of her 'spinster days'. As the day ended and the singing and dancing spilled out onto the Square, to everyone's surprise, bar Lacey, O'Hare and Gibney, a fine coach and pair came in through the gates to whisk the pair off to a local Hotel. Even in their absence the celebrations continued until the twelfth stroke of midnight. It had been, altogether, a most splendid wedding!

***

The following day, with his command unsurprisingly quiet, Lacey and O'Hare once more set out for Ilchester. Both were in the best of moods; perhaps it was the signs of an early Spring, perhaps it was this moment of freedom from command, or perhaps it was the two letters secure inside Lacey's tunic. He regularly felt inside to touch the smooth vellum, just to make sure that they were still there, and perhaps this added extra relish to what was about to transpire. The journey seemed to pass quickly and, once there, they stabled their mounts and entered the inn, to be confronted by a young potboy.

"Morning Sirs. How can I help?"

Lacey was in the lead.

"Good morning to you, young fellow, but no help needed, thank you. We know where we need to go. Been here before, you see."

O'Hare, following up, gave the lad a coin anyway. They came to the door and, before knocking, shared a wicked grim. Lacey knocked in a rhythm of three.

"Enter!"

They did so and found the room arranged as before and again Lacey re-arranged it again, as he had done before. Perry's face darkened and he spoke, before they had sat down.

"My order to you. To divide into two wings."

Lacey looked at him calmly, but Perry was impatient.

"Well?"

A pause.

"You understood my orders?"

488

Lacey noted the challenging tone, then reached into his tunic to extract the two letters, which he spread with his right hand, as though displaying two playing cards.

"Yes Sir, but I rather think that events have moved on and that you need to take into account the content of these two letters, which I have here."

He carefully selected one.

"This one came direct from Horse Guards. It says that we are to ready ourselves to join Wellesley's army to sail sometime in April back to Portugal. General Wellesley has specifically asked for us. It says that our District General, that being you Sir, has been informed regarding bringing us up to strength from the Somerset Militia."

Lacey watched the change spread over Perry's face, from secure impatience to insecure doubt.

"Have you not received anything on that, Sir? You should know something about it. By now, that is."

The response was the doubt turning to embarassment. Perry acknowledged within himself that he had lately been neglecting his duties, but Lacey continued, having placed the first letter, still folded, on the table.

"This other letter is, well, from our Colonel in Chief, Sir. That being the Prince of Wales."

This letter Lacey then carefully opened and turned so that it lay on the table the right way up for Perry, revealing the gorgeous letterhead; the lion, the unicorn and the shield, all occupying a quarter of the page. The look of annoyance returned, but somewhat concerned, worried, even, for an affront to the Prince of Wales was career suicide. Seeing that Perry was not going to read it, Lacey took his hand away, allowing the heavy vellum to re-curl itself.

"It says that I am to send a Colour Party to Bath, Sir. For our old Colours to be laid up and new ones presented. Somehow, I mean perhaps, his Highness has used his influence, because we are to be given three new battle honours."

Lacey paused and re-opened the letter to point half way down.

"There Sir. Rolica, Vimeiro and Corunna."

He took his finger away, because Perry had made no move towards it.

"Did you not know about that, either, Sir?"

489

This time he saw anger, but Perry's jaw was merely working, not speaking; however his eyes matched his jaw, darting feverishly hither and thither. Lacey continued, looking puzzled and sounding it.

"I'm not sure how that sits with your orders from our previous visit? Sir."

Perry seized the Horse Guards' letter and began to read it, his hand shaking. He then looked directly at Lacey, his eyes fierce and angry.

"This reinforcements thing. We'll have to see. The 14th and the 40th have a prior claim, especially if they are also to join Wellesley."

He folded both letters and carefully placed one upon the other, as though being folded and piled would reduce their significance. However, the letter from the Prince of Wales, being of the thickest vellum, disobediently unfolded itself to tip the Horse Guards letter back onto the table. Both were now visible, joint participants to the defeat of his cherished ambition. Lacey nodded.

"As you say, Sir."

Lacey then placed both hands on the edge of the table, as though about to rise.

"Well, Sir. If there is nothing further then, please, I have things to attend to, as listed there."

He pointed to the Horse Guards' letter and paused to look fully at Perry, who nodded curtly, but Lacey leaned forward to pick up the letters but not remove them, simply to look directly at Perry in order to fully make the point.

"I would like to take these with me, Sir. For the Battalion records, you understand. I mean they could add to our Regimental chronicle. I'm sure your own copies must be somewhere."

Perry's jaws came heavily together. The remark, supposedly innocent enough, was not lost on Perry, the implication being that the 105th would be a Regiment in existence long enough to build up their own history and traditions. Such was anathema in the eyes of the General, but Lacey was twisting the knife. There was no question of Perry taking the letters, they were not addressed to him, but he was not allowed to dwell on it, for the two now stood and Lacey brought the letters back to his side of the desk. He then stood and saluted, very formally. Perry returned with an even briefer nod than previously and the merest irritated twitch of his left hand. Formalities now observed, more in their absence than in their observance, Lacey and O'Hare turned and left.

***

Now lodged in the comfort of an Inn in town, Carravoy and D'Villiers were catching up with the daily newspapers, not so much reading, as rapidly turning the pages to find the article, or perhaps articles, that most involved themselves, these being any that covered Moore's campaign. Several, over the days since they had returned, had carried articles on the army's record in Spain and their state of the army when they landed. It was Carravoy who had The Times, amongst others, and was plainly the most eager to find criticism of their now dead Commander. His remaining as a Captain still rankled and added to the bile within him. He shuffled through each newssheet where he had found an article, and picked up that which most chimed with his own opinions.

"At last it seems some sense and judgment is emerging as a lesson from that shambles we've just escaped from. Here, listen. "Moore must be criticised for placing such over reliance on Spanish allies whose record against the French was utterly abysmal. Why did he think that they would give priority to his needs, when they were repeatedly failing so badly to manage any level of strategical support for their own forces?"

He crushed the paper together and slammed it down on the table.

"There! Finally! An informed and proper judgment. Justly condemned!"

D'Villiers did not respond for some moments, he was reading an article of his own.

"Well, that may be so, but it does seem wrong to speak so ill of the dead, who is not here to defend himself and I've got one here that speaks of the failures of the Commissariat and I think they have a point. Had we been fed sufficiently from, let's say Astorga to Lugo, on the retreat, then we'd have come through well enough."

He put down the paper.

"I think that's true."

Carravoy's eyebrows came together darkening his brow and his mouth became a thin line, but D'Villiers continued.

491

"You cannot argue against Moore having no choice but to retreat. He'd have lost the whole army otherwise, but had we been better supplied we'd have come out of it looking a whole lot better than the collection of scarecrows they are describing us as."

He waved another article he had read at Carravoy.

"This is a case for my point. In one of the first papers issued after we came back, they talk of the returning army being nothing more than "lice ridden scarecrows". Nothing about our fighting record. That's a bit thick! The Reserve Division beat the French back at every turn and then we threw back their whole army, come the end. What do they think an army looks like, after a battle and two weeks of retreat with nothing to eat?"

He was building up a level of heat himself.

"They should look to who was running the Commissariat before they start with their criticism of Moore and his men. Them being us!"

Carravoy ground his teeth before speaking.

"Moore got it wrong. That's the up and down of it. And what about the loss of the Dispatch and the Smallbridge, both wrecked with a loss of nearly 300 men!"

D'Villiers had raised his paper to read on further, but then lowered it. That also was 'a bit thick'.

"Moore can't be held responsible for shipwreck! After he was dead! No-one's saying he did not get stuff wrong. He got a lot wrong, but so did many others, some back here at Horse Guards. That's the full up and down of it!"

D'Villiers' opinions would not have been out of place in a carriage which at that moment passing the ruins at Glastonbury, containing The Colour Party, plus the three Senior Officers. O'Hare was doing his best to appease Barnaby Rushby who would have strongly agreed with D'Villiers.

"Now don't take it all so personal! Much as I dislike it myself, the army is never far from politics and one General or another is either a favourite of the Whigs or a favourite of the Tories. Moore was a 'child of the Whigs', as some put it. What I don't like is him being kicked about. His friends, if that's the correct word, using it all to hit our present Tory Government for incompetence, whilst the Tories blame Moore for the same."

He paused and was pleased that Rushby's face had become more thoughtful.

"But that's the game we're in, and we have to face up to it or give it all up."

He paused again, to assemble his final thoughts. No-one spoke; they knew that there was more to come.

"We got kicked out! That's a defeat and that sets off the politics. One thing I know. Moore drew off a whole French army that could have gone marching on into Portugal and taken over the whole shebang! Because of that, we're still there, holding Lisbon and that gives us a fighting chance. He should be given credit for that and, although no-one else seems to be doing that, then I certainly will."

The coach lurched to a stop outside the coaching inn, The George and Pilgrim, but before they got out, O'Hare had one more thing to say.

"So! This night there's a drink on me, and the toast is to John Moore and the other good lads we left behind in Spain."

The other four Officers laughed, there could be no more argument against a sentiment such as that, bar one comment from Lacey.

"I'll agree to anything you say, if it puts a drink in my hand!"

More laughter, then O'Hare alighted first, followed by Ensigns Rushby and Neape, then Brevet Major Carr and finally Lacey himself. Colour Sergeants Bennet and Deakin climbed down from the roof, carrying their own baggage, which included The Colours. They then caught the baggage of their Officers, thrown down to them.

That evening, whilst Deakin and Bennet kept their own company in a back room, the five sat to enjoy a good meal in a private room on the first floor. Initially, the good food cheered them all and the wine also sparked up all spirits, but it was not long before the subconscious depression that lay upon them all, manifested itself again. It began with Neape, the youngest and the least experienced. This was his first taste of 'post campaign' and, despite O'Hare's wise words earlier, resentment had built within him, to be finally released as the wine loosened his tongue.

"Are all homecomings like this? I mean, so, so, controversial!"

The other four looked at him quizzically, saying nothing, which prompted him to say more.

"Well, I mean, it's all been about what went wrong! We bested the French three times in set piece. Has all that been forgotten?"

This time it was Lacey who sat back and formed an answer.

"We hold the line!"

He paused whilst all turned to him, more than slightly puzzled.

"And that's what we did. Anyone who thinks Napoleon will be beat in one campaign is a fool. My money's on eight years, perhaps ten. He's beating all that Europe can send against him, bar us, and what were we, but a small event at the far end of the continent? Wars are attrition, slowly he will be worn down. The whole of Europe is against him, so six years if he makes a mistake, eight if he doesn't. For me, we did the best in Spain that could be expected, and the French now know that they have a foe at least their equal, probably better. We held the line, by which I mean that we're still in the fight. We weren't all captured, as we could have been. Back here, home, people like big, decisive victories, the sort that ends wars. Well, this is just the beginning; the grind. People are disappointed if that is not what is served up. It confirms the coming of years more of war. We know what we've done. The rest of the country may not acknowledge it as any form of progress, but we know that it is, and, as O'Hare here says, as soon as the politicians get involved, then it all goes to Hell in a handcart! It was their work that left us kicking our heels in Lisbon after Vimeiro."

He swallowed from hisdrink, whilst the others remained silent, but he had almost finished.

"So! You're still here, we're still here, and we're going back, this time under Wellesley. So there, we've held the line, we're still in it! And we're getting new Colours to prove it!"

All around, his companions nodded sagely, but the mood was not lifted and the meal ended quietly. However, this was in stark contrast to that pertaining in the back room, where Deakin lifted another quart of beer to Bennet.

"Here's to your legs, here's to your arms, here's to your head. They're all still where we'd wish for 'em to be!"

The two tankards clashed together and both drank deeply.

The following day, modestly clear for early February, saw the much ladened coach climbing the Mendips, affording a good view back of the Somerset Levels and so all Officers took the opportunity to look back at the curious feature of The Tor. Deakin and Bennet, although it was early morning, dozed on the back seat of the roof, somewhat worse for wear from the night before. Both were also influenced by the swaying coach and the sound of the wheels on the good turnpike road, lulling them back to slumber, but mostly they slept because there was nothing else to do. Thus, for them, the journey was broken into periods asleep and periods awake and soon enough, at least as they experienced it, the spires and spectacle of Bath came into view at the top of the final slope off the Mendips. The coach rolled into the courtyard of a very substantial coaching inn and all alighted, but Lacey went immediately to Deakin and Bennet.

"Get settled in. We could be here for some days."

The reason he did not give, this being that The Prince of Wales was notorious for failing to appear at appointed occasions, often being days late, being far more inclined to indulge his latest "fad", whatever that was, at that moment in time, but Lacey had no way of knowing the extent of any likely delay. So settle in they did, although the ceremony was scheduled for the very next day. O'Hare gave the odds as 'five to four against" and he was right. The next day at 11.00 am, all were about to leave the inn by the front door, when a messenger arrived to say, 'postponed for the following day, the Prince is indisposed'. As they climbed the stairs back to their room, Lacey gave his verdict.

"Mrs Fitzherbert has a headache, we presume!"

O'Hare, perhaps more worldly than even Lacey, volunteered another.

"Or perhaps not!"

Lacey looked at O'Hare, both amused and shocked at the inference, but the day was now their own. O'Hare took Deakin and Bennet off to the nearby racecourse where there was a meeting, Lacey wandered relaxed, taking in the sights, whilst Carr, Rushby and Neape decided on the Pumproom. Having decided on the premier social spot in the city, each ensured that the other was thoroughly spruced before leaving the inn and they wandered down the main street to turn into the square with the Abbey set before them and the Pump Rooms on their right. Once inside the interior with its pastel pink and cream décor, colours influenced by the Prince himself, so most believed, Carr took the pair to the fountain for the "waters" and prepared a cup for each of the Ensigns and one for himself.

"Now you've got to drink this, although it tastes like nothing on earth. The received wisdom is, that it's very good for you, so you must drink it all. Every drop!"

He handed out the cups and both drank. The expression on their faces matched his own feelings, but they had only sipped, which Carr judged to be wholly insufficient.

"All of it! There are many who drink at least one cup a day and swear that it lengthens their lives! Or improves it, in some fashion".

He drank from his own, the water tepid and earthy, a taste straight from the soil, much less than pleasant.

"Mind you, there's plenty more that think the sheer misery of drinking this stuff knocks off years!"

The two Ensigns laughed, then drank to the bottom and Rushby looked into his cup.

"We drank worse on the retreat, Sir, and would have welcomed anything this warm! Especially at Corunna, when everything we drank came from that vile stream."

On hearing the word 'Corunna', several well-dressed gentlemen turned to look at them, but the look of condemnation that soured their faces told all. Carr returned the look with a weary expression of his own, then gathered the cups.

"Come on! There's a good coffee room upstairs nextdoor."

However, they did not get very far than to exit one door, then enter another. At the bottom of the stairs mentioned they saw three young ladies coming down and it was plain to Carr that they recognised one of his Ensigns. It was soon evident that this was not Rushby, as two spoke in unison, the other merely beamed.

"Trenton!"

As they were evidently about to be joined by young ladies they retreated back down the stairs and there Neape made the introductions.

"Major Carr and Ensign Barnaby Rushby, may I introduce my cousins Natalie and Abigail and their friend Josephine Ambleside-Smith, Josie to us all?"

Carr came to the attention and saluted, which Rushby, very unsure of what to do next, gratefully copied, whilst Carr spoke his greetings.

"A pleasure to make your acquaintance, ladies. I trust I find you well?"

All three girls curtsied and smiled, their eyes soon regarding the floor, but one, the not-a-cousin, did manage the correct reply.

"We are quite well, Major. Thank you for your concern."

Carr grinned, suddenly in quite buoyant mood, significantly cheered by the happy scene of the two Ensigns, especially Rushby, and the three girls, chatting happily, so he stood politely by as they questioned Neape and Rushby. However, it soon became plain that both these were wholly at the centre of their attentions and that he was very much at the periphery. Rushby they plainly found very interesting, so Carr smiled to himself and then made his excuses to ascend the stairs once more. At the top he looked for a table that was free, but none were. However, a cup of hot coffee, in fact more than one, thoroughly appealed and so he took himself to the serving bar, where he could stand and obtain his desires, even if he could not sit. He made his way off to the side, not wishing to attempt a passage through the crowded tables, with seats and tables crowded with gentlewomen, gentlemen, top hats perched on canes, bonnets and trailing dresses. He found a space at the damp and stained mahogany of the bar and placed his order, which quickly came. He was soon halfway through the large cup's dark contents and feeling very much at peace, so he turned from gazing out of the window to look across the crowd of patrons. It took a while for the thought to hit him that this was the first time that he had been alone since he could not remember, with no-one in need of either his help or decisions. He smiled at the luxurious notion, that he was a free agent; at least for a few minutes, or perhaps even hours, Neape and Rushby were plainly in safe hands. He ordered another cup and drank it indulgently, perhaps with too much sugar, but he also indulged equally in the sound of the happy chatter that came from all around him. With the second cup finished he placed two sixpences on the bar and levered himself away from it, to follow his path of earlier, back to the head of the stairs.

He was absentmindedly descending, still in faraway mode and not particularly looking where he was going, when he heard his name.

"Carr!"

He looked to his right to see Lucius Tavender now ascending, but he had stopped just below Carr's own level. Carr looked over but did not speak, for alongside Tavender was a face he'd hoped never to see again, but Tavender was making the introductions.

"Captain Carr, have you met Lord Frederick Templemere?"

Carr made no reply. He did know Templemere, he was a dark figure from Carr's past, and relations were anything but good. They had fought a duel which Carr had finished with an illegal blow and the last words that Carr had spoken to Templemere were a threat to kill him at the end of a lone fight in a wood! Carr held his peace and allowed Templemere to speak first, which he did.

"Yes. The Captain is known to me."

He looked Carr full in the face.

"Carr!"

Carr nodded a reply.

"My Lord."

Tavender may have detected the chill between the two, but he did not show it.

"How are you, Carr? Why are you here?"

"I am well, thank you."

Carr thought it best to use no names, but continued as pleasantly as he felt able. It was always best to be circumspect with the likes of Lucius Tavender, so he replied neutrally.

"Summons from the Prince! Him whose name we carry, but how are you? Did you get back well enough? You weren't involved in either of those two sinkings, were you?"

Tavender's face remained cold. Was he resentful of the help he received from Carr back in the Peninsula, or was this the normal him? Or was it perhaps that Carr had practically ignored the second question. Carr smiled slightly, awaiting a response, which came from Tavender.

"No, I was not. In fact we got out early. We saw the smoke as we sailed; of your battle."

Carr waited for an enquiry on that subject, but none came, so he replied, he hoped civilly.

"Well, I'm pleased you got away, even though we could have done with your help. Your presence would have been welcome, but, having said that, it was a rearguard action and we both had our orders. It fell to us to hold back the French. Someone had to. The fortunes of war!"

Again no reaction in their faces that could be called warmth, but at least Tavender was speaking, repeating his earlier question.

"So, why are you here?"

"I'm here for a ceremony. We are here to receive new Colours, hopefully soon, hopefully tomorrow, from the Prince of Wales."

Tavender at last showed some genuine reaction.

"We? New Colours?"

Carr nodded.

"Yes. Three new battle honours for us, the 105th."

Carr kept it brief, he wanted this to end, but now Templemere spoke up, his voice still with the bullying, sarcastic edge that Carr remembered.

"Somewhat of a high level job, for such as a Captain? I'd say."

Carr smiled indulgently.

"I'm now a Major. Brevet after Corunna; only Brevet, but needed here to make the numbers up to the proper form."

Tavender's eyes just flickered, that was a surprise, but plainly he wanted nothing more of that topic, because no 'congratulations' were forthcoming. He changed the subject.

"Are you going back?"

Carr nodded. He knew it meant the Peninsula.

"So rumour has it."

At last Tavender smiled.

"So are we."

Carr looked puzzled. The word "we" could mean anything, but he was soon enlightened.

"We! Myself and Lord Fred here. He's purchased a Commission in my Regiment. 20th Light Dragoons. A Captaincy. Like myself."

Carr nodded but did not smile. The thought of Templemere leading men on a battlefield brought only horror, but politeness prevailed.

"The Regiment you charged with at Vimeiro! Well, good, I wish you both well."

He managed another smile, but Tavender retained his cold look, it bordering on dislike.

"But we won't be doing much charging. We're to join General Bentinck's Staff. He's preparing a Brigade to join Wellesley."

Despite the look, Carr felt the words to be engaging enough and replied in similar vein.

"Bentinck! Well good, you've landed on your feet there! We were part of his Brigade for some of the retreat and for Corunna itself."

The question immediately entered Carr's head regarding how they had obtained so soft a posting as that; influence somewhere, no doubt, but Templemere now joined in, his voice full of sarcastic disdain.

"We?"

Carr looked full at him, annoyed at the possibility of the slightest slur. His head lifted and his shoulders drew back.

"Yes we! The 105th Foot. The Prince of Wales Own Wessex Regiment!"

He took a pause to draw breath.

"Maida, Rolica, Vimeiro and Corunna."

Templemere voice remained unchanged.

"Corunna! You'll get few plaudits for that!"

That said Templemere's eyes and mouth conveyed pure malevolence; accompanied by a half smile that carried pure derision. Carr had had enough.

"That's as may be, but I can tell you that Johnny looks upon it somewhat differently to what's arrived in the newsheets."

He drew breath, making a slight pause.

"However, that's as may be. Right now, I must be away. I have two Ensigns to find."

He nodded in their direction and continued his descent.

"My best to you, gentlemen."

He did not look back, even when he heard Templemere's voice for the final time.

"I hear you have designs on a local girl, Carr. Well, good luck with that!"

Carr merely lifted his right hand into the air and continued on, failing to look back. The heavy sarcasm in Templemere's voice meant that he was wishing him anything other than good luck, but the likes of him would say anything to create strife and anxiety.

The Ensigns were not found, in fact Carr did not look for them, but returned straight to the inn and there he indulged in a glass of spirits to perhaps raise his own. The meeting with the two had left him in low mood. In Tavender he was disappointed, he felt that he was owed a debt of honour from him, enough for good regard at least, for twice he had helped him out of a situation almost irretrievable. On top was the re-acquaintance with Templemere, which could be nothing other than distasteful; stomach churning, even. However, he comforted himself with the logic that meeting Templemere could always happen, he was not dead and was part of the local social round. He then further lifted his mood at the thought of Templemere being faced with some crack French cavalry, Polish lancers or Cuirassiers, or somesuch. That made him smile, which made O'Hare think that Carr was in a good mood.

"Henry! Enjoying your freedom, I see. Have another, on me."

He waved at the barman, who brought them a bottle.

"We've another appointment with the Prince, tomorrow. And this time, we are assured, the Prince will keep it. It's in the morning and he wants back to London in the afternoon. Perhaps the waters don't agree with him, or he's got an appointment with his latest interest, the Beefsteak Club!"

Carr picked up the bottle and poured two more.

"On that subject, the waters I mean, I can only say that he has a very good point!"

***

One would have to go a long way to better the scene of Bath Abbey. There are superior religious buildings; higher, longer, more ornate, perhaps better designed, but for the intimacy of contact, of hand-in-hand sharing in the day to day business of the good inhabitants, Bath Abbey could not be bettered. For here the solid citizens of the noble city could walk unhindered within feet of its walls and its small, but personal, portico. Thus, their Abbey was part of the daily routine of many, it being right in the city centre and surrounded, almost to its very walls, with the buildings that gave character and shelter to the inhabitants of the premier city, excepting London, of Georgian England.

Thus it stood waiting and at 09.59 am, in thick drizzle, five Colour Parties formed up at the high end of Union Street, with a lone drummer before. The five were from the main British combatants of the Moore's army at Corunna, these being those who defended the ridge; the 4th, the 42nd, the 59th, the 81st and, finally, as the least senior, 105th. The Guards would receive theirs in London. As the town clocks all around struck 10.00, the drummer beat a roll, then, with single taps, he set the pace for the march down to the Abbey. The beating drum cleared the way as much as the four Parish Constables in the lead, but the rain depressed any uplift of the spirits that could have been aroused at the sight of five sets of Regimental Colours parading through the streets, these adding the only cheer to the drab scene. Most men removed their hats and almost all stood still in respect, but there was no cheering, no huzzas, the whole country remained in sombre mood. There was enough wind and also enough in the motion of the march to draw out The Colours to show the battle damage in each, but the impact of its significance on the watchers was impossible to gauge. The rainwater dripped down the facades of the yellow stone buildings on either side, over the pavement and into the gutters. The sound added a sad backdrop to the marching feet and the lone beating drum. Behind the 105th, the daily business of Bath closed behind them, as it would for the passing of a plain brewers dray.

Lacey was at the front of their party, sword drawn, followed on either side, but slightly behind, came O'Hare and Carr, followed by The Colour Party itself; Deakin, Rushby, Neape and Bennet. They traversed the ancient crossroads of Westgate and Cheap Street, to soon turn left into the Abbey Square, and march on towards the imposing, if not towering, West Front. At the entrance, small in proportion to the gracious stonework around, the drummer stood to one side and the five parties broke up, to enter as they may. Then to reform in the aisle, in a queue, for this was to be done quickly and efficiently, more as though they were new recruits being issued with new muskets at a quartermaster's store, instead of being issued with the very symbols of their existence.

The Prince was late, but not too late, so they were not stood waiting for more than ten minutes. During that time, Carr, stood closest to the pews on his right hand side, was soon aware of the three young ladies from the previous day, all stood simpering and giggling some way up from them. He turned to look at the Colour Party, stood just behind him. Whilst the faces of Deakin and Bennet would not have been out of place as gargoyles on the outside walls, those of Rushby and Neape, if only in their eyes and barely with their mouths, returned the amusement. Sparse and brief the occasion may be, but its significance was not lost on Carr, especially now that they were stood waiting in the nave of so significant and emotional a building. He glowered at both.

"Watch your front!"

The levity transformed into stern solemnity and Carr noticed the girl's expression also become subdued, then he saw that the 4th were now marching back out, along the North side aisle. The 105th marched up one place and waited some more. Carr, without moving his head studied the superb stained glass above the altar that seemed to create a sunshine of its own, but then they were moving forward again. This was to be quickly done. The kilted 42nd were receiving their Colour and then, after what seemed but minutes they were stood before the Prince. At this point Carr noticed three women who would not have been out of place alongside Neape's tittering relations, all three viewing the performance of their Prince with great amusement, but then Carr noticed something that did much to change his view of the whole affair. The Prince was wearing the uniform of the Colonel-in-Chief of the 105th, the lurid green facings stark and clear and Carr felt his emotion rising as Lacey, unlike the previous four Colonels, was motioned forward. Carr was close enough to hear.

"Morning Colonel."

"Morning Sir."

"Lacey, isn't it?"

"Yes Sir."

The Prince leaned forward conspiratorially.

"Sorry about this, Lacey. Would have much preferred to do the thing with full show back on Horse Guards, full fig, as it were, but politics! Politics, you see. Love to have a chat, as we did last time, I hear you did well, but, well, new circumstances, political ones. Hope you understand?"

"I do Sir. We are but plain soldiers."

The Prince's face lit up, almost matching the window behind him.

"That's it! Plain soldiers! We are! We do our duty, what! Nothing more."

"Yes sir. There is nothing more!"

The Prince's face showed full agreement, even emotion, and he said no more, other than to nod his head, then turn to take the bright green Regimental Colour while Lacey stepped aside. This was duly handed to Neape and then Rushby stepped up. The Prince recognised him.

"Ensign! How's your wound? What you picked up in Italy?"

Rushby spoke through a mouth restricted by a tight shako strap, but it was intelligible.

"All healed now Sir. Thank you."

The Prince stepped back and the Colour Party reformed. Salutes were exchanged and Lacey led his men in a right wheel. Outside was a small gathering, each of the other Colour Parties, stood within their own group of followers, although all were 'followers' of a higher social status than the common, soldier's usage of that term. Rushby and Neape were soon ensconced with the girls, but it was not long before Lacey had had enough. He shook the hand of each of his fellow Colonels, each wishing the other well, then he looked for Carr and found him, easily enough, standing statuesque and alone.

"Get The Colours cased and let's get home!"

On the evening of the following day, their coach rolled back through their barrack gate. Lacey took himself straight to his office, followed by Neape and Rushby who placed their respective Colour into the iron holders behind Lacey's chair. They saluted their charges, now safe in their proper stations, then left.

***

Nathaniel Drake stood in thoughtful pose, mostly amused, yet also critical. He was watching their servant, Corporal Edward Morrison, give a final brushing to Carr's uniform, paying particular attention to the sides and shoulders. Morrison was just as much diverted and taken with the occasion as was Drake and could not resist a comment.

"This is it, then, Sir!"

Carr frowned in Morrison's direction and Drake looked critically at Carr, but for the final time, him being the main character, now polished, burnished and twice shaved. The only blemish, apart from the scars, was Carr's eye, now an odd shade of yellow. Drake then went to the door of their room and opened it, for Carr to leave and then lead the pair down the narrow stairs and into the street. For the next hundred yards or so, Carr spent most of the time trying to ease, with his chin, the tight strap of his lustrous shako, its brass frontplate gleaming like gold and matching all other hardware. He felt in a worse state of nervousness than he would be were he leading his men against a French bastion, but Drake was chattering on, Carr hearing no more than every third word.

"I've asked Cecily to bring her to Lady Constance's and then we two will slope off, somehow, leaving you alone. We've set a date, by the way, Cecily and I. Very provisional, but, well, we live in hope."

Carr was in too much of a nervous state of mind to make any kind of point that his own Brevetcy was far from certain and could totally scotch Drake's plans, but he could only glace sideways at him.

"Congratulations."

Drake looked at him.

"Thanks. You're to be Best Man, of course!"

504

That did nothing to alleviate Carr's deep state of anxiety, in fact it added another layer, but at least it took his mind off what was to come, albeit very temporarily.

They marched and wheeled through the wide streets of the more salubrious and affluent end of Taunton, until faced with the imposing façade of the residence of Lady Constance Fynings. The last time Carr had walked up the path had been to sing in the good Lady's choir, but it was at this very place, during the first rehearsal, that he had first met Jane. The memories came back immediately on seeing the imposing limestone frontage and also the heavy, medieval door that was thoroughly in keeping with the uncommonly large blocks of stone forming the surrounding arch. Drake raised the huge doorknocker to let it thunderously fall and, as Carr remembered, it was the same maid of still indeterminate age who opened the door. As usual Drake took charge, as engaging as ever, even with servants.

"Hello Maud, good to see you. Is Cecily here? And Jane Perry?"

Maud curtsied.

"Yes, Sir. Yes to both. They're in the day room."

Drake stood back and pushed Carr through, fearing that he might run off. With Carr in the hall, Drake enquired further of Maud, pointing at the same time down a dark corridor, dark from dark varnish on dark wood, although the light shone through a window at the far end.

"That's down here, isn't it. At the end, overlooking the garden?"

Maud nodded.

"Yes, Sir. I'll go down and announce you. Her Ladyship is there, also."

Drake held out his arm across her path.

"No! No need, Maud. We can announce ourselves, can't we Henry?"

Carr twisted his head to show a face with no expression, neither did he make any gesture. He remained stock-still. Drake removed his own shako and gave it to Maud, but it was a good three seconds before Carr could bring himself to concentrate and remove his own to pass across to Maud. Drake then led on, to knock at the final door and open it. He entered, followed by Carr, in line astern, and almost hidden by Drake, who was immediately proceeding with bountiful bonhomie towards the seated Lady Constance, she sitting in a large single armchair, opposite Jane and Cecily, who occupied a small settee.

"Lady Constance. How do you do? We trust we find you well, you certainly look in the best of health!"

Lady Constance was well used to Nathaniel Drake's flummery and countered with a disapproving look, bar her eyes, which showed that she was more than a little charmed by the warmth and exuberance of his, as usual, theatrical arrival She had a deep affection for the always cheerful and egregious Nathaniel Drake, but her voice was flat and knowing, as though she had long fathomed the charmer that was this high spirited young man.

"Nathaniel! Thank you, I am well."

She tilted her head sideways to acknowledge Carr, although she could see him well enough.

"Captain Carr. You are welcome."

However, regarding Carr, her eyes showed both surprise and concern. Drake looked thin and worn, but Carr was positively gaunt and the odd shade around his left eye did nothing to improve matters. However, he was stood tall and straight and that must be to the good. She extended her hand to him and he strode forward and bent over it.

"Lady Constance. It is very kind of you to remember me, from so long ago."

Lady Constance laughed.

"I remember you very well, Captain. A rather effective bass baritone, as I recall."

Carr laughed. From somewhere he was gaining control of himself.

"Again, you are too kind. I cannot recall achieving much more than a lower end rumble."

The two girls laughed, which set Carr further at his ease, but it was Cecily who spoke.

"You were better than that!"

She looked at Drake, very affectionately.

"Better than one certain party, who did, at least, manage to sing the right notes, but not always as dictated by any form of rhythm!"

Drake feigned astonishment.

"Not so! I was the very sheet anchor of the bottom registers! As I recall it was left to me, many a time, to carry the tune along."

Cecily stared back, plainly not convinced, but Drake was by now pulling up a chair, with Carr just behind carrying another. At that point, Lady Constance's motherly instincts surfaced at the sight of the

506

pair, both plainly having played a part in the awful campaign that she had just lately read about.

"You are to stay to dinner! I insist! We are having roast mutton."

Drake turned to Carr, to ask a question to which he well knew the answer, but to ask it outright would add to the happy atmosphere now building.

"Can we do that? We're not on guard or anything?"

Carr looked at him, then looked away and swallowed. Now sat there, in the circle, his nerves had returned. He cleared his throat.

"Yes, to the first, no, to the second."

He bowed towards Lady Constance.

"You are most kind, Lady Constance. It will be a pleasure."

At that moment the door swung open and in came Maud with a huge tray that was just about within the span of her arms. Lady Constance looked up.

"Ah, tea! Nathaniel, bring over the table, please."

Drake stood and dragged a low long table into the gathering, to enable Maud to lower the tray perfectly into the centre of the five seated around. Lady Constance stood up and arranged the cups and saucers, the fine china causing a superior tinkling that spoke of the highest quality. As she poured the tea, she looked at Carr.

"Captain Carr, could you please circulate with this plate of sandwiches?"

She passed him a large plate of carefully prepared, crusts cut off, white bread sandwiches, which seemed to contain some kind of paste filling. Carr took the plate and looked up, first towards Cecily and Jane, who were staring at him, wide eyed and lovely, but more important, bathing him in two deeply affectionate smiles. Thus encouraged, he rose with the large plate and collected two small ones, which he handed to both girls. He gave one to Cecily, quickly, then looked at Jane and the look he received back told him all he needed to know, a look that held him fixed in place for some seconds, this by Cecily's indulgence, for she still holding an empty plate. However, noticing Lady Constance staring over at events that were not happening, she decided to finally interrupt.

"Henry! I would like some of those sandwiches."

Her words cut deep into Carr's thorough drowning in Jane's brown eyes and he re-surfaced with some confusion, nearly tipping off the sandwiches with the speed that he thrust them in Cecily's direction. She delicately selected two, for Carr to then offer the plate to Jane, which gave more opportunity for deep eye-to-eye gazing, but this time much shorter. He returned to his seat to select his own of the sandwiches, then to notice a fine china cup placed closest to him on the table. Polite conversation then ensued in which Carr played little part, other than answering direct questions. He was not often given to deep philosophising, but the thought came uppermost into his mind and would not depart, regarding the deep contrast of where he was now, in a refined drawing room, drinking expensive tea from a wafer thin teacup and eating delicate food from a spotless fine china plate, to where he had been but days ago, the seemingly endless starvation and being barely able to place one foot before another. His sojourn was interrupted by another question from Lady Constance.

"Captain Carr! What are your hopes that your Brevetcy will be confirmed?"

Carr released all the breath from within him and placed his cup and plate on the table. He looked down at his boots before answering, his hands together, outstretched over his crossed legs.

"Well, ma'am. I can only hope. About evens, I'd say, perhaps a bit better."

Drake swallowed a sandwich, almost whole.

"I'll say! Better than evens. I'll say!"

He looked at Lady Constance, then at Jane and Cecily.

"Notwithstanding the deeds of our esteemed and respected Colonel and Senior Major, he carried the whole battalion. Didn't put a foot wrong! In three tough battles, our Light Company, led by him, thoroughly distinguished itself and brought nothing but honour and recognition to the Regiment. Odds on, I'd say, or should be, at least."

Carr screwed his mouth sideways.

"That's kind, Nathaniel. One can only hope, but it rests on so many other things."

With that, there came a long moment of silence, perhaps the others hoping for some expansion on the "many other things", but nothing was forthcoming. Therefore, with the tea drunk and the sandwiches consumed, Lady Constance decided that this chapter of the afternoon was at an end and, in addition, she was thoroughly aware herself of the paramount reason for the gathering. She stood, to be immediately followed by Carr and Drake rising to their feet.

"I have some letters to write. I will leave you young people to your chatter and whatnot. Please make yourselves very comfortable. Dinner will be 7.00 for 7.30."

Accompanied by bows from Drake and Carr, she left the room. Drake now stood up and looked at Cecily.

"And we have things to talk about! Boring bits about a wedding."

Cecily looked at him with mock anger.

"They are not boring bits! They are thoroughly engaging, being the proper arrangements for our wedding!"

The last sentence was spoken forcefully, her voice rising to the last word, but then she noticed that Drake's eyes were moving from her to the door that led into the garden and back again, so it finally dawned on her that this was the excuse to leave. She stood and Carr again had to, as she spoke.

"Yes, but you're right. Who wants to listen to someone else's guest list and seating plans?"

Drake chimed in.

"And choosing of hymns!"

"Yes, that too! The choosing of hymns."

She walked towards the French windows.

"We'll take a walk outside. It's hardly cold at all."

With that she pulled her shawl off the back of the settee and wrapped it tightly around herself. Drake opened the door, they both left and Drake closed it behind them.

The sound came to Carr like the Clap of Doom. Through the glass, he watched both depart, then he looked at Jane. He sat down, then he stood up, then he sat down again. His boots creaked and his collar felt tight and hot. She, on the other hand, looked at him levelly and calmly, a half smile on her face. It was as though she were examining every part of him, carefully weighing all that she had come to know about Captain Henry Carr. He was transfixed by the look, almost mesmerised, so it was Jane who broke the silence, being shatteringly forward.

"You haven't got all day, you know!"

Then she laughed, looked straight at him, then down at her hands, then back up, her hands now together in her lap. Carr laughed himself and was still laughing when he crossed the room to kneel at her feet. He had crossed the distance so quickly that his knees skidded on the carpet, almost colliding with her. He needed both to retain his balance, whilst but one was the proper form, but holding to that would have deposited him in a sprawling heap at her feet. He was still laughing when he managed the correct words.

"Jane Perry, will you do me the honour of agreeing to marry me?"

He thought he should say more. Typically, he had given the thing no previous thought.

"And thus become my wife."

She giggled again at the clumsy collection, then looked down, then back up, directly at him.

"Yes. I will. I accept."

Carr took both her hands in his, which then gave vent to another burst of clumsy merriment from him, something between a hiccup and a chuckle. He was reaching up to kiss her, but at that moment Maud came in to clear the table, to see him now on one knee before Miss Perry, both chortling like rainspouts in a thunderstorm. Carr looked around at the treasured servant, himself now carried away on a wave of euphoria.

"Hello Maud. We're going to get married!"

"Oh Sir! That's lovely. I'm glad, right glad!"

***

Chapter Twelve

A Question of Letters

Carr stood before the door, summoning up the kind of spirit and determination that was more the requirement for storming a breach than to approach a prospective Father-in-Law. Jane was standingto one side, looking at him encouragingly; smiling, nodding and then going through the motions of knocking on a door. Carr, on the other hand, felt himself to be embarking on a 'Forlorn Hope', with little chance of success, merely performing what needed to be done, a 'going through the motions'. Jane had insisted that they ask her Father's permission, but Carr knew full well the dislike that the General harboured for the 105th and, as he had gathered from previous personal meetings with him, the dislike extended and intensified when applied to himself. Nevertheless, again dressed as though for Royalty, Carr did knock at the study door. A muffled "Enter" came from beyond the thick wood. Carr took one last look at Jane, she still nodding encouragingly, then he turned the handle and went in.

He was in the General's study. For any other Officer, Carr would have believed himself to be in the private room of a dedicated soldier, for all around were artefacts, pictures and maps, of wars and conflicts since Roman times. However, somehow the impression was of dust, age and slight neglect, although all surfaces shone in the weak daylight. Pride of place was given to the full set of armour of one of Cromwell's Ironsides, this actually found by Carr when out on manoeuvres in Somerset and donated by him and Drake to Perry himself, but he was not allowed to dwell on the thought nor even mention it. General Perry was sat behind a huge desk with a face like granite, granite going red from a bloody sunset beyond the tinted windows. He twirled a medieval poignard between the thumb and forefinger of his right hand, the point meeting the top of the index finger of his left. The whole was lowered to just below his chin as he sat regarding Carr, who came to the attention, his shako held regulation in his left hand against his ribs, whilst he saluted with his right. There was no acknowledgment other than the General dropping the weapon onto his desk. His voice came from way back in his throat, as though he were issuing a threat both dire and final.

"I know why you've come, Carr, and you can save your breath! Jane marrying you is out of the question. I've nothing further to say and neither have you. Leave, now, get out! That's an order!"

Carr opened his mouth as if to say something, but closed it again. Get out was the order and so he did exactly that, spinning on his right heel and leaving the room, closing the door behind him. He went immediately to Jane, whose demeanour had changed utterly, she was almost in tears. Carr took her right hand and kissed it.

"This changes nothing! We will be married. If not soon, then soon after that. My Brevetcy will be confirmed and you have your majority, you're over 21. With both, we'll get by."

He looked at her and smiled, a wide, encouraging grin. She responded, though her mouth was quivering and Carr laughed out loud.

"Take a midnight ride, if we have to!"

He grinned at her again, his teeth white in the gloom of the corridor.

"We'll make plans! But first things first. I become a Major, that means we can stand on our own feet, needing no-one!"

He kissed her hand.

"Don't worry. I love you! Everything will be fine."

He replaced his shako on his head and gave the top a cheerful tap, which made her grin, briefly, and then he left, seeing himself out. Jane dried her eyes, tidied her hair and went into her Father's room, without knocking. He, surprised, looked up and over the rims of sparse, steel rimmed glasses at his utterly unhappy only daughter, but her grief made no difference.

"It's out of the question. The man's unreliable, a poor Officer and wholly the wrong sort that I have in mind for you. Put it out of your head, I'm telling you, out of your head!"

Jane stood her ground, her eyes sparking defiance.

"No! It is Henry Carr that I want for my husband and he wants me for his wife!"

The vehemence of her outburst took Perry by surprise. Never had his daughter spoken to him in such a fashion; it was such a shock that he could think of but one reply, which was already lodged in his mind.

"I wish you to meet Lucius Tavender. I've made mention of you to him and he appeared quite receptive."

Jane exploded.

"Quite receptive! Do you think I'd settle for that? Tavender is an odious man that I'll have nothing to do with. As you reject Henry Carr, so I reject him!"

She calmed and took a deep breath. Perry himself remained silent and in a state of amazement.

"Father! I am over 21, which means I now have full control of the money of my own, left me by both Grandparents. I'd wish to marry with your consent, of course I would, but Henry Carr is my choice! I wish him for my husband! If I have to, I'll pack a bag and leave. We'll be married somewhere, and that'll be that. I'll be sorry that you'll not be there, but, but, ... so be it. If needs be, when he's a Major, I can join him in The Peninsula. That's where he's going and I'm prepared to join him."

She straightened her shoulders and lifted her head. Her defiance was complete. Perry, meanwhile had aroused to anger, both at his daughter's rejection of his wishes and the sheer rebelliousness of her words and attitude towards him. He rose from his desk to look at her, the steel glasses now thrown onto the desktop.

"Do that and I'll disown you! Do you hear? Disown you. You'll be no daughter of mine!"

Jane looked him full in the face. Her boldness and confidence growing by the second.

"So be it!"

She paused to allow that to sink in.

"If you want me, I'll be at Mother's grave."

Carr, meanwhile, had arrived at another door, Lacey's, which was open and Lacey could be seen at his desk. Carr knocked and stood in the doorway. Lacey looked up.

"Ah, Carr! Good day to you!"

Carr walked forward, searching for words, a hesitancy which Lacey noticed and he decided to help, if only to help Carr relax somewhat.

"Sit down. Now. What?"

Carr took a deep breath.

"Sir, I feel embarrassed to press the point, but my Brevetcy, Sir. Have you heard anything?"

Lacey's eyebrows accelerated up his forehead.

"I have, indeed I have."

He leaned over to get a sight of the door.

"Bryce!"

There was the sound of a chair scraping back, then the slightly rotund Sergeant Clerk stood in the door.

"Sir?"

"That letter from Horse Guards, about Captain Carr, here. Be so good as to fetch it!"

As Bryce disappeared, Lacey grinned encouragingly, looking directly at Carr, who felt his spirits rise more then a little.

Within a minute, Bryce returned and placed the letter, minus its cover, on Lacey's desk. Lacey picked it up and looked at Carr. Bryce had not departed and he was smiling also. Lacey began to scan down the page, speaking odd words.

"Promotions Board ...... normal procedure .... current hostilities ..... yes, yes....."

He looked at Carr.

"Perry received a copy of this!"

He continued scanning.

"Ah, here it is. "Regarding the Brevetcy of Captain Carr, the Board wishes to have confirmation from the Brigadier who commanded Captain Carr at Corunna, who can vouch for the worthiness of the aforesaid to receive his Majority. Or otherwise, as the case may be."

Lacey looked at Carr, grinning between licking his lips. Carr, on the other hand, felt mildly disappointed and tried not to show it.

"Thank you, Sir. That's encouraging, but who would that Brigadier be, do you think?"

Lacey leaned forward, suddenly enthusiastic.

"Why Bentinck! Who else? He spoke to you personally in that damnable farm you held, did he not, and Hope spoke very highly of you, to him, myself and Anson. Him of the Guards. Should be a shoo in, I'd say. We got a letter off to him, ..."

He turned to look at Bryce, who was also grinning paternally at Carr.

" .... when, Bryce?"

Bryce needed a second, before realising that he was being spoken to by his Colonel.

"From memory, Sir, four days ago."

Lacey turned to smile reassuringly at Carr.

"There, Carr. We can hope for a reply any day now."

Carr stood, now grinning from ear to ear himself, his spirits altogether much lifted.

"Thank you, Sir."

He paused to look at the smirking pair opposite.

"Well, we can but wait. Sir!"

He came to the attention and saluted, which was returned by both.

Whilst these events played out in battalion headquarters, General Perry, remaining in his study, had at last got around to reading his own correspondence and was reading his copy of the same letter from Horse Guards as read out by Lacey to Carr. His face reddened and his brow darkened, the idea of Carr becoming a Major, perhaps to command a battalion, appalled him. On top, his remaining a penuried Captain would do no harm to his own ambitions towards Jane. Tavender had very rich people; did she really want to live in some backstreet kennel somewhere? When faced with that as a future prospect she would change her mind, he felt certain. He may be too late, but Bentinck must be dissuaded. Immediately he reached for pen, ink and his stock of cheap paper, then he began to write. He got no further than:-

To: General William Bentinck – Plymouth Barracks
From: General Timothy Perry – Taunton

Perry then leaned back to gather his thoughts, but could not assemble too significant a collection, however, his inherent ill nature came to his rescue so that he then vented his pent up spleen on the specific notion of Carr becoming a Major and, also generally, on the subject of the 105th Foot. It stretched over three pages, but said nothing of particular substance other than Carr had resigned his Commission in order to fight a duel and had then rejoined as a Captain into what he described as a 'ragtag rabble of detachments'. With the last flourish of his signature, Perry felt much calmer, such that he was able to ring the servant's bell. In the space before the maid's arrival he sealed the letter and addressed a cover, to then place it on her tray, which she had brought, in the hope of clearing the several empty cups. He looked at her sternly, instantly making her very apprehensive.

"See that this gets posted today. Today! Am I clear?"

The maid curtsied and gratefully left the room.

Carr walked out of the barracks feeling both 'up and down', as he would put it, automatically acknowledging the 'present arms' of the sentries at the gate. He walked, almost without picking his way, the distance to their billet in the inn, his mind going in all directions, until he reached the room he shared with Drake, to find the door open. The reason for such was that he could hear a female voice coming from within, which in a second he recognised as Cecily's. He entered to find them sat on opposite sides of the small table, inevitably holding hands. Both looked up and both were torn between immediately giving their good news, or commiserating with Carr over his bad, which they had heard about, for Jane had left her Mother's grave to then call on Cecily. The result was a conflict, both speaking at the same time; from Drake, "Henry, we've excellent news" and from Cecily, "Henry, I'm so sorry!" Carr acknowledged first that which the lady had spoken.

"Never mind. We'll find a way, time is with us!"

She smiled weakly, then Carr looked at Drake.

"And your news is?"

Drake's face lit up with pure joy, much of which infected the visage of Cecily.

"That we are going to be married next month! We'll get the thing done before we depart again for Portugal."

He regained Cecily's hands in his own and both indulged in a deep exchange of affectionate eye gazing. Carr looked at the pair and debated for a moment whether or not he should break in on their amorous reverie, but it was Drake who broke his gaze away.

"Oh, I know, I said only when I get my Captaincy, but we can't see the point in waiting. A Captain's pay is not what we need, not any longer. Things have gone rather well at home, the iron ore quarries, so I'm told, and I now have enough money to purchase a Captaincy, even if my temporary is not confirmed. There are plenty of vacancies, what with the casualties from the retreat and all, and I can now describe myself as a veteran Officer, of Italy, Portugal and Spain!"

He turned to look at Cecily, who looked adoringly back. Carr could not decide whether to be glad or sad; to be happy for the pair, or dejected because they had the means to comfortably do what he could not, but good form prevailed.

"My sincerest congratulations to you both. I'm sure you'll be very happy."

Cecily rose up, ran over and kissed him.

"I'm sure you will, too. I'm sure things will come right for you."

In marked contrast, the next evening saw a major gathering in the almost squalid by comparison, barrack room of Jed Deakin, all gathered around the table close to the cribs of Jed and Bridie, and Nelly and Henry Nicholls. The gathering, as it usually did, included Miles, Davey, Byford, Saunders, and Fearnly. Joe Pike and Mary sat together, inevitably all touching that could. On this occasion Chaplain's Assistant Sedgwicke was present, awkward and saying little in such company, but glad of his place and their warm acceptance. The men were cleaning their muskets after musket drill with newcome Militia and the women were making and mending. It wasn't long before they descended into their usual squabbling, in this case complaints by Tom Miles to Zeke Saunders that he was using more than his share of the brickdust they had for polishing their muskets and other metal. Jed Deakin had soon had enough and went back into his crib to find a bag that he threw at Tom Miles.

"Thur! That's plenty. An' I expects to see a special shine on that bloody short-arse bundook that thee thinks so much of!"

Miles looked daggers at Deakin. His Baker was ten inches shorter that the standard musket, but he would stand no criticism of him nor anything of his. His voice rose petulantly.

"That's as maybe, but I can hit more with this than you can with that broomstick you carries!"

Deakin scowled back

"Ah, shut thee gob an' get on with that polishin'."

He craned his neck up.

"Seems to me that you've let that rifle get into none too good a state!"

Miles flared up and would have stood up, had Davey not passed him his own mug containing a good measure of rum.

"Here, have a drink! He's prodding of you."

Davey smiled good naturedly over at Deakin, before returning to the affairs of his own rifle.

"What'd you expect from a Sergeant?"

Deakin growled something but the tension had passed. Mary looked over at Bridie, mending the hole in a child's shirt, one from a French drummerboy at Vimeiro.

"Bridie, how's your foot?"

Bridie looked up.

"Oh fine, 'tis fine. The pitch've dropped off. Look, I'll show you."

She leaned back and brought her bare foot up onto the table. The gap from the missing centre toe stood out like that from a missing tooth, as all stopped their work to make an examination. Byford, somewhat more constructively, leaned over to examine the stump. First he nodded his head.

"Hmmm. Thoroughly healed! All thanks to our sawbones Sergeant!"

Reminded again of the nature of the operation, Bridie dealt a blow to the top of Deakin's arm.

"An' I've not forgot, neither, how the thing was done. What with an old chisel and a mallet! An' me drunk as Paddy's sow!"

Deakin looked askance at her.

"What you complainin' to me for? The thing's worked out fine! And if you want someone else to blame, go to him, him there!"

He pointed at Joe Pike.

"'Twas him as fetched the tools. Wouldn't have happened otherwise."

She hit him again.

"Now wasn't he just followin' orders. Orders from you!"

"Ah! Yes. An' I got my marchin' orders after, straight from you. That punch! I can't chew the same since. Army beef now gets swallowed a lot more whole than it used to."

At that all laughed, but Bridie had more.

"Serves you right! So it does."

More laughter, then Deakin looked across to Davey.

"Have you heard back from Molly?"

Davey's face lit up.

"Yes. I got a letter this week gone. Parson here read it out. She can't come as she's too pregnant."

Miles looked over, his face astonished.

"Pregnant! You've already got one and now another's due. You'm knockin' childers out like pots from a factory!"

Much amusement passed around the table, then Nelly spoke. Her memories of Molly were not of the best, but the news of a family mellowed her thoughts considerably.

"So how is she?"

Davey brightened again.

518

"Fine! Just fine. They've turned the smallholding over to layin' birds of most sorts and the eggs is sellin' well. An' Tilly's at school. That's right, Parson, b'ain't it?"

Sedgwicke nodded and grinned.

"Yes, she is. In fact it was Tilly that wrote the letter!"

The sounds of satisfaction came from all round, until Tom Miles had a more restraining thought, which he just had to give voice to.

"That's all to the good, but don't think you'll be joinin' of 'em too soon. The army'll get shot of you when you'm either dead or they don't want to pay for your keep no more!"

Nelly Nicholls dropped her hands full of sewing down onto the table.

"Well now, is that not just like you, Tom Miles? Spreadin' full misery when all others is sat lookin' on the bright side of things, at the good in it all."

Battle lines drew up, but at that moment the bugle notes sounded for lights out. All stood and returned their appointed places, some to their crib, some to another room. The last act of the evening was for Tom Miles and Nelly Nicholls to exchange a look of full-grown enmity; hostilities to be resumed at the next opportunity.

***

In even greater contrast again, as 'lights out' was being enforced throughout the barracks, lights were being replenished at the residence of Lady Constance Fynings, as the candles for the early part of the evening had by now burnt down. With the light restored, the activities of the soiree resumed; cards, small talk, someone on the piano and a game of backgammon. Carr was part of the latter, making up a foursome, and playing for pennies. Carr was a very accomplished player, but he did not show his dismay at the appalling mistakes made by his partner, such that they were about to lose their whole stock of copper. Besides, Carr was making crass errors himself, his mind often wandering elsewhere to where it wanted to be, rather than concentrating on the brightly coloured board, to thereby determine future strategy.

The company was more than pleasant, the guests being Carr himself, Drake and Cecily, Jane, and a few more from Lady Fyning's choir. The game was now lost and the winners graciously returned all the brown coins won, as Carr's partner, a local Magistrate, spoke across the board.

"Constance hopes to reform the choir again, Henry. Are you game?"

Carr nodded and smiled.

"Of course, but I fear for a mere month. We sail again soon."

The opponent on the right, looked across to him.

"Where?"

Carr sat up with a jolt. Strictly, that was a state secret. On the other hand, British forces were already in Lisbon and the planned expedition seemed common knowledge. He compromised.

"The Peninsula, again, I feel sure, although I am not privy to the final deliberations of their Lordships at Horse Guards. But I suspect back there, to one part or another."

The final member of the four assembled the dice and counters and returned them to the polished mahogany box. Of all those sat around the table he was the least inclined to be agreeable and the most likely to be argumentative.

"With the same result?"

Carr sat back.

"One can only hope not. Events conspired against us, I feel justified in saying. Moore felt honour bound to advance into Spain, when the Spanish decided to defend Madrid, an advance at some risk, especially with Napoleon in attendance. We all know the result, but General Wellesley I feel to be more likely to prioritise the safety of his own army, giving poor second to any pleas from the Spanish. I, for one, feel thoroughly content to be led by Wellesley. Under him we fought two battles and won both, the last inside two hours. He cleared Portugal of all French. Led by him we all feel we have the beating of any French army, even outnumbered."

The man tapped the box lid. He was not appeased.

"Then he signed the Convention of Cintra!"

The Magistrate now rejoined the debate.

"Ordered to, Lawrence! Ordered to! That's now thoroughly established."

He looked across, but Lady Constance, as an accomplished hostess, had noticed that the game had finished and also the serious tone of the conversation.

"Game finished? Good! Perfect timing. We are to test our memories, and voices, in the choir, albeit much depleted. Around the piano, now, all of you."

There was no arguing and the four stood to walk the yards of Persian carpet and assemble around the gleaming instrument. Lady Constance placed all according to their registers, meaning that Carr had to sing bass, he was the only one even close to capable. The same girl that Carr remembered from previous rehearsals long ago, arrived at the keyboard and Lady Constance distributed the sheet music. At the sight of the notes so low on the stave, Carr groaned inwardly, but they were starting with a good tune, the newly penned "Billy Boy" and all cheerily sang with gusto through the discussion of "Is she fit to be your wife?" and "Can she cook an Irish stew?" The lively ditty enthused everyone and several tunes more were sung, some from their past choir and the performance was judged "Passable and encouraging" by Lady Constance.

It was time to go and all exited to the hall for Maud to distribute the cloaks, capes, bonnets and gloves. Carr and Jane had naturally gravitated together and they were surprised to find Lady Constance suddenly stood just behind and between them. She leaned forward and spoke quietly.

"Go back into the drawing room. I want a word."

With that she sailed forth towards the front door and bid all her other guests a good goodnight. Drake, ever curious, looked back.

"Jane and Henry?"

Lady Constance gave him a firm look, one that said, "there is more to be done."

"They are remaining with me for a little while longer. You go on in your own carriage, mine is standing ready to take them home."

This fact told Drake all he needed to know, that this was all pre-planned and he had best keep out. He offered Cecily his arm and they left.

In the drawing room Carr and Jane were seated, still wearing their heavy outdoor clothing. Lady Constance lowered herself into a chair opposite and looked at the pair, quickly from one to the other.

"You have my blessing, in every way, for your hopes to be married!"

The sudden outburst took both of her audience by surprise and they exchanged a surprised but delighted look. Carr smiled warmly back at Lady Constance.

"Well, that is very kind of you, Lady Constance and very much appreciated. Things have not exactly gone our way ...."

Carr was about to finish with ".... of late", but Lady Constance had held up her hand. Carr remained quiet.

"Which is why I wish to speak to you both, now."

She looked at Jane.

"I have no wish to speak ill of your Father to you, Jane. In fact I have held my opinion back over all these years, but your Mother was a dear friend of mine and I watched the joy and happiness that was in her nature, gradually drain out of her over the years of her marriage to your Father. From that, you can gather my opinion of him and so I'll say no more on that subject."

She paused, as though gathering impetus for what she really wanted to say.

"I've had an argument with your Father, just today. He called this morning. He knows that I have some influence with you and he required me, yes, that's the word, required me, to persuade you of the wisdom of your marrying Lucius Tavender. I told him frankly that I would do no such thing! He left in high dudgeon! Relations between us are now thoroughly soured, never having been so very sweet in the first place."

She looked at Jane, who broke off towards her from sharing a serious look with Carr.

"Jane! I understand that you have some money of your own?"

Jane nodded. Lady Constance took a deep breath.

"Jane, my house is yours to use if ever relations between yourself and your Father deteriorate to such a level as to require it. My niece and myself bounce around in here like peas inside a rattle. If ever you need a roof over your head, for whatever reason......."

She paused, the implication being obvious.

"You may live here and welcome, to remain attached to each other until such time as you are able to be married."

She drew another breath.

"On the subject of money, my man of business will invest your sum for you to create an income, if you so wish. Any losses to the capital, I will make good myself."

She paused again.

"I am offering you my house and my help, such as I can give. This, so that you can pursue your hopes towards each other, and one day be married.

The interim was a stunned silence, broken only by the loud ticking of the ornate clock behind her Ladyship. Carr looked at Jane, who was showing a hesitant smile on and off, then he rose and went over to Lady Constance to take her hand and kiss it.

"Lady Constance, you are more than kind. That Jane should be safe, well, and cared for whilst I am abroad has been my constant worry. Your kindness has set me thoroughly at ease. We can now wait, for my Majority, whenever it comes and then be married. I hate to speak so openly of money, but with that promotion my pay will be such that we can buy our own house together and call it our home. I, too, have a little sum of my own."

He looked over at Jane, who was now grinning openly, almost laughing and looking all around the room in her happiness and confused surprise. Lady Constance stood up.

"My carriage is waiting for you outside, to take you home. Both of you."

Jane rose and the three took themselves to the front door. Before leaving, Jane threw her arms around her Ladyship. No words were spoken, except finally from Lady Constance.

"Go now, both of you. And God Bless you both!"

In the carriage, Carr, installed at last, naturally sat, as convention dictated, opposite Jane, but she soon changed that by bringing herself over to his side, seizing his right arm and draping it around her shoulders, holding two fingers of his right hand in each of her hands. Then she snuggled in under his armpit. Carr was in another place, seeing nothing, feeling everything!

"I told you things would move our way!"

The reply was a squeeze of his fingers, so he reached across to take her left hand in his.

"You are my love!"

The result was her head moving, in the dark, but it was light enough for him to know that she was looking up at him and smiling, then her head fell against his chest. His left hand came back to hold the handle beside the door, to steady them both from the coach swaying on the uneven road, but the coach rattled on, the dark world beyond the door and the polished glass completely irrelevant.

***

The following afternoon, exactly eighty miles away, William Bentinck sat in his office, this being in the highest part of a tower, for he worked not in the relative luxury of a common barracks, instead he spent his days in a fortress and it gave frequent reminder of that function, being bleak, bare and formed from large, unaesthetic, blocks of stone. Bentinck was a man of the country, therefore fresh air was a priority requirement and all windows were open to allow in both the sea air and the sounds of the preparation of an army, or at least his Brigade of it. The sounds came in of marching men, screaming instructors and squabbling seagulls. On his desk, and trying his patience, for he was necessarily a busy man, were two letters, both concerning the remembered Captain Carr, but one asking for support for his Majority, the other condemning the idea utterly. However, the first gave clear explanation of support, whilst the second contained little more than ill-tempered invective, this from a General Perry, of whom Bentinck had no knowledge. The first, from Colonel Lacey, whom he remembered, gave details that he could well recall, prominent being the fight to hold the farmhouse above Elvina. This same letter also made the very reasonable point that Lacey needed to establish his Officer Corps soon, as they were ordered to be part of the army being despatched back to the Peninsula.

His impatience was not improved by the fact that, to deal with this, required three replies, one to Horse Guards, a second to Lacey and a third, a courtesy reply, to General Perry. He sighed, this was irksome on a busy day amongst other busy days, but there was one simple way out. He raised his head and looked through the door.

"Tavender!"

A voice came from the outer office.

"Sir?"

"Are there any clerks there?"

"No Sir. They are taking their midday meal. Its just me and Captain Templemere."

"You'll have to do. Bring pen and paper."

The two appeared and came in, both resplendent in their Light Cavalry uniforms, their outlandish decoration, which Bentinck always found to be somewhat irksome. They spoke in unison.

"Sir?"

"Bring up a chair, opposite, here. We have letters to write."

He pointed to the space before his desk. With the two seated, Bentinck pointed at Tavender.

"Yours is a copy to Colonel Lacey of the 105th Foot at Taunton."

He pointed to Templemere.

"Yours is a copy to General Perry, Regional General Western Counties at Taunton."

He watched as the two wrote their headings.

"Mine is to Horse Guards."

He began writing his own heading, then he looked up.

"Now, have you written your recipients?"

Both nodded and again in unison.

"Sir."

"Very good, so, as I dictate, so you write."

He dipped his pen again in the ink and shook off the surplus.

"Sirs. Concerning the Brevetcy of Captain Henry Carr of the 105th Foot, Prince of Wales Own Wessex Regiment. This has my full support, therefore it is my wish that his current rank should now be confirmed and Captain Carr promoted to full Major. In support I would list the following."

He reached over for Lacey's letter, which gave the details that he wished to add of Carr's record in the Peninsula; 'Why take the trouble to compose my own words?' he thought. Whilst he did so, Tavender and Templemere exchanged a disturbed look, one that conveyed their shared distaste with what they were now part of. By now, Bentinck had Lacey's letter in front of him and he called out, for all three to write, the details as given. That done, Bentinck composed his final sentence, which he again dictated.

"I consider this record to provide ample justification for Captain Carr's Majority to be confirmed."

He signed his own, then signed the two copies.

"Wax!"

Tavender stood and fetched the small pot of sealing wax suspended above a small candle, whilst Bentinck folded all three. The wax was poured and the letters sealed. The wax dried quickly, then Bentinck wrote above each seal the recipient and then pushed all three across to Templemere.

"Get covers addressed for each and sent off. Today. Dismiss."

As the pair departed, Bentinck reached across for the next bundle of orders, bound in red tape, congratulating himself that a nuisance job had been quickly dealt with.

Outside in the Office, Tavender and Templemere looked at each other, the letters now on a desk beside them. Tavender looked at the three, looked again at Templemere and then selected the letter destined for General Perry. He picked up a pen, inked it, and wrote above the seal "assume other copies lost in the post". He signed his own name, then he folded a cover, sealed it around the letter and wrote on the address. He picked up the remaining two, crossed to their blazing fire and casually threw them in, then he looked at Templemere for both to share a satisfied grin.

"Shocking state the post is in these days! Must be the war!"

The red wax melted and spluttered as the thick paper was consumed by the flames.

***

A week passed, then two, both of routine, both filled with orders and scheduled duty. As each day passed Carr grew more apprehensive, whilst Perry sat ensconced in his office, gloating that "no news was good news", and that soon Carr would be gone abroad, for the whole affair to be buried under bundles of paper. Carr himself called into Lacey's office, twice daily, to ask of Bryce the same question and the conversation was very repetitive.

"Anything?"

"No Sir. Nothing today."

Cecily and Drake were hastening on with arrangements for their marriage, which thoroughly occupied all the spare time of Drake, leaving Carr to wrestle alone with his own worries, one at least being that Jane remained still at her family home. Relations were frosty, to say the least, but, at least, there was no resumption of outright hostilities over Carr or Tavender, such that when Carr saw Jane, everyday, in a variety of meeting places, the troubles of both seemed to melt in the company of each.

The issue was also troubling Lacey and O'Hare, both sat in Lacey's Office. It had been a subject of discussion for more then one occasion and now Lacey raised it again at their morning meeting, voicing all the concerns that both had identified.

"We need to know who our Officers are! We cannot set out again to take on the French with only one and a half Majors. What if a refusal for Carr comes whilst we are out there? If it's refused, then we need another Major, from where, God Knows! Could be Horse Guards send one, or could be a purchase, but I want someone whose heard the sound of gunfire more often than a few times and from more than hearing distance. If there is someone of such, then Carr goes back to Captain. If he gets it, then it's a new Captain we need!"

O'Hare looked up over the fingers of his hands poised thoughtfully together.

"Easier to find a Captain than an experienced Major!"

Lacey nodded.

"This needs to be resolved! Send another letter?"

O'Hare raised his eyebrows and tilted his head sideways, his shoulders and hands gesturing in common; a "what else?" gesture.

Lacey looked out through the open door.

"Bryce!"

The reply was instant.

"Sir?"

"In here, if you please."

Bryce rumbled in to come to the attention.

"That letter which you sent to General Bentinck. Do you have a copy?"

Bryce nodded.

"Yes Sir. We always keep a copy."

"Right. Re-copy and send again."

Concern came over the face of the old Sergeant.

"To General Bentinck, Sir?"

Lacey looked puzzled.

"Yes! Who else?"

The concern increased.

"Well Sir. I've just been looking through the latest Movements and General Orders and I read that General Bentinck and his Brigade have been ordered to Lisbon, Sir. Reinforcements for General Cradock, Sir. They sail day after tomorrow, Thursday."

Lacey looked at O'Hare.

"We'll have to send someone. Deliver the thing by hand! Who?" O'Hare's eyes widened.

"Who else, but Carr? And that's a serious ride, here to Plymouth in two days, but the man deserves his chance. To get it out of Bentinck, either a yay or a nay!"

Lacey nodded, then looked up at Bryce.

"Send for Carr."

Bryce saluted, confident that Carr would not be too far away; him awaiting the morning delivery. He was correct. Carr came in, just as Bryce was leaving to look for him.

"Sir. I'm afraid there is nothing again, Sir, but the Colonel is writing another letter, as we speak and he would like you to take it, in person, Sir. To General Bentinck, to see that it gets there."

Carr was nodding vigorously, all the while, but then came the vital question.

"Where is Bentinck?"

"Plymouth, Sir."

The nodding lessened.

"Plymouth, right, not too arduous. I'll see about a horse and come back. The letter's being written now?"

"Yes Sir, but there's one further thing."

Carr turned back. Bryce's tone spoke of problems.

"General Bentinck sails day after tomorrow, Sir, to the Peninsula. You've not much time."

Alarm crossed Carr's face and he turned away, calling back as he went on his hurried way.

"I'll be back, soon as I can."

He ran, almost all the way, making his calculations. He knew the maths, that on horseback thirty miles a day was an achievement, which meant two and a half days to Plymouth, when he had but two, and only from that moment. Once in the billet of himself and Drake, he found, exceptionally, Drake alone and reading a newspaper.

"Nat! I need a horse, a good one, one that will get me to Plymouth come morning of Thursday."

The newspaper hit the table.

"Thursday! Plymouth! Is it vital?"

"Yes, I'm to carry another letter to Bentinck and get a reply, one way or the other, about my Majority. Now I need a horse, one that will get me there and not give up halfway. Any ideas?"

Drake exhaled all the air out of his body in both thought and exasperation.

"The Colonel and O'Hare, do they have decent mounts?"

Carr shook his head.

"Neither! I know that for a fact. Neither has yet replaced what they lost in the retreat."

Drake looked at his newspaper, but saw nothing, then some light came into his face.

"I recall, even before we left for Sicily, Royston D'Villiers sounding off about his horse. It threw him during a hunt or a point to point, or something, but he was cracking on that it had a lot of racehorse in him. Perhaps he still has him, and perhaps he will ……… you know, give you the loan. It'll take something like him to get you there on time."

Carr was galvanised.

"Where is he? Is he in his room?"

"Yes, I heard the pair of them stomping down the corridor, not an hour ago."

"Definitely coming?"

"Yes. Definitely not going."

"Tell Morrison to pack me a bag for four days."

"Four days! That means you'll miss the wedding!"

Apprehensive both regarding that fact and what he was about to ask; and looking it, Carr left the room and paced along the corridor. He came to the appropriate door and stopped, to listen to Carravoy's voice, complaining about the standard of the latest tranche of recruits. There was no halt in the invective, so he knocked anyway, his anxiety for his own affairs paramount.

"Enter!"

It was D'Villier's voice. Carr went in and found both in a state of thorough relaxation, especially D'Villiers, examining with some glee a newly purchased brace of pistols, definitely new, for both box, green beige and pistols were wholly unblemished. Carravoy, on the other hand was eating pastries, but, were it not for the presence of Carr, an observer would think that he had just bitten his tongue. Carr felt moved to speak first.

"Good afternoon to you both and my apologies for the disturbance."

He drew a deep breath, this could be make or break.

"Royston, I have a great favour to ask of you."

D'Villiers returned a pistol to its place in the box and gave Carr his full attention, but nevertheless he looked puzzled as Carr continued.

"I need to ride to Plymouth to be there on Thursday. It's vital, but I have no horse, not of any kind, never mind one that could undertake such a journey."

Both D'Villiers and Carravoy sat forward, as Carr continued.

"I seem to remember you talking about a horse of yours. One that had some quality, some 'racehorse' in him as I recall."

D'Villiers sat back; it was obvious what was coming next, but he said nothing.

"May I borrow him? If there's a question of payment for hire, I will understand and pay what you think, but it is vital that I get to Plymouth close to dawn on Thursday."

Carravoy also sat back, somewhat languidly. This was good, Carr clearly in some difficulty and only they could help. This should be explored, for which he used his most aristocratic tone.

"And the reason why you have such need of such an animal?"

At this point D'Villiers sat bolt upright and interjected before Carr could begin his explanation, which he was about to give.

"Take him! No payment required, nor explanation. Take him. He's down below in the stables. A good run will do him good."

He picked up a pistol again.

"His name's Junot, by the way. I renamed him after our opponent at Vimeiro!"

Carr looked at him, a smile of gratitude clear on his face.

"Thank you, Royston. I am in your debt."

Carr took a deep breath of relief; the first part of the plan was in place.

"May I take him now?"

"Of course, things seem urgent with you."

"They are, and thank you once again."

With that Carr was gone from the room, but as D'Villiers sighted down the barrel, he could not fail to notice the displeasure that was plain on the face of Carravoy, at which point something stirred inside the good D'Villiers, possibly brought on by the joy of his latest

530

acquirement, or possibly his new and growing inclination to speak and do as he saw fit.

"He's never been anything other than straight with me!"

Carravoy could only stare straight back, as D'Villiers continued.

"I count him as a decent comrade and a brother Officer!"

Carravoy, pastries now consumed, reached for a newssheet and lifted it, to cut the space between them.

***

Carr at last was on the road. Precious time to him, for every minute was precious, had been taken up with tearing the bags away from the fussing hands of Morrison and obtaining the letter from Bryce, but now Junot was out on the good turnpike from Taunton to Wellington, this being the first part of the main Bristol to Exeter road. Two bags were bouncing behind the saddle and, under his right leg, in the holster of the dragoon saddle, was a huge dragoon pistol, placed there by Morrison as Carr mounted.

"You never know who's about, Sir."

Carr had realised instantly that he was mounted upon a special quality of horse. Junot, immediately on seeing the open road, sprang into a loping canter that began to eat up the miles, but the sun was inching into a space in the first break in the cloud to the East as it moved inexorably in the Heavens to quit the thankless task of illuminating an overcast late February day. Carr allowed Junot his head, allowed the reigns to go slack and left Junot to simply take the road as he would. Carr had now some thinking to do, now that he had the time to do it. Holding Junot back to conserve him, they could ride through the evening and night, but a stop was inevitable, for him, if not for Carr. If they rode through the night, it meant a stop during the day, when neither would gain satisfactory rest. So, it made some sense to stop in Wellington, give Junot rest and feed, then push on with the daylight, but Junot would have thirty miles stored in him, or thereabouts, whilst Wellington was barely ten. Therefore, Carr decided to take the chance and push on into the late evening and hope for some kind of inn with a stable. Depending on progress, he would decide what was to happen on the following night.

So, they rode on, through the growing gloom that soon became full dark. Carr pulled out his watch, and he could just make out that the hands were set for just after 7.00. However, even if he could see little of the road ahead, Junot could and soon they were passing between the candle-lit windows of Wellington High Street. Back out on the open road again and for what seemed a good hour Junot cantered on, then he slowed, then stopped. Carr peered ahead; Junot had stopped at a turnpike gate. Carr leaned over and rang the bell for the keeper, but it was more than three minutes before he appeared and ill-naturedly took the four pence, double the day rate, and opened the gate, to close it behind what he hoped was the last traveller of what was now a full, dark night. Carr enquired the time, but the door closed upon any answer, even if there was one. However, the keeper had re-lit the gatelamp and Carr used it to discover from his own watch, that it was well past 8 o' clock, pushing up for 9. Carr patted the horse's neck.

"We'll need somewhere soon, boy, but I'd say that you've just about done your thirty."

To confirm, Junot had slowed to a trot, which Carr allowed. He was well content with progress so far. Carr judged an hour, but nothing came that even resembled an Inn, and he was hungry. Morrison had included little more than bread, cheese and apples, but Junot needed rest, so Carr halted in a small stand of trees and decided to make a camp. He unharnessed Junot, wiped him down, and then replaced the thick horse blanket, tying it under his body. The horse, on a long tether, wandered off to munch some grass, which was the only sound that Carr heard as he consumed his own food. He sat for a short while before rolling into his own blanket, when straight away he heard a sound, which could only be described as a loud whisper which sounded as though it included "up in the trees." Carr froze to listen more carefully, then came a rustling sound as though someone was approaching, He rolled out of his blanket, pulled out the pistol from the holster and cocked it, then he looked for human shapes. He soon saw two and up came the pistol.

"Stop! Stop there and come no further, or I'll blow your bloody head off!"

Then, amazingly, came a female voice.

"'Tis alright, Jim, that don't sound like Father!"

Carr kept the pistol erect, but certainly he felt more secure.

"That's right, Jim, I'm not Father but come to me slowly and no-one gets hurt."

Out of the gloom came a young couple, holding hands, but she with her free hand was reaching across to take the muscle of 'Jim's' nearest arm. Even in the gloom, Jim saw the yawning muzzle of the pistol and felt an immediate need to say something.

"There's no need for that gun, Sir, 'tis just me an' Bethan here, we bin out, ... ah ..... courtin', an' now we'n movin' back on our way home. Least back to Bethan's."

Carr lowered the pistol and released the hammer, but Bethan was more than curious and not the least afraid.

"But what be you doin' here, Sir? You'm miles from any place as'll give you shelter."

Carr peered at the girl. Even in the dark he could see she was quite pretty.

"That's your answer. So I'm making the best of it here, for the night."

Perhaps she was the more intelligent of the two.

"You must be on some urgent journey, to come so far past Wellington on such a dark night."

Carr smiled and so did they, when they saw his white teeth through the gloom.

"It is, you're right. I need to get to Plymouth by Thursday morning."

He decided to appeal to their sympathies as kindred spirits.

"If I get there on time, then I can be married!"

Carr saw her move in the dark

"We hopes to get married soon, don't we Jim, but my Father's determined to keep us apart, till then, strict Methodist, you see, Sir."

Then Jim's voice.

"Which is why you finds us out yer, Sir, at such a time."

Carr could see Bethan's head turn towards her swain, after she had elbowed him in the ribs.

"You've got a decent barn. Warm hay and some oats for the pony, there."

Jim's head nodded.

"We have, Sir, an' you can spend a night there. 'Tis better than here, for both of 'ee."

Carr laughed.

"You are very kind, but only if I'll not cause any problem between you both and Bethan's Father."

533

"No fear of that, Sir. Our barn is but two doors down from Bethan's an' after we gets her back in, then I'll get the doors open for you."

Carr smiled in the dark, sincerely moved at their kindness. Five minutes later he was leading Junot down a track where stood a rank of buildings of various shapes, but Bethan was still curious.

"You'm a soldier, Sir. An Officer?"

"Correct."

"Well, we've just 'ad one from our village come back, without half his leg. Says he lost it at a place called Vimeero."

"I was at that battle."

"You was, Sir, what Regiment? His was the One Hundred and Fifth."

"My Regiment."

"Well, that's amazin', Sir. He come back last Autumn, an' is always goin' on about his Regiment what beat two French attacks and tipped the balance against a third."

Carr laughed in the dark.

"That's putting it a bit high, but we were involved in all of that, certainly. What's his name?"

"Spencer. Bill Spencer."

"I'm afraid I didn't know him."

Then came Jim's voice.

"He came back with all sorts of stuff, as fetched a tidy sum of coin."

Instantly, from Bethan.

"Now don't you get no ideas, Jim Darley, of takin' the shillin', 'specially with a baby likely sooner not later, the way you carries on!"

Carr smiled in the darkness and tactfully changed the subject.

"What's Spencer doing now?"

Bethan continued.

"He's a carter. Says his new foot gives him better balance on the cart seat!"

Carr looked across, even in the dark, and laughed.

"I'm glad. So many end up begging on the street."

"You 'ere that, Jim. Runnin' the farm's a better bet!"

No answer, perhaps because they had stopped. Tying up Junot, Carr helped Jim insert Bethan through a downstairs window and

534

averted his eyes as the couple shared a last fond kiss. Before the window closed, Bethan's head re-appeared to talk in a hushed whisper.

"Make sure he gets somethin' to eat, Jim, afore he goes on."

With that, the window closed and soon after Carr found himself in a small, but well appointed barn. Junot was given some oats in a bucket and water alongside, whilst Carr spread his blanket on the hay. Jim went out, but soon re-appeared with a cloth bundle.

"'Tis a fruit pie, Sir. Take it an' welcome. Mother always makes a batch, for sellin' on, like."

Carr stood and felt inside his pocket. He felt the thick shape of a guinea and handed it to his host.

"Here's for your kindness. Buy Bethan something special."

Jim had bit on the coin and found it very satisfactory.

"I will, Sir, and the very best to you on your quest."

"And when you see your Bill Spencer, tell him that Captain Carr of the Light Company wishes him the very best of fortune."

"Captain Carr. Light Company. I'll do that, Sir. Good night."

The door closed. Carr took three bites of the pie, then settled to a sound sleep. He was awoken in the early dawn, either by the noisiest cockerel he had ever heard, or Junot pulling at his halter, to bring the weight noisily crashing against the ring-stop on the stall side post. It may well have been both. He fed Junot the rest of the oats and washed himself in the trough outside. Within ten minutes he was back in the saddle and on the open road, eating the rest of the pie to the regular rhythm of Junot's long striding canter.

The day passed according to the hands of his watch. Two hours in the saddle, fifteen walking, with every four hours for a ten-minute stop, for water and grazing. He was hungry again, but, just in time, the church spires and then the central tower of Exeter Cathedral came into sight. He debated stopping for a meal, but Junot was still fresh, tossing his head and working the ground with his hoof, so he bought pies, pasties and fruit and set Junot to the Plymouth road, again paying tuppence for a turnpike. From that point was an endless climb up onto the heights of Dartmoor and the sight sent a chill through his memory of barren, cheerless Spanish uplands. It was but a short step of the imagination to paint onto the austere landscape before him, the picture of the endless column and the harrowing sights at the side of the road, almost every yard. He thrust the image back and set about the concerns of his own present. The day was passing and it was beginning to rain.

Junot had slowed to a fast walk and Carr had only a vague idea of how far he was from any destination, this being hopefully a place to spend the night. The rain became heavier and so he donned his heavy cloak and rescued his blanket from the rain by draping its roll over the pommel of the saddle where it would be protected. It was plain that both he and the horse were weary, but the road stretched on, bleak, grey and stony, puddles forming by the minute.

The dying of the day lasted an hour and then Carr was forced to follow not the road ahead, but the stone road markers passing to his left. Suddenly came a series that were whitewashed and he looked up and ahead to see dim lights through the mist and rain. One set proved to be an Inn, not well appointed, but, on enquiry, large enough to have a stable and rooms for rent. He saw Junot well fed and stabled, then he returned to the main bar, where it soon became obvious that he was the centre of attention for those within, bar the Innkeeper, who stood like one of the stone waymarkers behind his bar, drying his beer pots, studiously ignoring Carr's arrival. Carr looked at him directly and did not like what he saw, but there was no other choice.

"Please, may I know the name of this village?"

There was no pause in the polishing hands, and the reply was a flat, staccato single word.

"Moretonhampstead."

The brevity of the reply created a pause before Carr's next question.

"How far to Plymouth, please."

The reply was equally brief.

"Twenty five mile. Give or take."

Carr studied him.

"Give or take what?"

The Innkeeper grinned showing canine yellow teeth.

"Give or take what gets in your way."

Laughter emerged from around the bar, but the Innkeeper continued, on a slightly different tack.

"We don't much like King's men around here."

Carr nodded.

"No King's man, me, at least no Revenuer. Your business is your own, mine's in Plymouth. I just want to get there."

The slightest change came over the man, as Carr continued.

"I'm just a soldier, just back from Spain."

No comment, so Carr continued.

"Can you provide me with a meal, please?"

At least this time a nod and a clear answer.

"'Tis just stew!"

Carr smiled, in the hope of making some measure of friendly connection.

"Stew is perfect. Hot and warming!"

The Innkeeper nodded and went through a side door to reappear with a large bowl of the stew, although it looked to be mostly vegetables. Carr picked up the spoon, but did not yet eat; instead he looked at his watch. It said 6.30 exactly.

"Can I be woken at 2 o' clock in the morning? I need to be at Plymouth by daybreak."

The answer was a series of head nodding as the Innkeeper disappeared back into the kitchen. Then Carr's attention was turned to a wizened old man sat but two yards away.

"Be this the King's business you be on, or your own?"

Carr looked at him before dipping in the spoon.

"Both!"

"Soldier's business?"

"In part. The rest, my own."

The man nodded and returned to his drink, but spoke whilst looking into it.

"My son were taken by the Press. Down Plymouth."

Carr swallowed the mouthful.

"I'm sorry! Have you heard from him?"

This time the Innkeeper answered, having re-appeared.

"Heard what? How can you hear word of a pressed man? You only knows they've been taken when they don't come back at day's end, and you only knows they'n still alive when they walks back through the door, years later. You don't even get the word, when they gets DD."

Carr looked at the Innkeeper.

"DD?"

"Discharged dead!"

A pause.

"You'll find few with a likin' for King George round 'ere!"

There was finality to the last words and Carr did not pursue the discussion any further.

"I'll pay you now. You'll not want to be woken, yourself, at two, I'd imagine."

A faint nod again.

"Two shillin' 'll cover it."

Carr fished in his pocket for change and brought some out. The required two were included there and Carr placed both on the bar, each with its own click. Then he added a twopence piece.

"A drink for this man!"

The response was a glower from the Innkeeper, but a thank you from the man. Carr finished his stew and once again had cause to trouble the Innkeeper.

"I'd like to go to my room now, please."

There was no change in the Innkeeper's expression as he gestured with his head towards the corner of the bar.

"Up the stairs, there. Your's is on the left."

Carr nodded a response; he felt no inclination to give thanks. He picked up the two portmanteaus and was glad that the butt of the horse pistol was protruding clearly from one of them. At the top of the stairs was a candle on a small shelf, which he took into the room. Inside he considered his options. He did not like the company he was in, he did not trust them. Evidently he was thoroughly disliked and that made him a target. That meant risk and Carr never ignored any risk. He felt sure that they would not make any attempt on him, not armed as he was with sword and pistol, but Junot in his stable was another question. The only solution was to sleep with Junot in the barn, but the only way to it was back through the bar, where all could see his movements. He opened the window and looked out. It was still raining, heavier, but he was looking down at only a ten-foot drop from the window sill. He knotted two sheets together, ran the result loose through the handles of both portmanteaus and lowered them down. The sheet came back up when he released one end, then he tied that end to the bed and slid down himself, taking the candle. His last act was to make a sizeable knot in the end of the sheets and throw them back in through the window. This he did at a second attempt. The window had to remain open, but no matter. In the barn, Junot greeted him with a toss of his head.

"Hello boy."

Carr patted and stroked his neck.

"Seems like you and I are sharing the same bed.

He unfastened the blanket from his saddle and settled himself into the hay store, instructing his mental reveille that he was on guard in three hours. Soon he was asleep.

He was not sure what woke him, perhaps Junot restless in his stall, him now stamping and whinnying. Carr lay still and listened. Something was pushing against the door, he could hear the regular creak of the hinges, but Carr had put the inside bar in place. He rolled back his blanket, took the pistol in his left hand and drew his sword with his right, then he marched forcefully to the door, whilst shouting loudly a pirate threat, remembered from his boyhood reading.

"If anyone's out there, I'll see the colour of his insides, see if I don't!

He threw up the bar, yanked open the door and stepped into the space. Nothing! He stepped outside and looked both ways and again to the front. Again nothing. He looked down for footprints in the mud and saw only pools of water. He shook his head and re-closed the door, then somewhat wearily slotted home the bar. Returning to his bed, he sheathed his sword and dropped the pistol onto a portmanteau. Once back under his blanket, he examined his suspicions. Were they unjustified? These were people who were getting the rough end of the war with France; did they not have a right to voice their discontent? He shook his head and returned to a fitful sleep, often disturbed by Junot's shenanigans in his stall. After a spell of sleep that both came and went, he came fully awake. It was still dark. He found his tinderbox and lit the candle, to then lead Junot out of his stall.

"Come on, boy. We've had enough of this!"

He found a sack, which he stuffed full of oats and this he added to his own portmanteaus behind the saddle, before placing another twopence piece in an obvious place for payment, then led Junot out into a dry night and, encouragingly, found a good moon dodging in and out of scurrying clouds. Soon they were on the road and Junot was into his ground-eating stride.

***

The first person he saw was a fisherman, obvious from him hauling his catch of fish in a wide basket onto a landing stage from his small boat, which he had sculled up a tributary of the estuary, Carr assumed as he Carr stopped on the bridge.

539

"Morning!"

The man nodded, but seeing the uniform he said no more.

"How far to Plymouth, please?"

The man pointed.

"Not far, now, just some beyond that hill."

Then a thought came to Carr.

"There's a fleet of transports, for soldiers, off to Spain. When will they sail, can you tell me?"

For the first time the man smiled.

"Well, Sir, they'n still thur, and the tide's only just now come on the make, so twill be a good while afore there's water over the bar!"

Carr took that to mean that, at that moment, there was low water and the tide had just turned to come in. For the first time he genuinely smiled and felt in his pocket to find a shilling. He held it up for the man to see, then tossed it for him to catch.

"That for your children. Buy them all a sweetmeat! I'm sure you have several."

The man raised a finger to his forehead, as Carr spurred Junot on.

'Beyond that hill' proved to be miles of road, but he could now see Plymouth and the gathering of masts behind the congregation of church spires and towers. Junot cantered on and in, for them soon to be swallowed into the narrow streets. Carr was hungry, but more urgent was to find Bentinck. The part of town near the harbour was a maze, the widest street little more than a London alley, but Carr wanted the quayside and so, following logic, he kept turning downhill, following the widest of the choices of route presented between the polyglot of buildings.

This proved to be wise. Soon, at the end of a street he had chosen, could be seen no more buildings, only the expanse of the harbour. A few more strides from Junot and they were on the quayside, for Carr to see transports, lined up bows to stern, with files of soldiers going aboard. An Army Captain was stood, idly observing proceedings, particularly the guns being hoisted aboard, suspended from a ship's yardarm. Carr approached.

"Good morning!"

The Captain turned. He looked very young.

"Morning!"

"Can you help me, please? I have an urgent letter for General Bentinck."

The youngster raised his eyebrows.

"Well, Headquarters is up there."

He pointed both up and along the quay.

"But they're about to embark. The place was emptying rapidly, last time I was there. They're probably on their way down."

Carr paused.

"Which road will they use?"

The young face contorted from strained thought.

"I know the road, but not it's name."

Carr looked sternly at him.

"Describe it."

"Well. It's cobbled and has two buildings, one at each corner, but ...."

More cudgelling of his memory.

"One is an Inn. On the right, I think."

Carr nodded, somewhat exasperated.

"My thanks."

He remounted and Junot trotted on. At the first street that met the description, Carr turned up. He could see a gloomy fortress at the top and, thus encouraged, he spurred Junot on. There was a sentry at the gate, who immediately presented arms. Carr dismounted.

"At ease."

The musket came down.

"General Bentinck. Is he still within?"

Nervousness came into the face below the shako.

"No Sir. Just gone this minute, Sir, and his Staff with him."

Carr stepped closer.

"Which street?"

The sentry pointed.

"Why that one, Sir. 'Tis the quickest down."

It was not the one that Carr had come up. He remounted and pulled Junot's head around. Almost at a gallop they entered the narrow entrance, but there were twists and turns and the cobbles steepened, so Carr reined the horse back for safety's sake, however, around a turn, he saw red uniforms and beyond them was the harbour. Somehow, Junot also felt the urgency and he broke into a canter for the last hundred yards. On reaching the first red uniform at the rear, Carr dismounted and led Junot past them all and up to the head of the group. There he saw Bentinck, whom he recognised, and it was not long before Bentinck recognised the Captain stood rigidly to attention before him. Bentinck peered forward.

"Carr, isn't it? I remember you from that farmhouse at Corunna. What are you doing here? Have you transferred out of the One hundred and Fifth?"

"No, Sir. I am still with them, and I have a letter for you, here, Sir, from Colonel Lacey."

Carr reached inside his tunic to extract the letter and offer it to Bentinck, who took it, but did not open it.

"What does it say?"

"That they've received no reply to their enquiry to you regarding my Brevetcy, Sir, and they need some decision if they are to finalise their Officer Corps, Sir."

Total puzzlement came across Bentinck's face.

"But I replied to that, two weeks ago."

The puzzlement remained, then he turned to look at his Staff, halted behind.

"Tavender! Templemere!"

The names hit Carr like a hammer, but he did not show it, forcing every part of himself to remain stock-still. The two emerged from the group behind and saluted, Templemere stealing a look in Carr's direction. Bentinck was in a hurry and in no mood for detailed questioning.

"Those letters about Carr's brevetcy. They haven't arrived. Why?"

There was silence for a long second, before Tavender spoke, but there was just a flicker of anxiety in his eyes.

"Can't say, Sir. We sent them off."

Bentinck turned again to Carr.

"And you've heard nothing from Horse Guards?"

"No Sir."

Bentinck turned again to the pair.

"So, we can assume that that letter did not arrive, either!"

He stood foursquare before them.

"Two! Two vital letters gone astray. And no explanation!"

The last word was shouted, but the reply was mumbled.

"No Sir."

He turned again to Carr.

"So you've ridden yourself to catch me?"

"Yes, Sir. I left last Tuesday Noon."

Bentinck nodded before turning to glower again at the pair before him.

"Right. You'll not go back empty handed."

He looked back through his Staff.

"Rogerson!"

A Clerk Sergeant hurried forward.

"Break out pen, ink and paper."

Bentinck turned into a shop doorway.

"Follow me!"

Rogerson flicked open the deep leather bag, but extracted nothing. He now required to follow Bentinck, who turned at the doorway to look at Tavender and Templemere.

"And you two!"

He then looked at Carr, his face quickly becoming less irate.

"You too, Carr, if you please."

The four entered the shop, where Bentinck stood before the shop counter behind which was a very anxious shopkeeper, anxious at the sudden influx of soldiery.

"My apologies, but may I make use of your counter? I have some urgent letters to write."

The shopkeeper nodded vigorously and then equally vigorously brushed the already spotless surface with a cloth. Bentinck looked at Rogerson.

"Enough for three letters."

Very efficiently Rogerson found paper, pens and a travelling inkpot, which was arranged on the wide surface. Bentinck looked back at his two trailing Captains.

"You two up here. You're the third, Rogerson."

The three placed themselves as ordered, Rogerson, businesslike, Tavender and Templemere plainly reluctantly. It needed a fierce look from Bentinck to persuade them to pick up the pens. He pointed to each scribe as he allocated their recipients to them and waited whilst each wrote the required name.

"Now, write as I say."

The pens were poised.

"Captain Carr's Brevetcy is to be confirmed. It has my full support, justified by events which I witnessed myself and reports from other Officers who took part in the recent Spanish campaign, Officers who enjoy my complete confidence."

He paused.

"Captain Carr, of the ……"

He looked at Carr, the question obvious and the answer came.

"The 105th Foot. The Prince of Wales Own Wessex Regiment."

"Just so, get that down all of you, …… has shown himself to be both a courageous and resourceful Officer, well deserving of his Majority."

There was something about the hunched shoulders of Tavender and Templemere that caused Carr to smile and it was still there when the two filed past, murder in the eyes of Templemere, but perhaps a hint of embarrassment in those of Tavender. Meanwhile, Bentinck was signing the three and Rogerson preparing their covers. Within a minute, all three were in Carr's hand. Bentinck was the last to leave.

"Well. Good luck to you Carr. Perhaps we'll see you once again down amongst the Dons!"

"Thank you, Sir, and good luck to you also, if I may."

A brief nod and Bentinck was out and in the street, to hurry on to the quayside. Junot was stood obedient, his reins trailing on the damp cobbles. Carr gathered them up.

"Right, boy. We've a wedding to get to, but first I fancy one of what that fellow in there is selling."

He tied up Junot and re-entered the shop.

***

The church was bathed in early Spring sunshine, the stained glass windows adding their own colour to the early daffodils and snowdrops. The doors were then flung open, by two red-coated Officers, which added again to the colour, and then the sunshine was also augmented by the glowing happiness of those emerging, Bride and Groom first, both in a state of unrestrained rapture. Soon, they were joined by the rest of the congregation and their tumbling laughter drowned out the chirruping of the Spring birds, but not the ringing of the single joyful bell. Soon there was but the Vicar left standing in his

544

doorway, hands folded across his stomach, a gesture that matched the look of satisfied contentment spread across his face.

Soon, the happy couple were progressing down the path to the Lychgate and rice was flying through the air, the grains held momentarily in the bright sunlight. All were in a state of sincere happiness and it showed on all smiling faces. All bar one. For Lieutenant Richard Shakeshaft what was coming next was going to be particularly dire. It was one thing to stand beside the Groom and hand him a ring, it was quite another to make a speech at a Wedding Breakfast! His face registered the passage of several phrases, which may, or most likely may not; meet the requirements of his role to come. Thus, he hung back towards the end of the procession, hurriedly scribbling notes and then, exasperated and frustrated, crossing many of them out. He had the greatest of respect for his Company Commander, but at that moment he could have wished him in Jericho, or preferably not, preferably right there!

The procession wended its short way into the largest hotel that the town could offer. In its echoing main hall, the formalities were observed; the Bride, Groom, Parents and Shakeshaft greeting their guests, the last of those who were dispensing the greetings coping with a pulse rate approaching danger level. Then, into the dining room to take their places and Shakeshaft took himself to the top table, his nerves now affecting his ability to walk. Gratefully he sat down beside the Chief Bridesmaid, Jane Perry. The courses came and went. Shakeshaft could not taste, nor barely swallow, because between unknown mouthfuls he continued to scribble, head down. Then he heard a voice that he would have supposed came from Heaven, had he not known it so well.

"That's fine, Richard. I can take over from here."

Anyone observing his face would think that he had just been beatified as Shakeshaft stood to allow Carr to take his place. Carr sat and his first act was to place his elbows on the table, the better to support his chin as he looked knowingly at Jane Perry, a half smile on his face. She looked back, her face a combination of delight and disapproval.

"Take your elbows off the table!"

General Perry was sat far back, too far back to speak to his daughter, yet well placed to see her. He had noticed Carr's arrival and now, this Officer and his daughter were evidently conspiring in some way, but he could hear nothing. Then, Carr placed his hand on the table in the space between himself and Jane, then he took it away. Jane looked down; obviously something had been left there. She picked it up and studied it in the palm of her hand for half a minute, often changing the angle, as though there was something special to see or even read, but, to Perry's relief, it was passed immediately up to Cecily, now Mrs. Drake. It was definitely small, for she also examined it, again as though there was something to be read, but he had no idea what it was. Perry felt relief, it must be some wedding gift, then, to his horror, it was passed back and Jane took it below the table. At that moment, a spoon was tapped against a glass to begin the speeches. Carr stood up, with a sheaf of papers in his hand. Jane looked up at him, smiling, with a warmth in her eyes that even Perry could recognise. She raised her left hand before her, absorbed with her fingers, but Perry could not see. She then sat forward, the hand still before her face and in plain view. Perry felt his breath catch, for on the centre finger was a posy ring. Jane sat back and looked at it, but it must have felt tight, for she then moved it to her third finger, where it fitted perfectly.

***

Footnotes

During the retreat, the 28th were, indeed, ordered to throw the treasure of Moore's army over a cliff. Many camp followers froze to death trying to find some of the coin. One women drowned when she fell into the harbour during the evacuation with her pockets full of coins.

Sergeant Ellis' role at Betanzos portrays the achievement of Sergeant William Newman of the 43rd (Monmouthshire) Regiment of Foot. On finding a large group of stragglers defending themselves against cavalry, he organised the group into two sections and they retreated alternately down the slope and into the town, beating off several cavalry attacks. He was rewarded with an Ensign's Commission in the 1st West India Regiment. How he felt about this is not known, for serving in the West Indies was almost a death sentence, mostly from disease.

I have taken somewhat of a liberty with events at Corunna. The 50th defended the slope immediately above Elvina and defeated the first French attack to appear up from it, which had come from Mermet's Division. The 50th then took Elvina to be immediately ejected by another French column following on. In disorder, they fell back, for their place to be taken by The Guards, who held this second attack. Then events become more of a mystery. Two more French assaults were made on the village of Elvina and the hill behind it. One came from a column of Merle's Division (to the French right of Mermet) that was engaged by The Guards, amongst other Battalions, and defeated. However, the other came from Mermet throwing his last divisional reserves into Elvina to reinforce his men still holding the village and perhaps breakout. The question is - how was this fresh attack by Mermet prevented from advancing further into what was a gap in the British line to the right of The Guards? They were only partially covering Elvina and to their right, the position is weak, only a gentle slope that the 50th had previously retreated from. There are no accounts that I can find, not even in Napier, to answer this, and so I can only assume that they were held by the 50th returning on The Guards' right flank and holding there until Merle's column was defeated.

547

Lightning Source UK Ltd.
Milton Keynes UK
UKOW03f1030031213

222277UK00002B/111/P